Climate Change and Agriculture in India: Impact and Adaptation

Syed Sheraz Mahdi
Editor

Climate Change and Agriculture in India: Impact and Adaptation

Springer

Editor
Syed Sheraz Mahdi
Mountain Research Centre for Field Crops
Sher-e-Kashmir University of Agricultural Sciences and
Technology of Kashmir
Khudwani, Anantnag, Kashmir, J & K, India

ISBN 978-3-319-90085-8 ISBN 978-3-319-90086-5 (eBook)
https://doi.org/10.1007/978-3-319-90086-5

Library of Congress Control Number: 2018950537

Printed on acid-free paper

This Springer imprint is published by the registered company Springer Nature Switzerland AG
The registered company address is: Gewerbestrasse 11, 6330 Cham, Switzerland

In memory of my beloved late father, Syed Mahdi Shah, to whom my desire to serve utmost remained a desire only

Preface

Climate change is perhaps the most serious environmental threat to fight against hunger, malnutrition, disease and poverty, mainly through its impact on agricultural productivity. Agriculture is a climate-sensitive sector and is also a sector that provides livelihood for more than 60% of Indians. Warming due to climate change is now reality as evident from the significant increase in the CO_2 concentration (410.88 ppm as on June, 2018) which has caused most of the warming and has contributed the most to climate change. Yet again, year 2016 and 2017 set a global heat record for the third year in a row. A record El Nino lasting from 2015 into 2016 played a role in further pushing the planet's temperature higher. The rising temperatures will adversely affect the world's food production and India would be the hardest hit. There are reports of shifting in the sowing time and length of growing seasons geographically, which would alter planting and harvesting dates of crops and varieties currently used in a particular area. It is estimated that crop production loss in India by 2100 AD could be 10–40% despite the beneficial effects of higher CO_2 on crop growth. The impact of climate change on water availability will be particularly severe for India. About 54% of India's groundwater wells are decreasing with 16% of them decreasing by more than one meter per year. Dynamic of pests and diseases will be significantly altered leading to the emergence of new patterns of pests and diseases which will affect crops yield.

No doubt, Indian farmers have evolved many coping mechanisms over the years, but these have been fallen short of an effective response strategy in dealing with recurrent and intense forms of extreme climatic events on the one hand and gradual changes in climate like rise in surface temperatures, changes in rainfall patterns, increases in evapo-transpiration rates and degrading soil moisture conditions on the other. Region wise climate change adaptation and mitigation options have been identified as important strategies to safeguard food production.

To this end, this book ***Climate Change and Agriculture in India: Impact and Adaptation*** provides the most recent understanding about climate change and its effects on agriculture in India. Further in-depth research is showcased regarding important allied sectors such as horticulture and fisheries and examines the effect of climate change on different cereal crops. The individual chapters discuss the different mitigation strategies for climate change impacts and detail abiotic and biotic stresses in relation to climate change. The book provides an insight into environmentally safe and modern technologies approaches such as nanotechnology, utilization of underutilized crops and breeding climate resilient crops under a changing climate. This book provides a solid foundation for the discussion of climate resilience in agricultural systems under both temperate and subtropical conditions of India and the requirements to keep improving agricultural production.

I am grateful to Dr. A.K. Singh, Vice Chancellor, BAU, Sabour, for providing necessary support, guidance and encouragement in compiling and editing this first edition of the book. The encouragement, support and guidance received from Dr. R.K. Sohane, Director Agriculture Extension Education, BAU, Sabour, and Dr. R.P. Sharma, Associate Dean-cum-Principal, BAC, Sabour, is sincerely acknowledged. I express sincere thanks to all the authors who had contributed in time and sharing their knowledge.

I firmly believe that this publication will be highly useful, in one way or other, for researchers, academicians, extension workers, policy makers, planners, officials in development institutions/agencies, producers, farmers and students of the agriculture and allied sciences.

Khudwani, Anantnag, Kashmir, J & K, India
June 12, 2018

Editor

S. Sheraz Mahdi

Introduction

Due to increase in anthropogenic activities, global temperatures have shown a warming trend of 0.87°Cover the period 1880–2015. Annual surface air temperatures over India also have shown increasing trends of similar magnitude during the period 1901–2016, making 2016 the warmest year in the period of instrumental data. Warmer temperature during the monsoon season (June-September, +0.72°C above average) and the post monsoon season (October-December, +1.1°C above average) mainly contributed to the warmer annual temperature. Climate change is now reality as evident from the significant increase in the CO_2 concentration (408.84 ppm as on June 2017) which has caused most of the warming and has contributed the most to climate change. Two drought years in south central India created panic and for the first time in the history and special trains were put in place to provide drinking water. Climate Change and its extremes are increasingly one of the most serious national security threats, which will have significant impacts on agriculture, natural resources, ecosystem and biodiversity. At the same time, it is likely to trigger food insecurity, human migration, economic and social depression, environmental and political crisis, thereby affecting national development. Although, scientific reports have amply proved that future food production is highly vulnerable to climate change. But, an important source of uncertainty in anticipating the effects of climate change on agriculture is limited understanding of crop responses to extremely weather events. This uncertainty partly reflects the relative lack of observations of crop behavior in farmers' field under extreme heat or cold. Crop yield increases have been slowed and may go into decline as the region runs out of natural resources. Therefore, regular assessments to understand the science of earth's changing climate, and its consequences; primarily driven by global warming, which in turn is highly extensive, complicated, and uncertain, is a scientific challenge of enormous importance to society. The summary of the recent IPCC's fifth assessment report (IPCC 2015) has again a stark warning on how climate change is threatening the South Asia, but report has also shown the way out to combat rampant climate change. India needs to develop a regional strategy for adapting to climate change and its variability in order to ensure food and ecological sustainability. Recognizing the importance of science issues that need to be addressed to deal with climate change, new approaches and policy interventions are desperately needed to enable and encourage smallholder farmers to adopt new technologies and practices under more uncertain and extreme climatic conditions for a resilient agricultural production system. No doubt, Indian farmers have evolved many coping mechanisms over the years, but these have been fallen short of an effective response strategy in dealing with recurrent and intense forms of extreme climatic events on one hand and gradual changes in climate like rise in surface temperatures, changes in rainfall patterns, increases in evapo-transpiration rates and degrading soil moisture conditions on the other. Region wise climate change adaptation and mitigation options have been identified as important strategies to safeguard food production.

Through this book we have bought together a series of chapters that provide scientific insights to possible implications of climate changes for different important types of crop and fisheries systems, and a discussion of options for adaptive and mitigative management. The book provides recent understanding about the climate change and its effects on agriculture. It also includes information regarding important sector like horticulture and fisheries. It examines the effect of climate change on different cereal crops and discusses different mitigation strategies for climate change effects. Details about the abiotic and biotic stresses in relation to climate change are also discussed. It also presents environmentally safe and recent technologies approaches such as nanotechnology and utilization of underutilized crops under changing climate. The information about climate change trends and crop scenario of temperate and cold arid region of Ladakh, Kashmir has been also discussed in last two chapters. This volume provides a solid foundation for the discussion of climate resilience in agricultural systems and the requirements to keep improving agricultural production.

The book is an excellent resource for researchers, instructors, students in agriculture, horticulture, environmental science, other allied subjects and policymakers.

Mountain Research Centre for Field Crops
Sher-e-Kashmir University of Agricultural Sciences
and Technology of Kashmir
Khudwani, Anantnag, Kashmir, J&K, India

S. Sheraz Mahdi

Contents

List of Figures

List of Tables

The original version of this book was revised: The author group of the chapters 18 and 19 have now been restored as in the original submission. The correction to this chapter is available at https://doi.org/10.1007/978-3-319-90086-5_21

Chapter 1
Future Changes in Rainfall and Temperature Under Emission Scenarios over India for Rice and Wheat Production

P. Parth Sarthi

Abstract Indian Summer Monsoon (IMS) prevails during June–July-August-September (JJAS) and 80% of the annual precipitation is received during JJAS. The spatial and temporal variability of ISMR and surface temperature has been influencing agriculture and water resources. The changes in earth's surface temperature is affecting the patterns of weather and climate and influencing the agriculture. The Coupled Model Inter comparison Project 5 (CMIP5) models output data is generally of higher resolution with different emission experiments and therefore rainfall and surface temperature over India is analyzed under CMIP5 which may be used for agricultural purposes.

The Indian Summer Monsoon Rainfall (ISMR) in simulation of BCC-CSM1.1(m), CCSM4, CESM1(BGC), CESM1(CAM5), CESM1(FASTCHEM), CESM1(WACCM), and MPI-ESM-MR for the period of 2006–2050 under RCPs 4.5 and 8.5 at 99% confidence shows possibility of excessrainfall over homogeneous monsoon regions of NWI, NEI, WCI and PI, while deficit rainfall over NWI, NEI, WCI, CNI and PI. At 99% and 95% confidence levels, deficit rainfall is found over CNI, NWI and PI. The CMIP5 model GISS-E2-H, BCC-CSM1.1 m and GISS-E2-H-CC for Tmax; GFDL-CM3, MRI-CGM3 and MRI-ESM1 for Tmin; and CESM1 (CAM5) for T under Representative Concentration Pathways (RCPs) 4.5 and 8.5 for the period of 2021–2055 shows possible significant warming of 0.5°C–0.7°C at 99% confidence level over homogeneous temperature regions of NC, NW, and WC. The warming of 0.2°C–0.5°C might be possible at other locations.

These future projections may be used in crop simulation models which may assist adaptation to climate change-through changes in farming practices, cropping patterns, and use of new technologies.

Keywords Rainfall · Temperature · Future projections and GCMs

P. Parth Sarthi (✉)
Center for Environmental Sciences, Central University of South Bihar, Patna, Bihar, India
e-mail: ppsarthi@cub.ac.in

S. Sheraz Mahdi (ed.), *Climate Change and Agriculture in India: Impact and Adaptation*, https://doi.org/10.1007/978-3-319-90086-5_1

1.1 Introduction

The summer monsoon season in India lasts during June-July-August-September (JJAS) (Rao 1976) and 80% of the annual rainfall occurs during JJAS. The Indian Summer Monsoon Rainfall (ISMR) spatial and temporal variability has been largely influencing agriculture and water resources. As well as, rapid increase in earth's surface temperature is affecting the patterns of weather and consequently influencing the agriculture production.

Under warming conditions, ISMR simulated in different Global and Regional Climate Models have been studied by various researchers; however, uncertainties exist in the regional climate projections due to biasness in the global climate models (Meehl and Washington 1993; Lal et al. 1994, 1995; Rupa Kumar and Ashrit 2001; May 2004; Kripalani et al. 2003; Rupa Kumar et al. 2002, 2003). The variability of ISMR has great importance over the Gangetic plain for agriculture, but it is poorly simulated in many climate models (Lal et al. 2001; Rupa Kumar and Ashrit 2001; Rupa Kumar et al. 2003)., Kripalani et al. (2007a) suggested significant increase in mean monsoon precipitation of 8% and possible extension of the monsoon period, in doubling of CO_2 experiment of Coupled Model Inter Comparison Project Phase 3 (CMIP3). Further, Kripalani et al. (2007b) applied t test and F ratio and suggested possible significant changes in future rainfall from −0.6% for CNRM-CM3 to 14% for ECHO-G and UKMO HadCM3 for East Asian monsoon. Menon et al. (2013a, b) suggested increase in all-India summer monsoon rainfall (AISMR) per degree change in temperature of about 2.3% K − 1, which is similar to the projected increase in global mean precipitation per degree change in temperature in CMIP3. Sarthi et al. (2012) suggested, under A2, B1 and A1B experiments of CMIP3, a future projected change in spatial distribution of ISMR which shows deficit and excess of rainfall in Hadley Centre Global Environment Model version 1 (HadGEM1), European Centre Hamburg Model version 5 (ECHAM5), and Model for Interdisciplinary Research on Climate (MIROC) (Hires) over parts of western and eastern coast of India. Multi-models average of CMIP5 simulations, less uncertainty in CMIP5 projections of rainfall and temperature.

The global mean surface air temperature has been increased by 0.60°C in 20th century experiment of Intergovernmental Panel on Climate Change (IPCC)'s Third Assessment Report (AR3), while as per IPCC's Fourth Assessment Report (AR4), it is estimated to have increased by 0.74°C and could rise up to 1.1°C–6.4°C during twenty-first century depending on a range of possible scenarios (IPCC 2007). In IPCC's Fifth Assessment Report (AR5), increment in temperature are largely due to anthropogenic emissions (IPCC 2014). In IPCC (2014), increase in global mean surface temperatures for 2081–2100 relative to 1986–2005 is projected to increase in the range of 0.3°C to 1.7°C (RCP2.6), 1.1°C to 2.6°C (RCP4.5), 1.4°C to 3.1°C (RCP6.0), 2.6°C to 4.8°C (RCP8.5) in CMIP5. Lal et al. (2001), Rupa Kumar et al. (2003, 2013), and Pattnayak et al. (2015) have analyzed projected surface temperature in various coordinated climate models experiments such as CMIP3 and CMIP5.

Lal et al. (2001) has been suggested projected mean warming of 1.0°C–1.4°C and 2.2°C–2.8°C by 2020 and 2050, respectively over the Indian subcontinent (IPCC, 2014). In 21st century, increase in temperature is particularly conspicuous after the 2040s over India and suggested an increase in Tmin (up to 4°C) all over the country, which might be more in Northeast India (Rupa Kumar et al. 2003).Overall increase in surface temperature by 4.8°C, 3.6°C and 2.2°C in A2, A1B, and B1 emission scenarios, may be possible at the end of 21st century experiment of CMIP3. Extremes in Tmax and Tmin are also expected to increase over the West-Central India in different scenarios (Rupa Kumar et al. 2013). Pattnayak et al. (2015) have analyzed the six CMIP5 model's namely GFDL-CM3, GFDL-ESM2M, GFDL-ESM2G, HadGEM2-AO, HadGEM2-CC and HadGEM2-ES in RCPs 4.5 and 8.5, which are able to capture spatial distribution of temperature with an increasing trend over most of the regions over India. Chaturvedi et al. (2012) worked on multi-model and multiscenario temperature projections over India for the period of 1860–2099 using CMIP5 under RCPs 6.0 and 8.5 scenarios, it is found that mean warming in India is likely to be in the range of 1.7–2°C by 2030s and 3.3–8 4.8°C by 2080s relative to preindustrial times.

Under World Climate Research Programme (WCRP),Working Group on Coupled Modelling (WGCM), Climate models are integrtaed out for past, present and future climate under differerent emission scenarios in Coupled Model Inter comparison Project 5 (CMIP5) and simulated rainfall and surface temperature area analyzed for agriculture purposes in this paper Section 1.1 deals with Introduction and literature survey; data and methodology is in Sect. 1.2. Results and discussion is placed in Sect. 1.2 while conclusions are in Sect. 1.4.

1.2 Data and Models

The gridded observed rainfall of India Meteorological Department (IMD) at resolution of 1°×1° during 1961–1999 and observed rainfall of Global Precipitation Climatology Project (GPCP) at resolution of 2.5°× 2.5° during 1979–1999 is considered for validating the model's performance. Table 1.1 and 1.2 shows list of CMIP5 models considered under RCP 4.5 and 8.5 experiments for analysis of rainfall and surface temperature T, Tmax (for March, April, May) and Tmin (December, January, February). RCPs 4.5 and 8.5 experiments are very likely that world will follow these mild and high emission scenarios in future time periods. The simulation of a Historical experiment in CMIP5 is equivalent to 20th century experiment (20C3M) of CMIP3; models are integrated from 1850 to 2012 with external forcing changing with time. The external forcing includes GHGs, the solar constant, volcanic activity, ozone, and aerosols.

For rainfall, the period of historical experiment is 1961–2005 and for future project is 2006–2050; for surface temperature, the period for historical experiment is 1971–2005 and for future project is 2021–2055.

Table 1.1 List of CMIP5 models their surface resolution and available RCPs experiment for rainfall analysis

Serials No.	Models	Surface Resolution	RCP 4.5	RCP 8.5
1	BCCCSM 1.1(m)	320 × 160	√	√
2	CCSM4	288 × 192	√	√
3	CESM1(CAM5)	288 × 192	√	√
4	CESM1 (BGC)	288 × 192	√	√
5	CESM1 (WACCM)	144 × 96		√
6	MPI-ESM-MR	192 × 96	√	√
7	CESM1 (FASTCHEM)	288 × 192		

Table 1.2 List of CMIP5 models their surface resolution and available RCPs experiment for T, Tmax, and Tmin

	Models	Surface Resolution	RCP 4.5	RCP 8.5
T	BCCCSM 1.1(m)	320 × 160	√	√
Tmax	GISS-E2-H-CC	144 × 90	√	√
	BCCCSM1.1(m)	320 × 160	√	√
	GISS-E2-H-CC	144 × 90	√	√
Tmin	GFDL-CM3	144 × 90	√	√
	MRI-CGCM3	120 × 120	√	√
	MRI-ESM1	120 × 120	√	

1.3 Results and Discussions

Fig. 1.1 shows homogeneous monsoon regions namely North West India (NWI), Central Northeast India (CNI), North East India (NEI), West Central India (WCI), Peninsular India (PI) and Hilly Regions (HR). Fig. 1.2 depicts Temperature homogeneous regions namely North East (NE), North West (NW), North Central (NC), East Coast (EC), Peninsula India (IP), Western Himalaya (WH), West Central (WC) of India. To analyze the reduced uncertainty in future projection of rainfall and temperature, students T test at 99% and 95% confidence levels is applied on the projected values.

A large number of CMIP5 models under Historical experiment is evaluated with observed rainfall of IMD and GPCP (Fig. not shown) and only BCC-CSM1.1(m), CCSM4, CESM1(BGC), CESM1(CAM5), CESM1(FASTCHEM), CESM1 (WACCM), and MPI-ESM-MR performed well and therefore used for analyzing future projections of ISMR in June-July-August-September (JJAS) under RCPs 4.5 and 8.5 emission scenarios. The spatial distribution of future projected percentage changes in JJAS (mm/month) rainfall during 2006–2050 in RCPs of 4.5 and 8.5 of BCC-CSM1.1 (m),CCSM4, CESM1 (BGC), CESM1 (CAM5), CESM1 (WACCM), CESM1 (FASTCHEM) and MPI-ESM-MR with respect to Historical experiment (1961–2005) is shown in Figs. 1.3a–k at 99% and 95% confidence levels using student t-test. In Figs. 1.3a–b, an excess of 5–25% rainfall at 99% and 95%

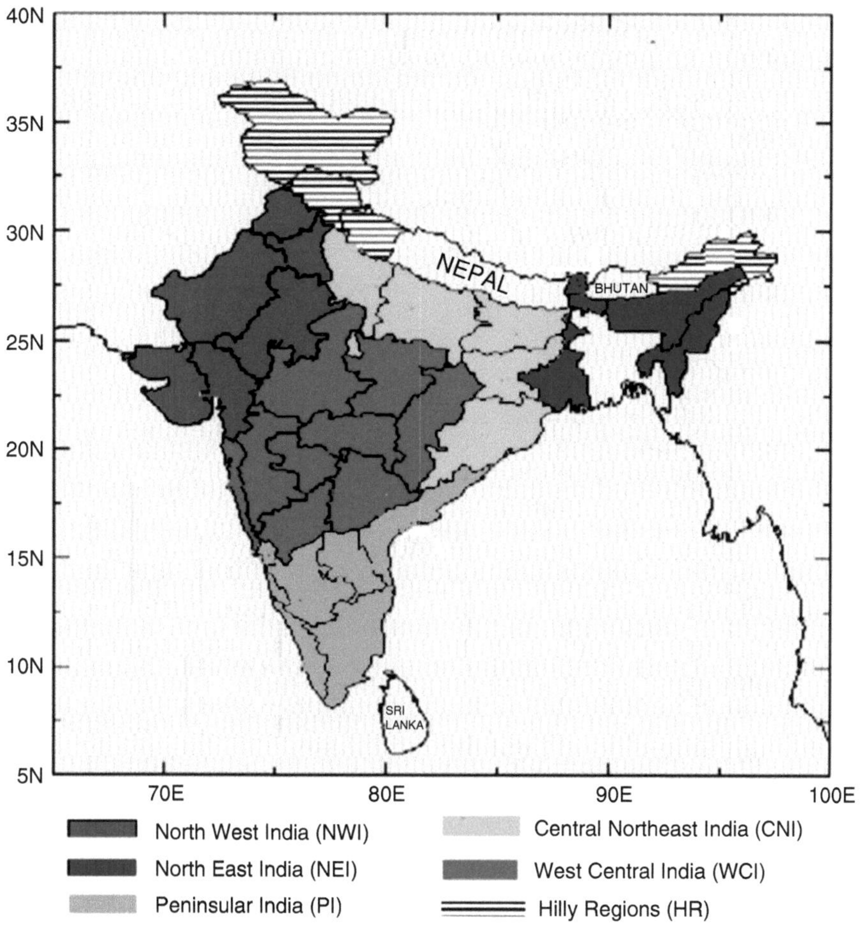

Fig. 1.1 Homogeneous monsoon regions of India (Source: India Institute of Tropical Meteorology, Pune, India)

confidence levels may be possible over the parts of NWI, Gangetic plain of CNI and PI. Figs. 1.3c–d shows possibility of 5–15% excess rainfall at 99% and 95% confidence levels over Western Ghat, parts of WCI and Gangetic plain of CNI, in simulations of CCSM4. In CESM1 (CAM5) simulations (Figs. 1.3e–f), 5–15% deficit rainfall at 99% and 95% confidence levels may be possible over the Gangetic plain of CNI and 5% deficit rainfall at 95% confidence over PI. Excess rainfall of 5–15% at 99% and 95% confidence levels may be possible over parts of CNI and PI in CESM1 (BGC) (Figs. 1.3g–h). 5–10% deficit rainfall at 99% confidence level over NWI is simulated in both RCPs. In MPI-ESM-MR simulations (Figs. 1.3i–j), 10–15% excess rainfall at 99% and 95% confidence levels may be possible over WCI, while 5–10% deficit rainfall over parts of NWI and CNI. In Fig. 1.3k, CESM1 (WACCM) shows, 10–15% excess rainfall at 99% confidence level over parts of

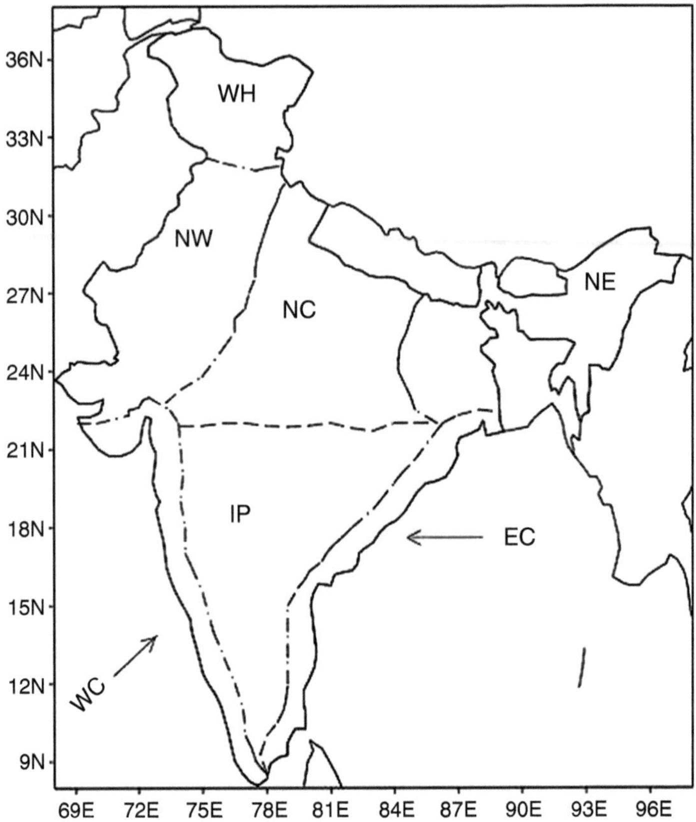

Fig. 1.2 Temperature Homogeneous Regions of India (Source: IITM, Pune, India)
NE – North East **NW** – North West
NC – North Central **EC** – East Coast
IP – Peninsula India **WH** – Western Himalaya **WC** – West Central

NEI and PI. Out of seven models, six models shows possibility of excess rainfall in JJAS at 99% and 95% confidence levels in future time periods over rice production regions of CNI and PI. If distribution of rainfall within JJAS is taking place equaly in months of june, july, august and september, then it would be favourable for production of rice. Further analysis of rainfall variability within rainy months is important over these regions.

In simulating surface temperature, models evaluation shows that model CESM1 (CAM5) for T, GISS-E2-H, BCC-CSM1.1 m, and GISS-E2-H-CC for Tmax and GFDL-CM3, MRI-CGCM3, and MRI-ESM1 for Tmin are relatively better performing than other models. Figs. 1.4a–n shows future projected changes in Tmax, Tmin and T during 2021–2055 under RCPs 4.5 and 8.5 with respect to Historical experiment (1971–2005). Student's t-test at 99% and 95% confidence levels is

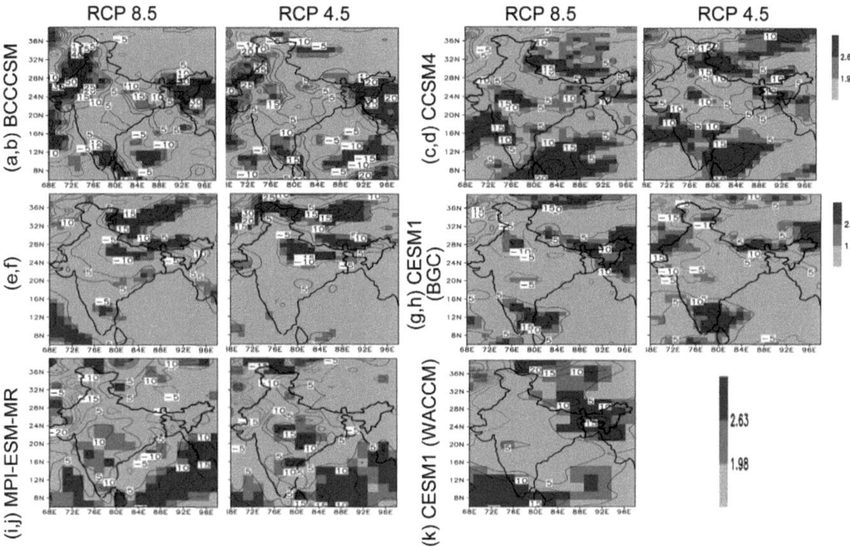

Fig. 1.3 **(a–k)** Percentage change in JJAS rainfall (mm/ day) for the period of 2006–2050 under RCPs 4.5 and 8.5 with respect to Historical experiment for the period of 1961–2005 at 99% confidence level (in dark blackish grey color) and at 95% confidence level (in masked with light grey color) at student's t test

applied to know the significance of future projected changes. In Figs. 1.4a–n, 99% and 95% confidence levels are masked with dark gray and medium gray color respectively, while non-significant area is masked with light gray color. GISS-E2-H, GISS-E2-H-CC projected Tmax (Figs. 1.4c–f) shows possibility of 0.3°C–0.4°C warming over NW and NC and 0.6°C–0.7°C over NW and the Gangetic plain under RCPs 4.5 and 8.5. The model BCC-CSM1.1 m (Figs. 1.4a–b) shows possible warming of 0.4°C–0.5°C at 99% confidence level over the entire region of India except over the central west region while 0.25°C–0.45°C over southern India at 99% confidence level. In Figs. 1.4g–l, GFDL-CM3, MRI-CGCM3, and MRI-ESM1 simulated Tmin in RCPs 4.5 and 8.5 shows possible warming of 0.3°C–0.7°C over major parts of India, but more warming is seen over northwest and southeast part at 99% confidence level while 0.4°C–0.6°C at 95% confidence level is depicted over whole India. In simulation of CESM1(CAM5), future projected T in RCPs 4.5 and 8.5 (Fig. 1.4m–n) shows a significant warming of 0.4°C–0.5°C over NC and the Gangetic plain at 99% confidence level, and the same magnitude of warming at 95% confidence level may be possible over whole India except WC and EC. The projected inrecase in Tmin over over noinfrthwest and other parts of India may not be supportive for wheat production and diseases may be possible in wheat grains. It would be important to understand the variability of Tmin within December, January and February and its relations to the wheat production.

Further, the significant projected change of rainfall and temperature in CMIP5 models may be used as inputs to Crop simulation models over the Gangetic plain

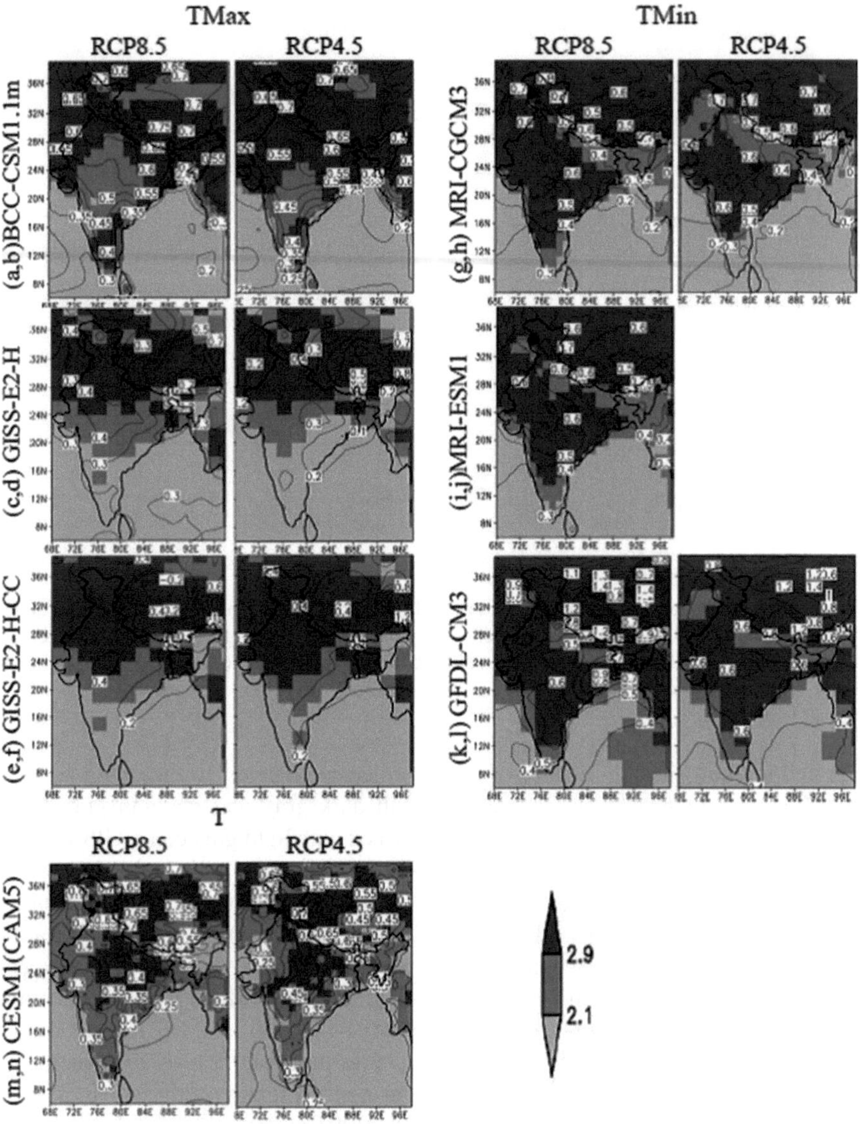

Fig. 1.4 **(a–n)** Future Projected Changes in Tmax, Tmin, and T (°C) for the period of 2021–2055 under RCP 4.5 and 8.5 with respect to Historical experiment for the period of 1971–2005. (Regions with statistically significant at 99% (cut off t value ≧ 2.9) confidence level (masked with dark grey color) and is 95% (cut off t value equals to 2.1) confidence level (masked with medium grey color) on two-tailed student's t test). The light grey shade shows the area with confidence lesser than 95%

which is more scientific approach to study the impact of climate change on agricultural production. Such study may help in framing adaptation to climate change-through changes in farming practices, cropping patterns, and use of new technologies which may reduce impacts of climate change.

1.4 Conclusions

These seven (7) models namely BCC-CSM1.1 (m), CCSM4, CESM1 (BGC), CESM1 (CAM5), CESM1 (FASTCHEM), CESM1 (WACCM), and MPI-ESM-MR simulated future projected percentage changes in JJAS rainfall for the period of 2006–2050 under RCPs 4.5 and 8.5 at 99% confidence level is analyzed which shows possibility of excess rainfall over homogeneous monsoon regions of NWI, NEI, WCI, CNI and PI. CMIP5 models GISS-E2-H, BCC-CSM1.1 m and GISS-E2-H-CC for Tmax; GFDL-CM3, MRI-CGM3 and MRI-ESM1 for Tmin; and CESM1 (CAM5) for T are close to observation for the period of 1961–2005. In RCPs 4.5 and 8.5 experiments for the period of 2021–2055 with respect to historical experiment for the period of 1961–2005, significant warming of 0.5°C–0.7°C at 99% confidence level may be possible over homogeneous temperature regions of NC, NW, and WC. The changes of 0.2°C–0.5°C might be possible at other locations.

These future projections of rainfall and Tmin may be used in assessment of rice and wheat production and assist adaptation to climate changethrough changes in farming practices, cropping patterns, and use of new technologies.

References

Chaturvedi, et al. (2012). Multi-model climate change projections for India under representative concentration pathways. *Current Science, 103*, 791–802.

Kripalani, R., Kulkarni, A., Sabade, S., & Khandekar, M. (2003). Indian monsoon variability in a global warming scenario. *Natural Hazards, 29*, 189–206.

Kripalani, R. H., et al. (2007a). Response of the East Asian summer monsoon to doubled atmosphericCO2: Coupled climate model simulations and projections under IPCC AR4. *Theoretical and Applied Climatology, 87*, 1–28.

Kripalani, R. H., et al. (2007b). South Asian summer monsoon precipitation variability: Coupled climate model simulations and projections under IPCC AR4. *Theoretical and Applied Climatology, 90*, 133–159.

Kumar, R., et al. (2002). Climate change in India: Observations and model projections. In P. R. Shukla et al. (Eds.), *Climate change and India: Issues, concerns and opportunities* (pp. 24–75). New Delhi: Tata McGraw-Hill Publishing Co Ltd..

Kumar, R., et al. (2003). Future climate scenarios. In P. R. Shukla et al. (Eds.), *Climate change and India: Vulnerability assessment and adaptation* (pp. 69–127). Hyderabad: Universities Press.

Kumar, R., et al. (2013). High-resolution climate change scenarios for India for the 21st century. *Current Science, 90*(3), 10.

Lal, M., et al. (1994). Effect of global warming on Indian monsoon simulated with a coupled ocean-atmosphere general circulation model. *Current Science, 66*, 430–438.

Lal, M., et al. (1995). Effect of transient increase in greenhouse gases and sulphate aerosols on monsoon climate. *Current Science, 69*, 752–763.

Lal, M., et al. (2001). Future climate change: Implications for Indian summer monsoon and its variability. *Current Science, 81*, 1196–1207.

May, W. (2004). Simulation of the variability and extremes of daily rainfall during the Indian summer monsoon for present and future times in a global time-slice experiment. *Climate Dynamics, 2*, 183–204.

Meehl, G. A., & Washington, W. M. (1993). South Asian summer monsoon variability in a model with doubled atmospheric carbon dioxide concentration. *Science, 260*, 1101–1104.

Menon, A., et al. (2013a). Enhanced future variability during India's rainy season. *Geophysical Research Letters, 40*(12), 3242–3247.

Menon, A., et al. (2013b). Consistent increase in Indian monsoon rainfall and its variability across CMIP-5 models. *Earth System Dynamics, 4*, 287–300.

Pattnayak, K. C., et al. (2015). Projections of rainfall and surface temperature from CMIP5 models under RCP4.5 and 8.5 over BIMSTEC countries. In *EGU general assembly conference abstracts* (Vol. 17, p. 556). EGU General Assembly, held 12–17 April, 2015 in Vienna, Austria. id.556.

Rupa Kumar, K., & Ashrit, R. G. (2001). Regional aspects of global climate change simulations: Validation and assessment of climate response over Indian monsoon region to transient increase of greenhouse gases and sulfate aerosols. *Mausam, Special Issue on Climate Change, 52*, 229–244.

Sarthi, P., et al. (2012). Possible changes in the characteristics of Indian summer monsoon under warmer climate. *Global and Planetary Change, 92–93*, 17–29.

Chapter 2
Impact of El-Nino and La-Nina on Indian Climate and Crop Production

Vyas Pandey, A. K. Misra, and S. B. Yadav

Abstract El-Nino refers to a large-scale ocean-atmosphere climate interaction associated with the episodic warming in sea surface temperatures (SST) across the central and east-central Equatorial Pacific. La Nina is an opposite event of El Niño which is termed as the episodic cooling of ocean SST in the central and east-central equatorial pacific. El Niño events are mostly associated with warm and dry conditions in southern and eastern inland areas of Australia, as well as Indonesia, Philippines, Malaysia and central Pacific islands such as Fiji, Tonga and Papua New Guinea. The inter-annual variability of Indian summer monsoon rainfall (ISMR) has been linked to variations of Sea Surface Temperatures (SST) over the equatorial Pacific and Indian Oceans. ENSO events have a profound impact on summer monsoonal rainfall across India and most of the major droughts have occurred during El Niño events. However, its reverse is not always true. Previously El Niño had a strong association with droughts in India but this relationship has been weekend in recent years. El Niño conditions mostly coincide with a period of weak monsoon and rising temperatures in India and thus the probability of drought occurrence surges during El Nino events that could be disturbing for Indian crop production and water supply. Moreover, El Niño resulting in deficit rainfall tends to lower the summer crops production such as rice, sugarcane, cotton and oilseeds and therefore the outcome might be seen in form of high inflation rates and lower GDP due to high contribution of agriculture sector in Indian economy. This paper describes the occurrence of El Niño events, its impact on climate in different parts of world with special reference to Indian monsoon and crop production.

Keywords El-Nino · La-Nina · ENSO · SST · ISMR · Agricultural production

---

V. Pandey (✉) · A. K. Misra · S. B. Yadav
Department of Agricultural Meteorology, B.A. College of Agriculture, Anand Agricultural University, Anand, Gujarat, India

S. Sheraz Mahdi (ed.), *Climate Change and Agriculture in India: Impact and Adaptation*, https://doi.org/10.1007/978-3-319-90086-5_2

2.1 Introduction

The term El Nino and La Nina both have their origin in Spanish language with their respective meanings are "Christ child" and "a little girl" respectively. The name El Nino was chosen since as El-Niño events starts around December near Christmas. El-Nino refers to a large-scale ocean-atmosphere climate interaction associated with the episodic warming in sea surface temperatures (SST) across the central and east-central Equatorial Pacific. Fishermen from the coast of South America recognized this event with the appearance of unusually warm water in the Pacific Ocean. La Nina is an opposite event of El Niño which is termed as the episodic cooling of ocean SST in the central and east-central equatorial pacific.

The fluctuations in SST during El Niño and La Niña events are associated with large fluctuations in air pressure which is termed as Southern Oscillation. The negative phase of the Southern Oscillation occurs during El Niño episodes, and refers to the situation when abnormally high air pressure covers Indonesia and the western tropical Pacific and abnormally low air pressure covers the eastern tropical Pacific. In contrast, the positive phase of the Southern Oscillation occurs during La Niña episodes, and refers to the situation when abnormally low air pressure covers Indonesia and the western tropical Pacific and abnormally high air pressure covers the eastern tropical Pacific (WMO 2014).

2.2 Oceanic Niño Index (ONI)

National Oceanic and Atmospheric Administration (NOAA) which is an American scientific agency has developed an index known as Oceanic Niño Index (ONI) that has become standard for identification of El Niño and La Niña events in the tropical Pacific. This index is calculated from running 3-months mean sea surface temperature (SST) anomaly for the Niño 3.4 region (i.e., 5 °N -5 °S, 120°- 170° W). Consequently, these events may be categorized as weak (with a 0.5 to 0.9 SST anomaly), moderate (1.0 to 1.4 SST anomaly), strong (1.5 to 1.9 SST anomaly) and very strong ($\geq$ 2.0SST anomaly) events. It is important to note here that for an event to be categorized as weak, moderate, strong or very strong it must have equaled or exceeded the threshold for at least 3 consecutive overlapping 3-month periods. On the basis of this index, the categorization of various El Niño and La Nina events since 1950 are presented in Table 2.1.

2.3 Impact of El Nino on Regional Climate

The direct effects of ENSO on regional climatic patterns have been found in areas near to the tropical Pacific. In general, Indonesia and surrounding areas in the western Pacific experience more precipitation as compared to the eastern Pacific e.g. as

Table 2.1 Historical El Nino and La Nina events and their severity based on Oceanic Niño Index

El Niño				La Niña		
Weak	Moderate	Strong	Very strong	Weak	Moderate	Strong
1951–52	1963–64	1957–58	1982–83	1950–51	1955–56	1973–74
1952–53	1986–87	1965–66	1997–98	1954–55	1970–71	1975–76
1953–54	1987–88	1972–73	2015–16	1964–65	1998–99	1988–89
1958–59	1991–92			1967–68	1999–00	
1968–69	2002–03			1971–72	2007–08	
1969–70	2009–10			1974–75	2010–11	
1976–77				1983–84		
1977–78				1984–85		
1979–80				1995–96		
1994–95				2000–01		
2004–05				2011–12		
2006–07						

Source: Golden Gate Weather Services. (2016)

Peru and Ecuador. But during El Niño events, warmer surface waters in the tropical Pacific move eastward due to of trade winds which results in to areas of low pressure and higher rainfall along the west coasts of North and South America. Conversely, the temperature of the waters in the western Pacific become cooler than normal, which leads to higher pressure and decreased rainfall there. Although many regions of the world can experience disasters in any year, certain locations tend to be particularly hard hit during El Nino events. The regional impacts due to El Nino are summarized below.

1. El Niño events are mostly associated with warm and dry conditions in southern and eastern inland areas of Australia, as well as Indonesia, Philippines, Malaysia and central Pacific islands such as Fiji, Tonga and Papua New Guinea.
2. During the northern hemisphere summer season, the Indian monsoon rainfall generally tends to be less than normal.
3. In northern hemisphere winter, drier than normal conditions are typically observed over south-eastern Africa and northern Brazil.
4. Wetter than normal conditions are observed along the Gulf Coast of the United States, the west coast of tropical South America (Colombia, Ecuador and Peru) and from southern Brazil to central Argentina.
5. Parts of eastern Africa usually receive above-normal rainfall.
6. Milder winters in north-western Canada and Alaska is experienced due to fewer cold air surges from the Arctic Warm
7. El Niño tends to lead to more tropical cyclone activity in the central and eastern Pacific basins and less in the Atlantic basin.
8. El Niño also suppresses Atlantic hurricane activity by increasing atmospheric stability.

2.4 Impact of El Nino and La Nina on Indian Monsoon

The inter-annual variability of Indian summer monsoon rainfall (ISMR) has been linked to variations of Sea Surface Temperatures (SST) over the equatorial Pacific and Indian Oceans. Several researchers have studied the relationship between El Nino events and Indian monsoon (Sikka 1980; Shukla and Mooley 1987). On the basis of All-India Rainfall Index data from 1856 to 2004, the relationship of ENSO events with Indian summer monsoon have been presented in Fig. 2.1. The plot indicates that ENSO events has a profound impact on summer monsoonal rainfall across India and most of the major droughts have occurred during El Niño events. Conversely, above average rainfall is shown to occur during La Niña events. Kumar et al. (2006) studied the rainfall trend for India for 132 years and concluded that severe droughts in India have always been accompanied by El Nino events. However its reverse is not always true and El Nino events with the warmest sea surface temperature (SST) anomalies in the central equatorial Pacific are more effective in focusing drought-producing subsidence over India than events with the warmest SSTs in the eastern equatorial Pacific. The reason behind El Nino having an inverse relationship with Indian Monsoon is warming of the Pacific Ocean due to El Nino results in weakening of the trade winds coming from Southern America during Southwest Monsoon. Consequently, moisture as well as heat content gets limited and results in reduction and uneven distribution of rainfall across the Indian sub-continent.

Gautam. et al. (2013) observed that during 1980–2014, all the six droughts faced by India were El Niño years, but still not all El Niño years led to drought in the country. Since 2000, there were four El Niño years (2002, 2004, 2006 and 2009),

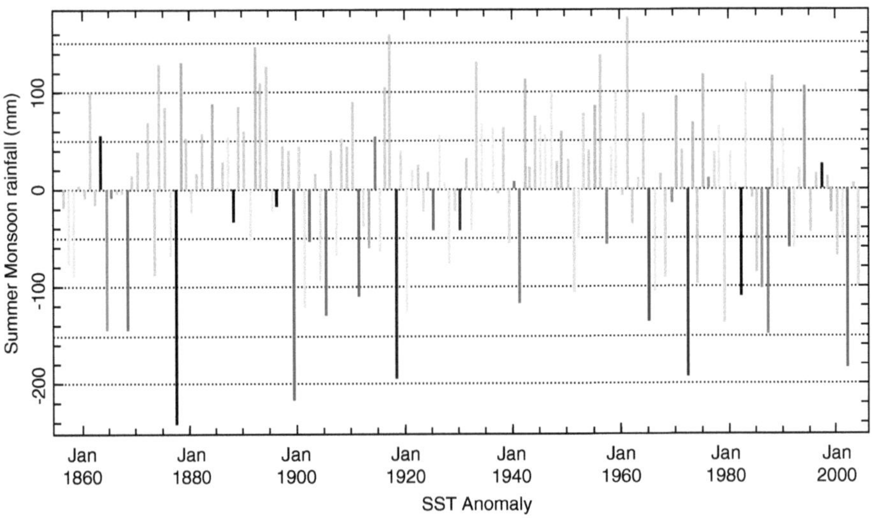

Fig. 2.1 Relationship between summer monsoonal rainfall in India and ENSO (Source: IRI Columbia)

and three of these (except 2006) resulted into drought years. The year 2006, which was an El Niño year, however, received normal monsoon rainfall. Similarly, year 1997–98 was a strong El Nino year but it resulted in excess rainfall in that year. Contrary to this, a moderate El Nino in 2002 resulted in one of the worst droughts of the country.

Saini and Gulati (2014) analyzed different global El Nino and Indian droughts during 1950–2000 and they found that there were 23 global El Nino years and 13 Indian drought years since 1950 (Table 2.2). It is interesting to note that out of the 13 + 1 drought years, 11 years were El Niño years. But of the 23 El Niño years, only 10 + 1 were drought years. Therefore, it is very clear that not all El Niño years converted into droughts for India and not all (but most) Indian droughts happened in El Niño years. This study also concluded that since the 1980s, the association between the two climatic phenomena appears to have strengthened. All the six drought years in the last 34 years were El Niño years, yet there were six El Niño years which did not convert into droughts for India. Contrary to expectation, the El Niño years of 1994 and 1983 saw India receive excess rains, exceeding 12 per cent of LPA.Since the 2000s, the association between the two phenomena seems to have strengthened yet further. Three of the four El Niño years converted into Indian droughts. However, India received normal rainfall in the El Niño year of 2006.

Rao et al. (2011) studied the El Nino effects on rainfall and thermal regime for Andhra Pradesh which concluded that the rainfall was below normal in 8 out of 10 El Niño and there are 80% less probability of getting above normal rainfall during the El Niño years. Furthermore, six of the deficit rainfall events associated with El Niño experienced a negative departure from normal indicating impending drought situations likely to occur in association with El Niño events. Another study con-

Table 2.2 Summary of the Global El Niño and Indian Drought Years during 1950–2010

Type of Events	Since 1950s (64 years)	Since 1980s (34 years)	Since 2000s (14 years)
Global El Niño	23 Years 1951, 1953, 1957, 1958, 1963, 1965, 1968, 1969, 1972, 1976, 1977, 1982, 1983, 1986, 1987, 1991, 1992, 1994, 1997, 2002, 2004, 2006, 2009	12 Years 1982, 1983, 1986, 1987, 1991, 1992, 1994, 1997, 2002, 2004, 2006, 2009	4 Years 2002, 2004, 2006, 2009
Indian droughts	13 + 1 Years 1951, 1965, 1966, 1968, 1972, 1974, 1979, 1982, 1986, 1987, 1991, 2002, 2004, 2009	6 + 1 Years 1982, 1986, 1987, 1991, 2002, 2004, 2009	3 Years 2002, 2004, 2009
Drought and El Niño	10 + 1 Years 1951, 1965, 1968, 1972, 1982, 1986, 1987, 1991, 2002, 2004, 2009	6 + 1 Years 1982, 1986, 1987, 1991, 2002, 2004, 2009	3 Years 2002, 2004, 2009
El Niño but not drought	12 years 1953, 1957, 1958, 1963, 1969, 1976, 1977, 1983, 1992, 1994, 1997, 2006	5 years 1983, 1992, 1994, 1997, 2006	1 year 2006
Drought but not El Niño	3 years 1966, 1974, 1979	None	None

ducted by Patel et al. (2014) for Gujarat region exhibited that annual as well asmonsoon seasonal rainfall during El Nino years was more as compared to non-El Niño years. Moreover, large spatial variability is noticed both in annual and seasonal rainfall during El Niño years in comparison to non-El Niño years. During weak and strong El Niño years, the rainfall towards the end of the monsoon season was found to be deficit at most of the study locations. However, in moderate El Niño years, the July month has a deficit rainfall indicating a break in monsoon activity.

Prasad et al. (2014), reported that the rainfall during El Niño years was likely to be less during southwest monsoon season as compared to non El Niño years in Himachal Pradesh. Furthermore, rainfall deficiency during SW monsoon increased with strength of the El Niño and subsequent winter season received higher rainfall. Rishma and Katpatal (2016) have also reported that the mean rainfall during El Niño years was less than that during La Niña and normal years. Manikandan et al. (2016) has found a negative departure of annual rainfall during El Nino years in the range of 1 to 10 percent for most of the districts in Chhattisgarh State. However, during strong El Niño years, negative deviation in rainfall went up to 21 per cent also. In the Tamil Nadu State which is located in the southern part of India, Geethalakshmi et al. (2009) found that during the summer monsoon, the relationship between ENSO and rainfall is negative and total rainfall is less than normal during El Niño years. In contrast, the relationship during the northeast monsoon is positive, i.e., rainfall is more than normal during El Niño years and less than normal during La Niña years.

Therefore, it is evident that there is an inverse relationship between the El Nino events and ISMR. Most of severe droughts over India have occurred in association with El Nino events. However, there is no one-to-one relationship. According to Kane (2006), previously El Niños had a strong association with droughts in India but this relationship has been weekend in recent years. In the 133-year data, only about 60% of the El Niño and La Niña are effective, and the magnitudes of the anomalies are not in good correlations with the strength of the ENSO.

2.5 Effects of El Nino on Indian Agriculture

The characterization of ENSO events and their impacts on agriculture is difficult to establish because there are several other factors also which have a profound impact on crop production and El Nino is just one of them. Furthermore, the impact of El Niño's on crop conditions and development is dependent on the sensitivity of the phenological phase of crops during the peak period of influence of the event, since the flowering and grain filling phases of cereal crops are more sensitive to water stress. It has also found that the areas where El Niño is most likely to negatively affect agriculture causing drought conditions, with ensuing reductions in agricultural production and potential food security implications (Rojas et al. 2014).

El Niño conditions mostly coincide with a period of weak monsoon and rising temperatures in India and thus the probability of drought occurrence surges during El Nino events that could be disturbing for Indian crop production and water supply. Moreover, El Niño resulting in deficit rainfall tends to lower the summer crops production such as rice, sugarcane, cotton and oilseeds and therefore the outcome might be seen in form of high inflation rates and lower GDP due to high contribution of agriculture sector in Indian economy. However, the relationship among Indian monsoon rainfall pattern and El Nino are not that strong as Australia and typically, only strong El Nino affects India negatively (Patel et al. 2014).

Selvaraju (2003) studied the variability of summer monsoon rainfall (SMR) with ENSO phases in major foodgrain producing sub-divisions of India for the period of during 1950–99 and concluded that warm ENSO years (El Nino years) reduces foodgrain production, as SMR is a critical input for both *Kharif* and *Rabi* season crops under intensive crop production systems.The percentage deviation of the foodgrain production variability has been depicted in Fig. 2.2 which clearly reflect that 12 out of 13 warm El Nino years, the total foodgrain production decreased in the range of 1.2 to 14.9%. On the other hand,10 out of 13 cold ENSO phases (La Nina years), the total foodgrain production increased from normal. There was a reduction in *kharif* season's foodgrain production in eight out of nine El Nino years and the impact was higher on *kharif* foodgrain production as compared to *rabi* season crops. In eight out of nine warm-phase years, the *Rabi* foodgrain production anomalies were negative and only in four out of ten cold-phase years was production less than normal.

Rao et al. (2011) has reported that the regions served by northeast monsoon rainfall are comparatively less vulnerable to decline in productivity during El Niño years in Andhra Pradesh. During the study period of 1981–2006, average production during El Niño years decreased by 42.7 per cent and the productivity decreased by 36.4 per cent for crops other than rice. Therefore, it is obvious that El Niño is exerting greater influence on productivity of rainfed crops as a result of decreasing tendency of southwest monsoon rainfall. It has been concluded by Prasad et al. (2014) that during El Niño years, there was a tendency in the reduction of the cereal grain production for major part in Himachal Pradesh except four ditricts viz., Una, Hamirpur, Bilaspur and Mandi districts.

The impact of El Nino events for different crops in Gujarat state (Patel et al. 2014) revealed that productivity for majority of the crops were adversely affected in different crop growing regions of the state. The crops such as rice, groundnut, mustard and wheat were found to be highly vulnerable due to El Niño episodes in all the major growing districts. However, these impacts were visible only during strong El Nino events for crops viz. maize, cotton, sugarcane and bajra.

The El Nino events have benefited the paddy production in some regions. Bhuvaneswari et al. (2013) reported that Tamilnadu state received more rainfall (with high interannual variability) during El Nino years and therefore the mean rice productivity was incresed in El Niño and normal years indicating the possibility of

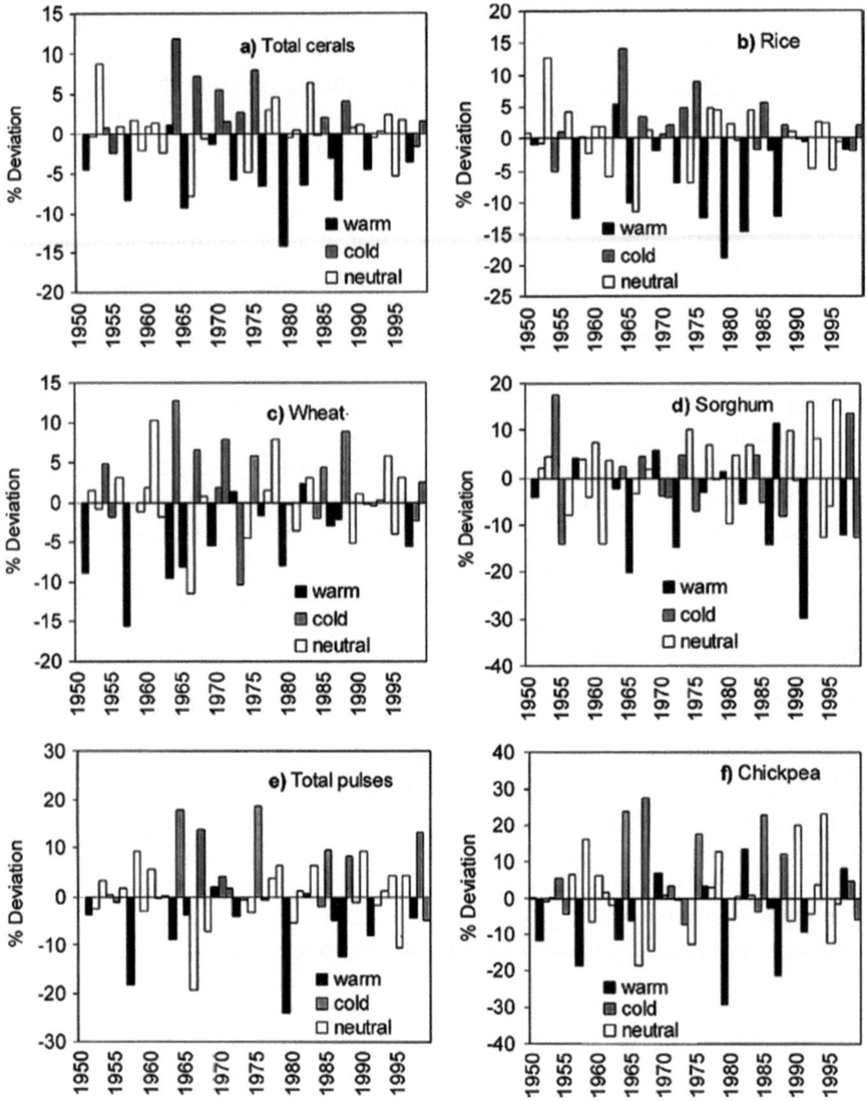

Fig. 2.2 Percentage deviation of Indian food-grain production anomalies (1950–99) from the normal under ENSO phases: (**a**) total cereals; (**b**) rice; (**c**) wheat; (**d**) sorghum; (**e**) total pulses; (**f**) chickpea production (Source: Selvaraju (2003))

getting more rice yields with less crop production risk compared to La Niña years. A regression model using monthly rainfall index and southern oscillation Index has been developed by Ramakrishna et al. (2003) which provides more accurate food grain production estimation as compared to all India summer monsoon rainfall alone.

2.6 Conclusion

It has been found that most of the El Nino years have either led to droughts or below normal rainfall and therefore it is clear that El Nino events adversely affect the monsoon rain with few exceptions. However, El Nino is just one of the several factors that influence monsoon rainfall and absence of an El Nino does not necessarily mean good rainfall. Furthermore, sensitivity of crops to El Niño and La Nina episodes is not uniform for all the regions and more rigorous study needs to be conducted to understand its mechanism in better way. In addition to this, to forecast an El Nino and La Nina events well in advance is another major challenge for Indian climatologists that needs to be addressed.

References

Bhuvaneswari, K., Geethalakshmi, V., Lakshmanan, A., Srinivasan, R., & Sekhar, N. U. (2013). The impact of El Nino/Southern oscillation on hydrology and Rice productivity in the Cauvery Basin, India: Application of the soil and water assessment tool. *Weather and Climate Extremes, 2*, 39–47.

Gautam, H. R., Bhardwaj, M. L., & Rohitashw, K. (2013). Climate change and its impacts on plant diseases. *Current Science, 105*(12), 1685–1691.

Geethalakshmi, V., Yatagai, A., Palanisamy, K., & Umetsu, C. (2009). Impact of ENSO and the Indian Ocean Dipole on the north-east monsoon rainfall of Tamil Nadu state in India. *Hydrological Processes, 23*(4), 633–647.

Golden Gate Weather Services (2016). El Niño and La Niña Years and Intensities, URL: http://ggweather.com/enso/oni.htm.

Kane, R. P. (2006). Unstable ENSO relationship with Indian regional rainfall. *International Journal of Climatology, 26*(6), 771–783.

Kumar, K. K., Rajagopalan, B., Hoerling, M., Bates, G., & Cane, M. (2006). Unraveling the mystery of Indian monsoon failure during El Niño. *Science, 314*(5796), 115–119.

Manikandan, N., Chaudhary, J. L., Khavse, R., & Rao, V. U. M. (2016). El-niño impact on rainfall and food grain production in Chhattisgarh. *Journal of Agrometeorology, 18*(1), 142–145.

Patel, H. R., Lunagaria, M. M., Pandey, V., Sharma, P. K., Rao, B. B. and Rao, V. U. M. (2014). "El Niño episodes and agricultural productivity in Gujarat" Techn. Report No. 01/2014-15. Deparment of Agril Meteorology, AAU, Anand, Gujarat.

Prasad, R., Rao, V. U. M. and Rao, B. B. (2014). El Niño-Its Impact on Rainfall and Crop Productivity: A Case Study for Himachal Pradesh, Res. Bull. No. 1/2014, CSKHPKV, Palampur, HP and CRIDA, Hyderabad

Ramakrishna, Y. S., Singh, H. P., & Rao, G. N. (2003). Weather based indices for forecasting foodgrain production in India. *Journal of Agrometeorology, 5*(1), 1–11.

Rao, V. U. M., Subba, A. V. M., Bapuji, R. B., Ramana, R. B. V, Sravani, R. C. and Venkateswarlu, B., (2011), El Niño effect on climatic variability and crop production : A case study for Andhra Pradesh, research Bulletin No. 2/2011. Hyderabad: Central Research Institute for Dryland Agriculture.

Rishma, C., & Katpatal, Y. B. (2016). Variability in rainfall and vegetation density as a response to ENSO events: A case study in Venna river basin of Central India. *Journal of Agro meteorology, 18*(2), 300–305.

Rojas, N., Yanyun Li, E., & Renato Cumani, N. (2014). *Understanding the drought impact of El Niño on the global agricultural areas: An assessment using FAO's agricultural stress index (ASI), Environment and natural resources management Series*. Rome: Food and Agriculture Organization of the United Nations.

Saini, S., & Gulati, A. (2014). *El Niño and Indian droughts-a scoping exercise, working paper 276, Indian Council for Research on international economic*. New Delhi: Relations.

Selvaraju, R. (2003). Impact of El Niño-southern oscillation on Indian foodgrain production. *International Journal of Climatology, 23*(2), 187–206.

Shukla, J., & Mooley, D. (1987). Empirical prediction of the summer monsoon rainfall over India. *Monthly Weather Review, 115*, 695–703.

Sikka, D. R. (1980). Some aspects of the large scale fluctuations of summer monsoon rainfall over India in relation to fluctuations in the planetary and regional scale circulation parameters. *Journal of Earth System Science, 89*(2), 179–195.

WMO. (2014). *El Niño/southern oscillation*. Geneva: World Meteorological Organization.

Chapter 3
Simulating the Impact of Climate Change and its Variability on Agriculture

Naveen Kalra and Manoj Kumar

Abstract Indian agriculture is mostly intensive and presently showing concerns of decline in productivity primarily due to deteriorating soil health (in terms of physical, chemical and biological), surface/ground waters concerns (quality and quantity), and newly emerging insects/pests). Climate change and climatic variability are of great concerns, especially in India. Probabilities of occurrence of extreme/episodic events have increased over the last three decades. There is need to sustain agricultural productivity and safeguard the environment under these climatic/episodic events. Agricultural productivity can be affected by climate change; directly, due to changes in temperature, precipitation or CO_2 levels and indirectly, and through changes in soil health and infestation by insects/pests. Our country has numerous and complex agro-ecologies and production environments, so there is a need to evaluate the impact of climate change on regional scale, by incorporating socio-economic and bio-physical drivers along with the climatic elements. Crop models (mainly dynamic/mechanistic) can effectively integrate these aspects for assessing the impact of future climate change as well suggest suitable options for suggesting suitable mitigation and adaptation strategies. Several simulation models viz. INFOCROP, WTGROWS, ORYZA, DSSAT, APSIM have been developed and widely used for resource and inputs management, plant ideo-typic designs, climate change/its variability impact evaluation, yield forecasting and addressing extreme/episodic events. By using these models, differential response of reduced crops yields to rising temperatures was evaluated at several locations. Interaction of changes in CO2 concentration and solar radiation with temperature was noticed through modifications in soil and crop processes and subsequent realization in grain yield. Crop models have been successfully employed for suggesting suitable mitigation and adaptation strategies (such as choice of crops/cropping systems, water/nutrients management options, adoption of suitable agronomic management) for reduction in GHGs emission and sustenance of agricultural production. Crop models have been successfully employed for assessing extreme/episodic events, viz.

N. Kalra (✉)
Division of Agricultural Physics, IARI, New Delhi, India

M. Kumar
Forest Informatics Division, Forest Research Institute, Dehradun, India

S. Sheraz Mahdi (ed.), *Climate Change and Agriculture in India: Impact and Adaptation*, https://doi.org/10.1007/978-3-319-90086-5_3

aerosol impact on crops as evaluated by using WTGROWS and DSSAT in the UNEP sponsored ABC project. The crop-pest-weather interaction and socio-economic components in the climate change impact evaluation, at present are relatively weak but are being continuously improved. We should develop agro-forestry models for effective land use planning. We also need to emphasize on development of inter- and intra- sectoral assessment models (IAM), for developing appropriate methods for sustaining systems' productivity under the prime concern of climatic variability/climate change.

Keywords Climate change · Climatic variability · Extreme climatic events · Crop simulation models · Mitigation and adaptation strategies · Sustainable agriculture

3.1 Concerns of Indian Agriculture

Indian agriculture is highly intensive, mostly small to marginal farmers, numerous agro-ecologies & production environments, several crops /cropping systems and presently showing concerns of yield stagnation or decline. The factor productivity of prime inputs is going down. Soil fertility, mostly secondary and micronutrients, is going down. There is a need to identify options for sustainable agriculture by ensuring safeguarding the environment. There exists a large yield gap (attainable minus actual yields) for important crops and we need to identify various biotic and abiotic stresses responsible for this gap. Newer insects/pests are appearing and we need to check the yield losses due to insects/pests through IPM options. Rapid land use and cover change is also the major concern, and we need to take care about the concerns of the small/marginal farmers. Enhanced industrial growth and increased population is causing concerns related to soil, air and water pollution. Land degradation viz. decline in soil fertility, increased soil salinity/sodicity/water-logging, decreased soil biological activity, increased solid wastes, are prime concerns and raises the need for proper management for agri-sustenance. Ground water table decline in most of the districts is noticed. We have to communicate appropriate agri-knowledge to the farmers for meeting the challenges of sustaining agricultural production. There is a need to sequester carbon content in our soils, could be through crop residue incorporation in soil, adoption of suitable land preparation/tillage options, organic farming. There is a need to develop regional platform for operational agri-knowledge dissemination and crop production monitoring and yield forecasting.

3.2 Climate Change and Its Variability

Crop growth and yield is influenced by several biotic and abiotic stresses. Among abiotic factors, inter-annual and intra-seasonal climatic variability is a concern. Occurrence of extreme climatic events (floods, drought, extreme high/low temperatures and aerosol) have increased in recent years, and farmers have to be trained for meeting these extreme situations through adoption of appropriate inputs and agronomic management options (Kalra et al. 2008). Climate change is a major concern, as noticed from historic long term historic-weather trends and through the future climate change scenarios as outputs from various GCMs. Future climate change scenarios indicate rise in temperatures, decreased solar radiation, increase in carbon dioxide concentration, mixed trend of change in rainfall (rainy days and intensity), which may adversely affect the growth and yield of crops. Impact of climate change on agriculture has been seen through delineating vulnerable regions and identification of suitable mitigation and adaptation strategies for sustained agricultural productivity.

Aerosol presence in the atmosphere has increased in the past, arising due to crop residue burning for saving delay in sowing, forest fires and rapid industrialization. Higher concentration may reduce the direct incoming solar radiation, whereas the diffused component reaching the earth's surface might increase. Overall, the event may cause reduction in total radiation, which may cause coupled weather phenomenon changed. Reduction in radiation may reduce net assimilation and subsequent reduction in yields to some extent.

Technical coefficients, as generated through quantification of inter-annual climatic variability through growth and yield of important crops, could be linked with regional climate change scenarios for assessing the impact for defining the vulnerable regions. Future climate change scenarios, although available on higher resolution for regional scales, but GCMs/RCMs needs to be perfectly validated for present day climate, and the other bio-physical and socio-economic aspects to be realistically integrated within the models for better future climate predictions.

3.3 Crop Simulation Approach to Assess Soil and Crop Processes

Growth of crops is a complex interaction of several biotic and abiotic stresses, and difficult in combining these factors to assess the growth and yield of crops/cropping systems. Crop models help to a great extent in integrating these factors, mechanistically as well dynamically. These models may be used as a tool for the grower to assist in decisions on management operations (e.g. in scheduling of irrigation, fertilizer application and crop protection), or to be used in process control (e.g. in climate control in greenhouses). A crop model may also be used for yield forecasting by also linking with remote sensing / satellite pictures. It may be applied in land use evaluation and planning e.g. to access the production potentials of new cropping areas in dependence of availability of water and fertilizer. Most of the time, these models have been used for inputs and resource management on point scales and expressed for regional analysis on crude scale, by not capture of spatio-temporal variations in inputs. There is a need to integrate relational layers of bio-physical and socio-economic aspects on regional scales with the decision tools, such as crop models, for realistic predictions.

Crop simulation tools (viz. WTGROWS, INFOCROP, CERES-DSSAT) help in evaluating growth and yield of crops and predicting the impact of climate change and its variability on soil and crop processes (Aggarwal et al. 1994; Aggarwal et al. 2006). These tools have successfully been used in the past on global scales, and have been time to time modified to cater the present needs of the research community. Most of the dynamic models include processes such as phenological development, water and nutrients balance components, assimilation & partitioning behavior, source-sink relationship & yield formation.

In the recent past, crop models have been linked with remote sensing and relational layers of bio-physical and socio-economic aspects for yield forecast, but on limited regional scales and there is a need to design the operational platform of yield forecast, well in advance, on national scale, may be on clusters operational/integrated on state levels. Presently a joint initiative by Indian Meteorological Department and ISRO, in association with SAUs and agricultural research institutes has been undertaken in this regard.

In terms of soil fertility and associated effects on growth and yield of crops, most of the crop models include nitrogen stress effects, with exception of DSSAT where N, P and K have been included in the present version, although need proper calibration and validation over regions for P and K. In the present day context, S, Zn, B, Fe, Mn and other nutrients have shown signs of depletion within the soils, in particular for south asia and south east asia. There is a need to integrate their associated effects for realistic estimate of growth and yield of crops and for suggesting proper nutrients' management. The possibility of their inclusion might be to address these nutrients and their interactions empirically, and also outside the main body of the program. Soil health indices, for addressing physical, chemical and biological condition of the soil and pedo-transfer functions for deriving moisture and nutrients availability from easily determinable soil parameters need to be developed and validated for various regions, before linking with the crop growth models for various applications. Processes related to problem soils (salinity/sodicity/waterlogging/acidic soils) need effective integration, within the dynamic window of the crop models. Socio-economic and insects-pests components are relatively weak in most of the existing crop models, and needs to be effectively integrated, raising the need of advanced computational methods, artificial intelligence tools and neural network for dealing with the complex agricultural systems.

3.4 Evaluating Climate Change and Its Variability Through Use of Crop Models

Climatic variability (inter- as well intra-) needs to be characterized for various production environments and to be linked with the soil and crop processes and subsequent growth and yield of crops and cropping systems. We need to learn for identifying options for agri-sustenance under extreme/episodic situations. Crop models effectively help in desiging the appropriate options (Singh and Kalra 2016).

Climate is not the only stress factor affecting agricultural production, there are other biotic as well abiotic stresses, which need to be included for evaluating the response and suggesting suitable technologies for maximizing crop production. The task becomes tedious, but need to be handled using advanced mathematical procedures. DSS platform to be designed by using relational layers and crop models.

Future climate change scenarios varies significantly amongst themselves in predicting the future climate change, needs regional validation to get the proper estimates of climate change for subsequent linkages with the crop models and other decision tools for impacts evaluation. Use of crop simulation models for climate change-agricultural impacts evaluation have been too encouraging, and reflects the vulnerable regions, addressing the shifting of the productivity centres in future. Along with climate change scenarios, the scenarios of other bio-physical and socio-economic aspects to be generated and linked. Socio-economic drivers of future change are rather poorly predicted for specifically for developing countries with intensive agriculture. We illustrate few of the research results on climate change impact analysis through use of crop simulation models.

Aggarwal et al. (2010) carried out simulation for rice and wheat crops in upper Ganga Basin using Info Crop and indicated that rice and wheat crops are likely to be affected more in the A2 scenario than in the B2 scenario for 2080 AD of PRECIS, IITM, Pune, India. Regions with relatively lower temperatures may be benefitted by 1 degree C rise.

Effect of sowing date on wheat yield in various meteorological-sub divisions was evaluated using WTGROWS-model (Rai et al. 2004). Results indicated that optimum date of sowing for wheat was spatially dependent, mainly due to prevailing climatic conditions. Simulation results indicated significant reduction over the optimal yields due to early or delay sowing of wheat from optimal dates (about 1% reduction in yield per day delay in sowing, (Aggarwal and Kalra 1994). Results also indicated that delayed sowing of wheat with increased temperature is one of the adaptation strategies for sustaining yield.

Effect of increase in seasonal temperature on grain yield of different crops, evaluated on the basis of district wise data in north-west India, indicated differential response to crops (Kalra et al. 2008). The extent of reduction in wheat yield due to temperature rise was highest (4.29q/ha/°C). Variation in amounts received in rains during winter in the Indo-Gangetic Alluvial Plains has pronounced effect on growth and yield of wheat crop, specially under rain fed and limited water supply situation. WTGROWS model was used to evaluate the interaction of temperature rise and variations in amount of winter-rains in various production environments and it could be concluded that the adverse-effect of temperature rise could be nullified through higher amounts of winter-rains received.

By running INFOCROP for wheat for New Delhi environment, it was found that with adequate moisture supply situation, N-input (>150 kg/ha) will provide benefits, whereas with limited amount of water availability, relatively lower amount of Nitrogen fertilizer is required. Potential yield of wheat in different growing regions was simulated by the INFOCROP-model using the historic weather datasets on meteorological sub-division scales. Results clearly showed the decrease in potential yield of wheat moving eastward and southward from Punjab, clearly indicating the dependence on prevailing temperature (Aggarwal and Kalra 1994).

The results of INFOCROP indicated that reduction in radiation upto 15% was not having much effect on wheat yield. With 40% decrease in the radiation, the yields reduced by around 10%. Diffused radiation may be more photo-synthetically active, and this feature has still to be included in most of the existing crop growth models, as the existing crop models do not differentiate between direct and diffused radiation (Kalra et al. 2006).

Studies conducted by Kalra et al. (2008), indicated significant interaction of radiation with temperature rise, by running Infocrop for wheat, it was noticed that increased radiation compensated the effects of temperature rise. Similar results have been reported in carbon dioxide and temperature interaction. With temperature rise, reduction in yield of crops is expected in most of the locations in south Asian environment. CO2 elevation may nullify the ill-effects of temperature rise to some extent. There is a need to evaluate the impact on regional basis by integrating crop models with relational layers of other biotic and abiotic stresses.

Agriculture offers promising opportunities for mitigating GHGs emissions through carbon sequestration, appropriate soil and land-use management, and biomass production. Most of the agricultural activities, carried out by the farming community to increase the sustainability of production systems, may qualify for the Kyoto Mechanisms and earn carbon credits. Policy should be formulated to encourage farmers for adopting the mitigation technologies with compromising production and income to some extent. The question of life cycle assessment of fertilisers/pesticides would aid in understanding the environmental impact on short/long term basis. Efficient methods, viz. Resource Conservation Technologies for enhanced output/input energy balance ratio will control the GHGs emission as well sustain the agricultural productivity through safeguarding the soil-plant-atmosphere health (Pathak and Aggarwal 2012).

Rice based cropping system is one of the major contributor to GHG emission, specially methane Mitigation options for control of GHGs emission to be effectively introduced. Models, as DSSAT, INFOCROP, DNDC, CENTURY evaluate GHGs emission under various land use types, and enables us to identify options for mitigating the climate change. And it has been established that alternate wetting and drying during growth of rice crop may reduce methane emission significantly when compared to flooded condition for rice growth. Choice of crops and cropping systems, fertilizers management play key roles in reducing the emissions.

Organic carbon in most of the South Asian soils is declining, possibly due to prevailing high temperature resulting in more degradation. There is a need to sequester more carbon within the root zone, through appropriate tillage management, following resource conservation technologies and through choice of crops and cropping systems. Lot of work has been done in this regard, but the process based models to compute the C – Sequestration although exist but needs extensive calibration and validation before talking to application in climate change studies. Newer emerging insects-pests, associated yield losses in relation to climate change and other stress attributing factors are addressed in models viz. Infocrop, DSSAT through forcing of percent infestation at a particular time, but there is a need to design the insect-pest population dynamics and associated crop losses as subroutines within the simulation models.

Identification of suitable adaptation strategies to combat climate change for agri-sustenance is effective through use of simulation models, some examples demonstrated in the works of Aggarwal and Kalra (1994), through altered date of sowing, choice of crops/cropping systems, altered water and nutrient management strategies. There is a need to on national scale, by integrating the simulation results from various agro-ecologies and production environments.

Scope of forestry- and agro-forestry based models to be explored in depth for expanding the sphere of application in climate change studies. In these sectors, the major elements could be land use and its cover change, net primary productivity, carbon sequestration, soil health aspects, water productivity, soil biotic population and associated insects-pests dynamics, which could be related to climate variability, climate change and other biotic and abiotic drivers. Various models in this regard are available, evaluating the impacts on short- and long-term basis, but need to be validated for South Asian regions extensively.

Integrated assessment modeling (including intra- and inter-sectors) is the present day need to evaluate the extent of land use and land cover change, drivers of change.

3.5 Conclusion

Climatic variability is a concern in tropical and sub-tropical regions, and we have to evaluate through growth and yield of crops and cropping systems. There is also need to identify suitable options for sustenance. Future climate change is noticed through historic weather analysis, and future climate change scenarios developed over the globe by use of GCMs and RCMs. Apart from climate change, there are other bio-physical and socio-economic drivers contributing to changes in system's productivity and health. Crop simulation models are effective in simulating the impacts of climate change and its variability on growth and yield of crops and cropping systems. Vulnerable regions arising due to climate change can be identified through use of crop modeling. Suitable adaptation and mitigation options are successfully demonstrated through use of crop models. There is a need to analyze the impacts on regional scales by integrating crop models with relational layers of bio-physical and socio-economic aspects.

References

Aggarwal, P. K., & Kalra, N. (1994). *Simulating the effect of climatic factors, genotype and management on productivity of wheat in India* (p. 156). New Delhi: IARI Publication.

Aggarwal, P. K., Kalra, N., Singh, A. K., & Sinha, S. K. (1994). Analyzing the limitations set by climatic factors, genotype, water and nitrogen availability on productivity of wheat. I. The model description, parameterization and validation. *Field Crops Research, 38*, 73–91.

Aggarwal, P. K., Kalra, N., Chander, S., & Pathak, H. (2006). Info crop: A dynamic simulation model for the assessment of crop yields, losses due to pests and environmental impact of agro-ecosystems in tropical environments. I. Model description. *Agricultural Systems, 89*, 1–25.

Aggarwal, P. K., Naresh Kumar, S., & Pathak, H. (2010). Impacts of climate change on growth and yield of rice and wheat in the upper Ganga Basin. *WWF India Studies*, 36.

Kalra, N., Chakraborty, D., Sahoo, R. N., Sehgal, V. K., & Singh, M. (2006). For assessing yields under extreme climatic events using crop simulation models: Aerosol layer effects on growth and yield of wheat, rice and sugarcane. In R. J. Kuligowski & J. S. Parihar (Eds.), *Agriculture and Hydrology Applications of Remote Sensing*. Genya Saito: Proceedings of SPIE. Vol. 6411–58.

Kalra, N., Chakraborty, D., Sharma, A., Rai, H. K., Jolly, M., Chander, S., Ramesh Kumar, P., Bhadraray, S., Barman, D., Lal, M., & Sehgal, M. (2008). Effect of increasing temperature on yield of some winter crops in north-west India. *Current Science, 94*(1), 82–88.

Pathak, H and Aggarwal, PK. (Eds.) (2012). Low carbon Technologies for Agriculture: A study on Rice and wheat Systems in the Indo-Gangetic Plains. *Indian Agricultural Research Institute.*, p. xvii + 78.

Rai, H. K., Sharma, A., Soni, U. A., Khan, S. A., Kumari, K., Chander, S., & Kalra, N. (2004). Simulating the impact of climate change on growth and yield of wheat. *Journal of Agricultural Meteorology, 6*(1), 1–8.

Singh, K. K., & Kalra, N. (2016). Simulating impact of climatic variability and extreme climatic events on crop production. *Mausam, 67*(1), 113–130.

Chapter 4
Are GCMs and Crop Model Capable to Provide useful Information for Decision Making?

Lalu Das, Hanuman Prasad, Jitendra Kumar Meher, Javed Akhter, and S. Sheraz Mahdi

Abstract Now-a-days simulations from Global Climate Models (GCMs) are extensively used for the purpose of impact assessment and decision making vis-à-vis crop simulation models are run to study the impact of climate change on agriculture sector. The present paper investigated how these models are able to provide useful information for policy making at sub-regional or local scale. The district level past rainfall over West Bengal has been estimated from the monthly district level IMD data through dividing the whole last century into four periods 1901–30, 1931–60, 1961–90 and 1991–2000 and percentage change of rainfall has been calculated with respect to the base period 1971–2000. The annual rainfall trend of 100 years data has revealed that all the districts except Nadia, Jalpaiguri, South Dinajpur and Bardhaman have shown a positive trend. Except Jalpaiguri, South Dinajpur and Nadia, all other districts have received an increased amount of rainfall during monsoon season. Apart from observed station data, how different GCMs from CMIP5 reproduces the observed increasing/decreasing rainfall trends has been judged through conventional statistics. Finally a group of better performing models have been identifed for North and South bengal seperately which can be used as an input for any decision making research. In addition, the future rainfall analysis using the better performing GCMs has revealed a significant increasing trend of seasonal (except winter) andannual rainfall at the end of 21st century . Analysis of percentage change of rainfall reveals that winter rainfall has shown 60–117% surplus in future for different RCP scenarios, whereas the post-monsoon has shown a 1–15% surplus of rainfall in future over North Bengal. On the other hand, 16-25%

L. Das (✉)
Department of agricultural Meteorology and Physics, Bidhan Chandra KrishiViswavidyalaya, Mohanpur, India

H. Prasad · J. K. Meher · J. Akhter
Department of Physics, Jadavpur University, Kolkata, India

S. Sheraz Mahdi
Mountain Research Centre for Field Crops, Sher-e-Kashmir University of Agricultural Sciences and Technology of Kashmir, Khudwani, Anantnag, Kashmir, J & K, India

S. Sheraz Mahdi (ed.), *Climate Change and Agriculture in India: Impact and Adaptation*, https://doi.org/10.1007/978-3-319-90086-5_4

deficit inannual rainfall has been projected over South bengal whereas there may be 3-24% surplus in postmonsoonrainfall. Such a change can alter the productivity of rice as assessed through Oryza 2000, a crop simulation model for rice. Gangetic West Bengal is expected to experience nominal change in the either sides as rice production increases by about 10% with 1 °C rise in temperature and falls by 30% with a rise of temperature by 2 °C.

Keywords West Bengal · Rainfall · Oryza2000 · Future projection · GCM · RCP

4.1 Introduction

Climate change, as a result of increased greenhouse gases in the atmosphere, is one of the important global environmental issues. Its regional assessment is highlighted in the different assessment reports particularly assessment report 4 and assessment report 5 of the Intergovernmental Panel on Climate Change (IPCC), as the vulnerabilities are expected to be more in the developing countries at sub-regional and local scales. On the basis of amount of rainfall received in different months there are four major seasons in India; Winter (Dec, Jan, Feb), Pre-monsoon (Mar, Apr, May), Monsoon (June, July, Aug, Sep) and Post-monsoon (Oct, Nov). Indian summer monsoon (ISM) is the major component of the Asian summer monsoon, which significantly affects the lives of more than 60% of the world's population. ISM provides more than 80% of the total annual rainfall in India and directly relates to water resources, agriculture, ecosystem, health, and food security.

Regional averaged rainfall analysis over the whole Indian subcontinent reveals that, rainfall in annual and monsoon has decreased while there is a sign of increment in the winter, pre-monsoon and post-monsoon season. In the context of West Bengal (WB) state of India, trend analysis of rainfall shows that regional averaged monsoon rainfall in the Southern part (South Bengal) has a significant increasing trend (Slope = +91.0 mm/100 year) and the Sub-Himalayan WB has a non-significant increasing trend (slope = 57.2 mm/100 year) during the period 1901–2003 (Guhathakurta and Rajeevan 2008). A similar type of study by Naidu et al. (2009) shows that the South Bengal has an increasing trend while Sub-Himalayan WB has as a decreasing trend of monsoon rainfall during the period 1871–2005. No long-term trends were found for the premonsoon rainfall over Gangetic WB and its neighbourhood during the period 1901–1992 (Sadhukhan et al. 2000), while during the period 1901–2003, Guhathakurta and Rajeevan (2008) reported a non-significant decreasing trend (slope = −6.7 mm/100 year) of premonsoon rainfall over South Bengal whereas Sub-Himalayan WB has a non-significant increasing trend (slope = +10.5 mm/100 year). During the winter and post monsoon season the same study has reported a non-significant decreasing trend of rainfall for Sub-Himalayan WB, where as South Bengal has shown increasing trend in post monsoon and decreasing trend in winter. The total annual rainfall over the state has an overall non-significant increasing trend.

Seasonal rainfall distribution over WB shows that, winter is the driest season of the year which contributes only 1–3.5% of the average annual rainfall. The hottest premonsoon season receives 8–17% of average annual rainfall, which is mostly associated with the premonsoon thunderstorm or 'Norwester' also popularly known as '*Kalbaisakhi*'.Bay of Bengal branch of the Indian Ocean monsoon brings 73–80% monsoon rain to the state from June to September. Monsoon rainfall has played a vital role in the last three decade as during 1970–2000 there has been a significant increase in productivity of food grains over the state of WB, and the state now has a surplus of food (Raychaudhuri and Das 2005). The post-monsoon season shares 5–15% of average annual rainfall.

Evaluation of the sensitivity of crop productivity to climate at regional and local level can be done with the help of crop models, in particular, crop-weather models. Crop-weather models are mathematical devices that can be used over a range of time scales to relate crop variables to meteorological variables. Two broad classes of crop-weather model can be distinguished. First, empirical statistical models, which relate a sample of crop, yield data to a sample of weather data for the same area and time period, using statistical techniques such as regression analysis. The second category is the crop simulation models, which express the dynamical relationship of crop growth such as photosynthesis, respiration, or translocation of assimilates and environmental and management factors such as climate, soils and cultivation practices. ORYZA 2000 (Bouman et al. 2001) model, which predict crop performance under different environment, have been extensively used in recent years to assess the potential impacts of greenhouse gas induced climate change on agricultural productivity.

There has been a plenty of studies based on the distribution of rainfall over the whole of WB and its neighborhood in different season. But, there is a scattered picture of exact rainfall distribution at district level. This study has been attempted to address the statistical distribution of rainfall over different districts of WB in annual and seasonal scale for a better understanding of the stakeholders, scientific community and for the general public. This study also address the simulations of rice yield under different conditions for a particular variety, IET-4784, using Oryza2000.

4.2 Materials and Methods

4.2.1 Description of the Study Area and Data Used

The present study has been considered over the state of WB. It is extending from 21.20^{0}N to 27.32^{0}N latitude and 85.50^0 to 89.52^0 E longitudes in the eastern part of India. Six districts from Northern part namely Darjeeling, Jalpaiguri, Cooch Behar, North Dinajpur, South Dinajpur and Malda have been collectively termed as North Bengal while rest of the thirteen districts of Southern part namely Birbhum, Murshidabad, Bardhaman, Nadia, Purulia, Bankura, Hooghly, North 24 Parganas, West Midnapore, Howrah, Kolkata, South 24 Parganas and East Midnapore have been together denoted as South Bengal. Monthly total rainfall data for 19 districts of WB state have been collected from India Meteorological Department (IMD), Pune during the period 1901–2000. This data has been used as reference observation for the present study.

For the purpose of impact study using Oryza2000 model, the six numbers of weather variables have been collected from Bidhan Chandra Krishi Viswavidyalaya's (BCKV) observatory, located at Mohanpur campus. Crop management data have also been collected from the experimental farm of BCKV. The experiment was laid out in a factorial randomized block design (Replication: 3; Number of treatment: 15; plot size: 5 m by 3 m; spacing: 20 cm by 10 cm; total number of plot: 45; Net plot size: 4 m by 2 m; seed rate: 15 kg/ha for hybrid i.e. for 1 seedling/hill and 60 kg/ha for H.Y.V i.e. 3 seedlings/hill; number of varieties: 5)

Time series analysis has been carried out for each district and their regional averaged (averaged over all available districts) both for North and South Bengal. Linear regression based trend test has been employed to find out monotonic trend in the rainfall time series.

4.3 Results and Discussion

4.3.1 Past Rainfall Trends

In general rainfall change trends during 1901–30 period have indicated a positive increasing trend of rainfall during winter over all the districts of North Bengal except CoochBehar while the South Bengal district like Birbhum, Murshidabad and Purulia districts have recorded positive trend and remaining 10 districts have shown the negative trend. The long-term linear trend analysis (1901–2000) of winter rainfall showed all districts of North Bengal and 6 districts of South Bengal recorded the positive change while the remaining 7 districts of South Bengal recorded a negative change.

In the South Bengal only one district Nadia (−38.14 mm) has shown decline trend whereas in Both North and South Bengal South 24 Parganas has recorded maximum increasing trend of rainfall (+487.47 mm).

On the other hand the annual rainfall during 1901–2000, in the North Bengal Jalpaiguri (−134.28 mm) and South Dinajpur (−59.54 mm) districts has shown a decreasing trend in rainfall whereas remaining 4 districts has recorded increasing trend. In the South Bengal, only Bardhaman (−52.18 mm) and Nadia (−161.60 mm) districts showed decline trend whereas in Both North and South Bengal South 24 Parganas recorded maximum increasing trend of rainfall (+671.73 mm).

4.3.2 Spatial Distribution of Rainfall Across North and South Bengal

The distribution of mean rainfall in different short-term and long-term periods is not uniform. As for example the monsoon rainfall (Fig. 4.1) during 1901–1930, shown a decreasing trend in Birbhum, Murshidabad, Nadia, some part of Darjeeling, Cooch Behar, and North Dinajpur districts. While the southern part of the Bengal

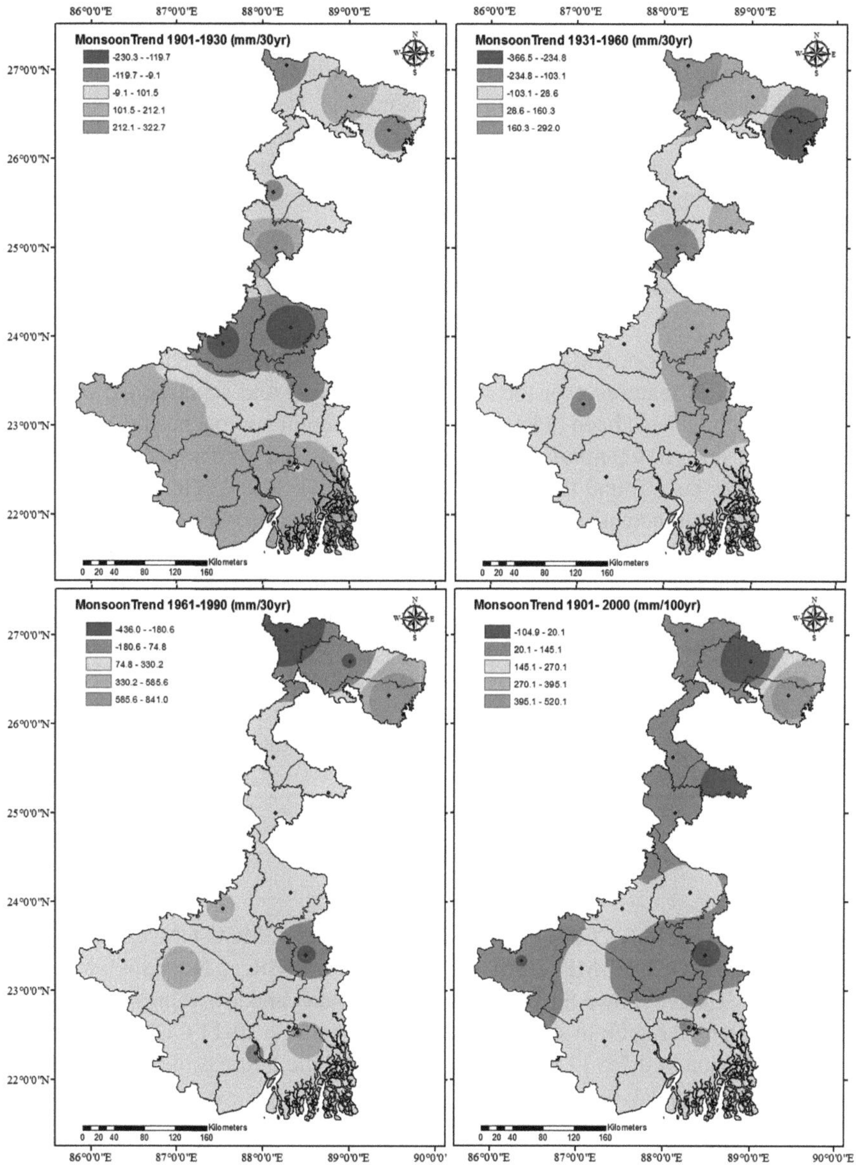

Fig. 4.1 Spatial variations of monsoon season mean rainfall in different time periods

like from Purulia to South 24 Parganas and some part of North 24 Parganas showed the increasing trend (101.5 to 212.1 mm/30 years). During the long-term period from 1901 to 2000 only two districts namely Cooch Behar and a very small part of the South 24 Parganas showed the highly increasing trend. In the Bengal during this time period, all districts showed the increasing trend except Jalpaiguri and some part of the South Dinajpur, Nadia, and Bardhaman districts. The increasing trend of West Midnapore, East Midnapore, Howrah and South 24 Parganas, some part of the North 24 Parganas, Hooghly, Birbhum, Murshidabad, Darjeeling and Jalpaiguri was from 164.1 to 326.6 mm/100 years. Similarly the spatial distributions of annual rainfall change trneds during different time periods are shown in Fig. 4.2.

4.3.3 Model Evaluation Over North and South Bengal

How the 42 numbers of CMIP5 GCMs (Table 4.1) have been able to reproduce the observed rainfall features over the whole West Bengal and separately over North and South Bengal have been judged using three different approaches i) Comparison of seasonal cycles ii) Comparison of long-term trends and iii) comparison of spatial patterns. It has been found that majority of the models have been unable to reproduce the observed increasing trends of annual and monsoon season adequately over north and south Bengal regions (Table 4.2–4.3). Based on long-term trends and values of statistical indices, we have seen that some models overestimated the mean observed seasonal cycles whereas some underestimate the same but all models underestimate the magnitudes of annual and monsoon rainfall compared to observation (Fig. 4.1 and 4.2). To select how models have been able to simulate seasonally observed rainfall, we have calculated the spatial correlation considering the 19 district values during the period 1901–2000 (results have not been shown here).

4.3.4 Better GCMs for North Bengal

11 models i.e. CCSM4, CESM1-BGC, CESM1-CAM5, CMCC-CM, CMCC-CMS, inmcm4, MIROC5, MPI-ESM-LR, MPI-ESM-MR, NorESM1-M and NorESM1-ME have shown the high value of agreement and low values error indices over North Bengal region in simulating annual rainfall. So all-11 GCMs or any suitable combination of these 11 models can be used for the future projection of rainfall over the North Bengal.

In case of monsoon season the 12 GCMs namely bcc-csm1-1, bcc-csm1-1-m, CCSM4, CESM1-BGC, CMCC-CM, CMCC-CMS, GFDL-CM3, inmcm4, MIROC5, MPI-ESM-MR, NorESM1-M and NorESM1-ME; in case of premonsoon 15 GCMs namely CESM1-CAM5, CSIRO-Mk3-6-0, FIO-ESM, GFDL-ESM2-M, GISS-E2-H-1, GISS-E2-R-1, GISS-E2-R-2, HadGEM2-AO, IPSL-CM5A-LR, MIROC5, MIROC-ESM, MIROC-ESM-CHEM, NorESM1-ME, MPI-ESM-LR and MPI-ESM-MR; in case of post monsoon season, 12 GCMs namely FIO-ESM, GFDL-CM3, GFDL-ESM2G, GISS-E2-H-2, HadGEM2-ES, IPSL-CM5A-LR, MIROC-ESM, MIROC-ESM-CHEM, MRI-CGCM3, NorESM-1, MPI-ESM-MR, IPSL-CM5B-LR; and finally in case of winter season, 4 GCMs

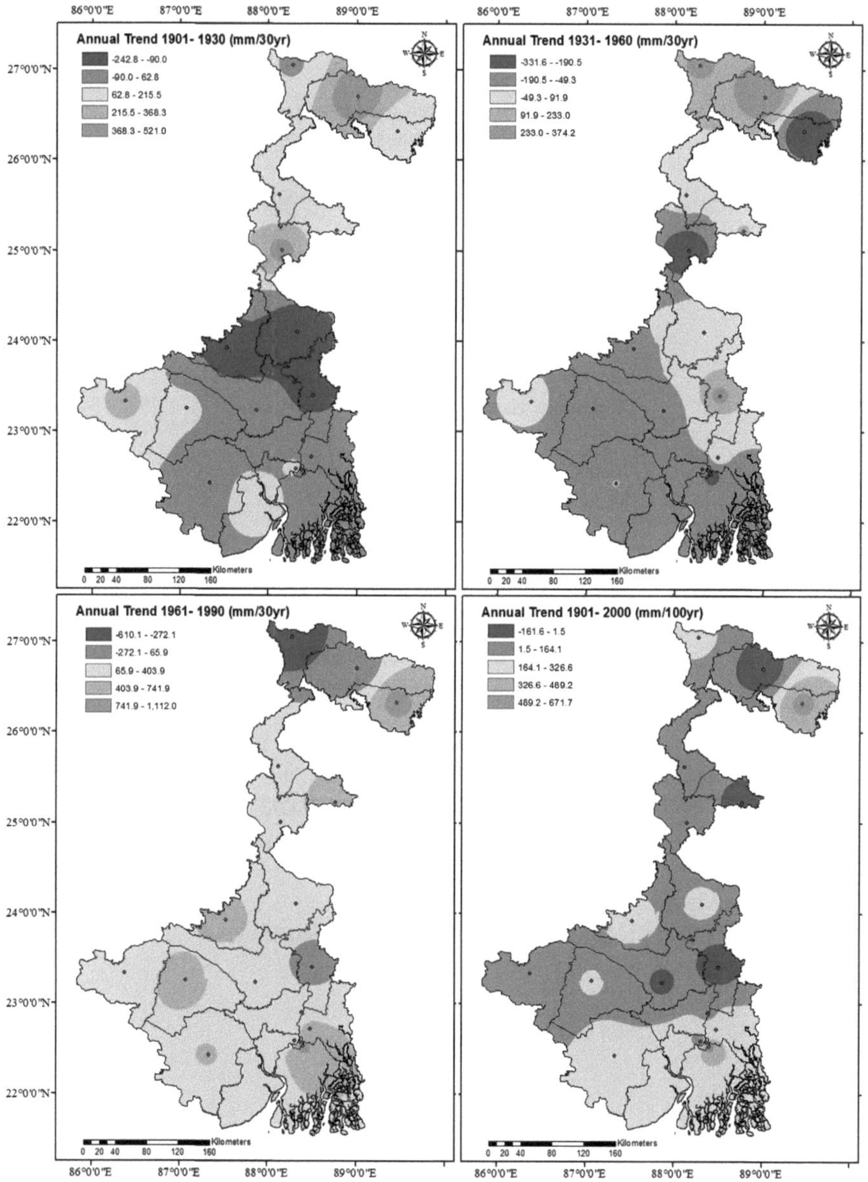

Fig. 4.2 Spatial variations of monsoon season mean rainfall in different time periods

namely CESM1-CAM5, GFDL-ESM-2G, GFDL-ESM-2 M and NorESM1-ME are selected for developing the future projection rainfall change information for any decision making and long-term developmental plan over the study area. For selecting those season-wise GCMs, more attention was put on the comparison of observed trends with GCMs simulated trends as well as values of spatial correlation.

Table 4.1 All CMIP5 GCMs taken in this study

S.N	Model	Institute	(Lon(degree) × Lat (degree))
1	ACCESS1-0	Commonwealth Scientific and Industrial Research Organization-Bureau of Meteorology (CSIRO-BOM), Australia	1.88 × 1.25
2	ACCESS1.3		
3	bcc-csm1-1	Beijing Climate Center, China Meteorological Administration	2.8 × 2.8
4	bcc-csm1-1-m		1.12 × 1.12
5	BNU-ESM	College of Global Change and Earth System Science, Beijing Normal University, China	2.8 × 2.8
6	CanESM2	Canadian Centre for Climate Modeling and Analysis	2.8 × 2.8
7	CCSM4	National Center of Atmospheric Research, USA	1.25 × 0.94
8	CESM1-BGC	Community Earth System Model Contributors	1.25 × 0.94
9	CESM1-CAM5		
10	CMCC-CESM	Centro Euro-MediterraneopelCambiamentIClima tici	3.75 × 3.71
11	CMCC-CM		0.75 × 0.75
12	CMCC-CMS		1.88 × 1.87
13	CNRM-CM5	National Centre of Meteorological Research, France	1.4 × 1.4
14	CNRM-CM5-2		
15	CSIRO-Mk3-6-0	Commonwealth Scientific and Industrial Research Organization Queensland Climate Change Centre of Excellence, Australia	1.8 × 1.8
16	EC-EARTH	EC-EARTH consortium	1.13 × 1.12
17	FGOALS-g2	LASG, Institute of Atmospheric Physics, Chinese Academy of Sciences	2.8 × 2.8
18	FGOALS-s2		2.8 × 1.7
19	FIO-ESM	The First Institute of Oceanography, SOA, China	2.81 × 2.79
20	GFDL-CM3	NOAA Geophysical Fluid Dynamics Laboratory, USA	2.5 × 2.0
21	GFDL-ESM2G		
22	GFDL-ESM2M		
23	GISS-E2-H	NASA Goddard Institute for Space Studies, USA	2.5 × 2.0
24	GISS-E2-H-CC		
25	GISS-E2-R		
26	HadCM3	Met Office Hadley Center, UK	3.75 × 2.5
27	HadGEM2-AO		1.88 × 1.25
28	HadGEM2-CC		1.88 × 1.25
29	HadGEM2-ES		1.88 × 1.25
30	INMCM4	Institute for Numerical Mathematics, Russia	2.0 × 1.5
31	IPSL-CM5A-LR	Institut Pierre Simon Laplace, France	3.75 × 1.8
32	IPSL-CM5A-MR		2.5 × 1.25
33	IPSL-CM5B-LR		3.75 × 1.8
34	MIROC5	Japan Agency for Marine-Earth Science and Technology, Atmosphere and Ocean Research Institute (The University of Tokyo), and National Institute for Environmental Studies	1.4 × 1.4
35	MIROC-ESM		2.8 × 2.8
36	MIROC-ESM-CHEM		2.8 × 2.8

(continued)

Table 4.1 (continued)

S.N	Model	Institute	(Lon(degree) × Lat (degree))
37	MPI-ESM-P	Max Planck Institute for Meteorology, Germany	1.88 × 1.87
38	MPI-ESM-LR		
39	MPI-ESM-MR		
40	MRI-CGCM3	Meteorological Research Institute, Japan	1.1 × 1.1
41	NorESM1-M	Norwegian Climate Center, Norway	2.5 × 1.9
42	NorESM1-ME		

Table 4.2 Statistical indices between the seasonal cycles of observation and model simulation along with the linear trends during the period of 1901–2000 over north Bengal

SL	Model Name	Indices from seasonal cycles				Trends (mm/100 year) during 1901–2000				
		d-index	Corr.	nrmse	pbias	Annual	JJAS	DJF	MAM	ON
	OBS	1	1	0	0	88.4	68.0	6.9	20.5	−6.0
1	ACCESS1-0	0.98	0.98	25.90	−7.60	−101.6	−101.0	−17.3	25.0	−9.0
2	ACCESS1-3	0.92	0.90	46.90	−13.10	−109.8	−144.6	−12.4	52.0	−5.6
3	bcc-csm1-1	0.77	0.85	73.60	−46.00	−6.0	1.0	−23.8	−28.4	43.0
4	bcc-csm1-1-m	0.67	0.97	81.00	−50.70	116.7	40.3	17.2	18.4	41.2
5	BNU-ESM	0.86	0.98	58.00	−40.00	−49.9	−48.2	18.0	−13.2	−5.5
6	CanESM2	0.86	0.95	61.30	−45.20	−130.1	−144.8	8.6	9.7	−2.8
7	CCSM4	0.94	0.97	39.20	−20.70	62.8	39.9	−7.1	−6.0	35.9
8	CESM1-BGC	0.95	0.96	38.40	−19.70	41.9	5.4	−12.8	65.2	−17.2
9	CESM1-CAM5	0.98	0.96	25.60	3.00	40.3	−44.8	16.8	49.2	20.1
10	CMCC-CM	0.96	0.95	36.20	−20.50	35.0	19.3	21.2	4.7	−8.8
11	CMCC-CMS	0.94	0.97	39.90	−25.20	84.3	53.0	−5.3	12.4	22.6
12	CNRM-CM5	0.92	0.99	44.90	−27.00	−201.4	−107.3	−11.0	−55.8	−27.3
13	CSIRO-Mk3-6-0	0.77	0.86	76.10	−52.10	−16.2	−23.0	−4.3	4.2	7.2
14	EC-EARTH	0.83	0.98	62.80	−42.50	135.3	101.6	2.3	9.7	22.0
15	FGOALS_g2	0.62	0.96	92.70	−63.60	106.5	42.7	12.8	35.0	15.7
16	FIO-ESM	0.86	0.94	59.80	−41.00	−21.8	−31.0	3.8	12.6	−6.0
17	GFDL-CM3	0.80	0.97	65.70	−42.00	−166.7	−115.7	14.6	−8.6	−56.2
18	GFDL-ESM2G	0.93	0.98	42.80	−30.20	39.4	73.9	13.1	−21.0	−28.5
19	GFDL-ESM2M	0.95	0.98	36.10	−23.60	−18.1	−62.7	5.6	20.5	18.8
20	GISS-E2-H-1	0.95	0.91	41.50	−4.00	−391.9	−402.5	2.6	4.4	4.6
21	GISS-E2-H-2	0.95	0.90	42.40	−8.80	−468.8	−434.3	−7.8	−4.4	−21.9
22	GISS-E2-H-3	0.93	0.92	59.00	21.70	−311.7	−283.1	−3.7	−22.2	−2.7
23	GISS-E2-H-CC	0.95	0.91	41.30	−12.60	−169.8	−154.2	2.1	−15.5	−4.5
24	GISS-E2-R-1	0.77	0.89	72.30	−46.90	−143.3	−159.1	1.7	3.7	9.9
25	GISS-E2-R-2	0.75	0.88	75.60	−49.20	−246.2	−227.8	4.3	1.0	−22.6
26	GISS-E2-R-3	0.88	0.90	56.80	−33.20	−77.0	−98.2	3.9	14.8	2.4
27	GISS-E2-R-CC	0.77	0.90	72.10	−46.90	−69.4	−122.3	4.9	22.9	25.8
28	HadGEM2-AO	0.96	0.98	33.40	−14.20	−94.4	−157.1	−2.2	42.2	24.4
29	HadGEM2-CC	0.97	0.98	30.30	−11.90	−251.8	−191.5	−4.5	−46.3	−6.3

(continued)

Table 4.2 (continued)

SL	Model Name	Indices from seasonal cycles				Trends (mm/100 year) during 1901–2000				
		d-index	Corr.	nrmse	pbias	Annual	JJAS	DJF	MAM	ON
30	HadGEM2-ES	0.96	0.98	32.40	−15.10	−179.5	−152.0	−7.4	−8.5	−12.3
31	inmcm4	0.93	0.98	41.20	−23.40	77.7	61.0	−3.5	−3.0	24.1
32	IPSL-CM5A-LR	0.63	0.79	91.80	−61.50	−29.9	−17.9	−7.7	9.4	−13.7
33	IPSL-CM5A-MR	0.63	0.93	92.00	−65.30	−72.5	−45.5	−6.7	−2.5	−19.5
34	IPSL-CM5B-LR	0.55	0.68	104.00	−67.90	−67.8	−102.4	21.4	18.0	−6.3
35	MIROC5	0.98	0.96	28.70	7.70	192.2	118.5	−2.2	60.4	15.4
36	MIROC-ESM	0.81	0.97	65.40	−42.40	−66.4	−48.7	−27.7	30.0	−19.3
37	MIROC-ESM-CHEM	0.80	0.96	66.10	−42.70	23.9	−9.1	0.8	50.2	−18.5
38	MPI-ESM-LR	0.99	0.99	21.80	−15.10	15.4	−37.9	8.0	16.7	28.4
39	MPI-ESM-MR	0.98	0.98	26.80	−18.80	31.2	37.5	−18.7	17.9	−4.4
40	MRI-CGCM3	0.62	0.86	93.90	−65.10	−97.8	−26.8	−22.9	−29.0	−20.5
41	NorESM1-M	0.98	0.97	27.50	−11.70	78.4	58.7	43.2	−8.7	−11.2
42	NorESM1-ME	0.98	0.97	26.90	−10.70	82.1	63.0	−3.1	10.9	15.2

Table 4.3 Statistical indices between the seasonal cycles of observation and model simulation along with the linear trends during the period of 1901–2000 over south Bengal

SL	Model name	Indices from seasonal cycles				Trends (mm/100 year) during 1901–2000				
		D	r	nrmse	pbias	Annual	JJAS	DJF	MAM	ON
	OBS	1	1	0	0	143.1	136.7	0.9	−13.6	22.6
1	ACCESS1–0	0.91	0.93	71.6	26.7	−10.5	−32.4	0.4	24.6	−4.6
2	ACCESS1–3	0.84	0.76	66.7	−23.4	−12.6	−110.1	38.1	51.6	5.1
3	bcc-csm1–1	0.79	0.90	71.0	−43.2	38.7	66.8	−51.2	−9.4	31.2
4	bcc-csm1–1-m	0.69	0.94	74.0	−33.2	117.1	18.4	41.2	6.7	49.7
5	BNU-ESM	0.83	0.98	63.6	−44.4	83.6	69.7	3.0	−13.8	25.8
6	CanESM2	0.96	0.98	34.1	−22.6	−35.0	−46.1	5.9	−3.1	8.3
7	CCSM4	0.98	0.99	24.2	−8.3	46.2	31.8	−0.1	−10.2	24.6
8	CESM1-BGC	0.98	0.99	25.0	−9.7	57.5	41.3	−11.3	19.2	6.4
9	CESM1-CAM5	0.96	0.98	34.5	−15.7	83.7	30.7	16.7	19.6	16.8
10	CMCC-CM	0.88	0.94	56.3	−37.9	−48.1	−25.2	5.9	−18.2	−10.4
11	CMCC-CMS	0.91	0.95	48.6	−29.6	13.9	6.4	−9.1	9.5	7.0
12	CNRM-CM5	0.97	0.98	27.7	−8.8	−104.9	−23.8	−22.4	−48.0	−11.8
13	CSIRO-Mk3–6-0	0.91	0.84	55.6	−13.2	20.1	22.6	−8.4	1.5	4.2
14	EC-EARTH	0.98	0.96	28.1	−3.6	19.1	−3.3	−0.9	−1.7	25.6
15	FGOALS_g2	0.78	0.96	72.6	−48.7	66.1	32.6	−0.6	19.3	14.4
16	FIO-ESM	0.82	0.97	65.0	−42.4	22.6	30.0	6.4	15.9	−30.0
17	NOAA1	0.99	0.98	17.3	−1.0	−127.2	−103.8	6.1	5.2	−34.1
18	NOAA2	0.95	0.97	40.9	−31.0	−25.4	−28.8	6.4	0.7	−5.9
19	NOAA3	0.92	0.97	48.1	−34.3	−52.7	−73.2	−7.9	4.4	24.1
20	GISS-E2-H-1	0.93	0.89	50.6	−21.7	−78.2	−106.3	3.3	17.9	7.2

(continued)

Table 4.3 (continued)

		Indices from seasonal cycles				Trends (mm/100 year) during 1901–2000				
SL	Model name	D	r	nrmse	pbias	Annual	JJAS	DJF	MAM	ON
21	GISS-E2-H-2	0.92	0.88	53.0	−23.6	−234.0	−223.1	−2.3	7.2	−15.4
22	GISS-E2-H-3	0.92	0.91	65.0	5.6	−131.4	−130.6	−1.4	−0.6	1.2
23	GISS-E2-H-CC	0.91	0.87	55.9	−28.2	−101.6	−109.8	0.7	−2.2	8.9
24	GISS-E2-R-1	0.80	0.84	75.6	−50.5	−95.1	−102.4	0.7	2.6	3.6
25	GISS-E2-R-2	0.78	0.82	78.4	−52.3	−131.8	−116.5	0.1	−0.7	−14.6
26	GISS-E2-R-3	0.89	0.87	59.2	−34.8	−19.4	−36.9	1.0	12.4	3.9
27	GISS-E2-R-CC	0.77	0.84	78.1	−52.6	−46.1	−76.2	−0.3	18.0	12.3
28	HadGEM2-AO	0.92	0.91	63.3	14.2	138.9	51.7	11.1	25.2	52.4
29	HadGEM2-CC	0.94	0.90	49.3	1.4	−39.0	19.3	−4.6	−38.6	−14.3
30	HadGEM2-ES	0.94	0.91	53.4	9.3	−43.1	−59.7	13.7	4.8	−2.4
31	inmcm4	0.97	0.94	32.4	−6.6	45.6	37.7	−11.4	−3.0	23.7
32	IPSL-CM5A-LR	0.74	0.89	80.9	−54.3	−55.8	−42.8	−3.6	3.1	−11.9
33	IPSL-CM5A-MR	0.92	0.92	48.7	−29.5	−87.1	−31.7	−16.0	−3.8	−37.6
34	IPSL-CM5B-LR	0.56	0.66	106.5	−68.6	−22.9	−14.2	1.3	−1.2	−8.7
35	MIROC5	0.90	0.93	72.9	36.4	−89.6	−119.6	2.8	8.7	19.2
36	MIROC-ESM	0.92	0.97	48.1	−35.1	−61.5	−34.5	−21.0	−16.0	9.8
37	MIROC-ESM-CHEM	0.91	0.96	50.7	−35.3	−111.7	−67.4	6.1	−26.4	−23.1
38	MPI-ESM-LR	0.91	0.96	50.4	−33.7	49.0	15.0	3.v5	5.1	25.0
39	MPI-ESM-MR	0.90	0.98	50.5	−35.2	4.1	24.5	−12.0	−2.8	−4.7
40	MRI-CGCM3	0.53	0.60	108.5	−67.6	2.7	75.1	−37.9	−9.9	−25.7
41	NorESM1-M	0.92	0.97	47.3	−30.4	47.0	27.9	40.2	15.7	−33.7
42	NorESM1-ME	0.92	0.97	45.6	−29.8	22.0	−18.7	5.3	14.6	19.5

4.3.5 Better GCMs for South Bengal

Based on the above mentioned different statistical measures, 9 model namely, CCSM4, CESM1-BGC, CESM1-CAM5, CMCC-CMS, EC-EARTH, inmcm4,MPI-ESM-LR, NorESM1-M and NorESM1-ME have shown better performances as they have shown the high values of agreement index, lower values of error indices. Therefore, either all the 9 models or any combination of these 9 model scan be used for creation of future rainfall change scenarios over the South Bengal apart from using those any impact research.

In case of monsoon season, the 10 GCMs namely bcc-csm1-1, bcc-csm1-1-m, CCSM4, CESM1-CAM5, CMCC-CMS, FIO-ESM, HadGEM2-AO, MRI-CGCM3, MPI-ESM-LR NorESM1-M; In case of pre-monsoon 10 GCMs namely bcc-csm1-1, CCSM4, CNRM-CM5, GISS-E2-H-3, GISS-E2-R-2, IPSL-CM5A-MR, MIROC-ESM, MIROC-ESM-CHEM, MPI-ESM-MR, MRI-CGCM3; in case of post monsoon season GCMs namely bcc-ccm1-1, CCSM4, CSM1-

CAM5 GFDL-ESM2M, HadGEM2-AO, MIROC5, MIROC-ESM, NorESM1-ME; in case of winter 7 GCMs namely CESM1-CAM5, GFDL-ESM-2G, GFDL-ESM-2 M, GISS-E2-H-1, MIROC5, MIROC-ESM-CHEM, NorESM1-ME are selected for future projection of GCMs on the basis of similar comparison of long-term trend and spatial correlation.

4.3.6 Selection of Suitable GCMs for Future Projection

As all the selected models in different seasons, do not provide simulation of all available RCPs, so a multi-model ensemble is proposed to construct the future climate change scenarios. In case of annual rainfall, the multi-model ensemble is prepared using simulations from CCSM4, CESM1-CAM5, MIROC5, NorESM1-M and NorESM1-ME GCMs. For monsoon rainfall are bcc-csm1-1, CCSM4, GFDL-ESM2G, MIROC5, NorESM1-M and NorESM1-ME; for pre-monsoon rainfall are CESM1-CAM5, CSIRO-Mk3-6-0, FIO-ESM, GFDL-ESM2M, GISS-E2-H1, GISS-E2-R1, GISS-E2-R2, HadGEM2-AO, IPSL-CM5A-LR, MIROC5, MIROC-ESM-CHEM, MIROC-ESM and NorESM1-ME; for post-monsoon rainfall are FIO-ESM, GFDL-CM3, GFDL-ESM2G, GISS-E2-R2, HadGEM2-ES, IPSL-CM5A-LR, MIROC-ESM, MIROC-ESM-CHEM, NorESM1-M; for winter rainfall are CESM1-CAM5, GFDL-ESM2G, GFDL-ESM-2 M and NorESM1-ME.All these mentioned GCMs in different seasons, the simulations for each model is commonly available under all the RCPs. Similar results can be seen over south Bengal which has not been displayed.

4.3.7 Ensemble Based Future Rainfall Change (in %)

The percentage of rainfall with respect to the climatological base period of 1971–2000 has been also calculated for different climatological periods for the 2020s, 2050s and 2080s and results are shown in Table 4.4 and 4.5. The annual rainfall for RCP4.5, RCP6.0 and RCP8.5 is shown increasing trends for 2050 and 2080s while a slightly decreasing also noticed for RCP2.6 over North Bengal. The winter rainfall has shown 60–117% surplus in future for different RCP scenarios, whereas the post-monsoon has shown a 1–15% surplus of rainfall in future over North Bengal.

Similarly, the annual rainfall over south Bengal has projected a deficient precipitation of 16 to 25% under different RCP scenarios and different climatological as well as long-term period while 3–24% surplus of rainfall is projected for different RCP scenarios in the post-monsoon over south Bengal.

Table 4.4 % change of rainfall during annual and monsoon season over North Bengal and south Bengal with respect to 1971–2000

	Annual				Monsoon			
Time period	RCP 2.6	RCP 4.5	RCP 6.0	RCP 8.5	RCP 2 6	RCP 4.5	RCP 6.0	RCP 8.5
North Bengal								
2001–2030	−6.0	−5.3	−4.0	−2.6	−18.9	−17.7	−17.5	−16.6
2031–2060	−5.1	−2.5	−1.3	−3.7	−19.1	−16.3	−16.7	−16.9
2061–2090	−2.5	2.0	−0.1	3.1	−17.3	−15.0	−15.1	−11.1
2091–2100	−1.7	2.1	5.5	10.1	−15.8	−15.0	−10.9	−3.7
2001–2100	−4.3	−1.5	−1.1	0.1	−18.2	−16.2	−15.9	−13.7
South Bengal								
2001–2030	−24.9	−23.5	−22.8	−24.0	−34.4	−33.0	−34.3	−34.1
2031–2060	−23.6	−23.8	−22.9	−23.7	−31.1	−30.9	−32.9	−32.2
2061–2090	−23.1	−19.4	−23.0	−17.8	−30.9	−29.8	−28.9	−25.2
2091–2100	−22.1	−17.0	−19.6	−15.9	−32.9	−28.1	−28.3	−22.9
2001–2100	−23.7	−21.7	−22.6	−21.3	−32.2	−30.9	−31.7	−29.7

Table 4.5 %Change of rainfall during winter and pre-monsoon season over North Bengal and south Bengal with respect to 1971–2000

	Winter				Pre monsoon			
Time period	RCP 2.6	RCP 4.5	RCP 6.0	RCP 8.5	RCP 2.6	RCP 4.5	RCP 6.0	RCP 8.5
North Bengal								
2001–2030	75.5	77.7	98.4	89.9	−30.2	−31.0	−32.8	−29.5
2031–2060	64.4	87.1	74.8	69.3	−26.8	−29.5	−31.1	−29.6
2061–2090	81.3	72.1	67.2	82.5	−28.0	−27.3	−31.5	−22.2
2091–2100	59.4	117.2	69.6	85.2	−28.9	−25.2	−28.7	−19.3
2001–2100	72.3	82.8	79.1	81.0	−28.4	−28.9	−31.5	−26.3
South Bengal								
2001–2030	−24.0	−26.6	−19.3	−16.8	−59.7	−57.9	−57.9	−59.9
2031–2060	−20.9	−22.2	−18.9	−22.6	−59.7	−61.3	−61.3	−57.5
2061–2090	−18.0	−24.2	−27.6	−27.3	−59.5	−54.1	−54.1	−52.8
2091–2100	−23.6	−19.5	−19.2	−23.7	−59.4	−51.2	−51.2	−51.4
2001–2100	−21.2	−23.8	−21.7	−22.4	−59.6	−57.1	−57.1	−56.2

4.3.8 Application of Oryza 2000: Validation and Impact Study

Potential as well as N_2 and water-limited situations are considered for two boro seasons (2000–01). The potential production (~10 t/ha) appears to be more than the highest reported value (~9 t/ha) in West Bengal conditions. Application of N_2 fertilizer under N_2 limited situation assures more yields (~8–9 t/ha) up to a limit of 140 kg/ha. Further increase of N_2 application has less impact on the production, indicating an optimized amount of nitrogen fertilizer as 120–140 kg/ha. On the

Table 4.6 Comparison between the observed and model simulation yields

Year	Field experiment Grain yield (t/ha)			Model simulation Final grain yield (t/ha)		
	N00	N120	N140	N00	N120	N140
1999–2000	2.1	3.5	4.1	4.6	8.6	8.9
2000–01	3.9	5.4	7.0	4.8	8.9	9.0

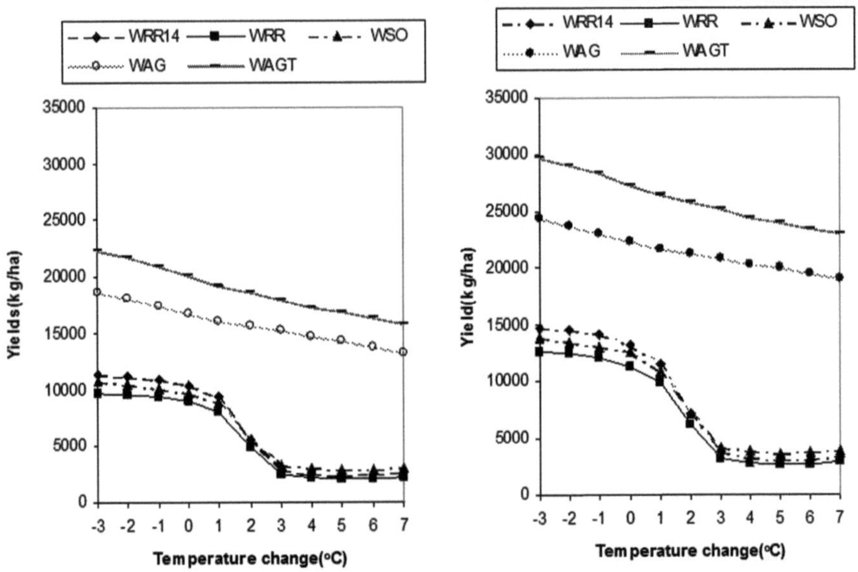

Fig. 4.3 Simulated yield components (kg/ha) for different changing temperature during the crop season 1999–2000 with 1 × CO_2 (left panel) and 2 × CO_2 (Right panel)

other hand, in water-limited situation, predicted yield components are less (~7 t/ha) (Table 4.6) even with the application of optimized amount of N_2 fertilizer.

Verification study indicates that the model estimation is more than the observed values. Keeping the overestimating nature of the model in view, the model has further been used to investigate the impact of climate changes on the production of yield. The variations of yield components under different temperature are shown in Fig. 4.3.

Simulations have been carried out under both the current value of CO_2 concentration and double CO_2 concentration. A part from the usual nature of overestimation by the model, the yield gets reduced with the increase in temperature and vice-versa. The simulated yield is enhanced by about 10% in case of doubled CO_2 concentration and 1 °C temperature increase when compared with the yield at current CO_2 concentration. This is particularly important when a temperature increase of 0.5–2.0 °C has been estimated for the period 2010–39 over the study area.

4.4 Conclusion

In general increasing trends have been found over the districts in South Bengal. There have been noticable decreasing trend(for example 134mm/100 year Jalpaiguri) as well as significant. increasing trend (for example 420mm/100 year) over North Bengal. Model evaluation has indicated that none of the CMIP5 model has been not able to reproduce mean seasonal cycles over north or south Bengal adequately but multi-model ensemble using the better performing models has been able to simulate the observed seasonal cycles quite satisfactorily. Almost 50% of the total 42 model have been able to simulate the increasing rainfall trends in annual, monsoon, winter and pre-monsoon season over North Bengal district while the majority of the models have indicated decreasing trends over south Bengal in annual and monsoon season. It is expected that annual rainfall over north Bengal will increase by 140-348 mm at the end twenty-first century as all RCPs while winter rainfall is projected to be decreased by 1–20 mm at the same time. Similarly annual rainfall will increase by 29–167 mm at the end of twenty-first century over south Bengal. In terms of percentage change of rainfall with respect to the climatological base period of 1971–2000, the winter, pre monsoon and monsoon rainfall will increase both for north and south Bengal as per all RCPs.

Further, sensitivity study has indicated that optimum use of N_2 is 140 kg/ha in this climatic condition with ample supply of water, which assures yield of about 8–9 t/ha. Verification analysis has shown overestimation of the yield production by the model. It has been noticed that production of yield gets reduced with the increase of temperature and vice-versa in the current as well as doubling CO_2 conditions. Increased CO_2 in the atmosphere has indicated higher yield. Combined effect of CO_2 and temperature has indicated less change of rice yield over the study area.

It may be concluded that the present generation GCMs is not provide useful information for impact research or any policy making issues at district or block level due to its courser scale resolution which is unable to capture the details complexity of earth atmosphere system but all these information can provide a realization about the past and future climate scenarios. These GCMs simulation can be more reliably use for local scale decision making after developing down scaled scenarios from the raw GCMs simulation. On the other hand crop model still today cannot be used as operational purpose for day-to-day agricultural practice as it has its inherent biases and limitations to reproduce the crop phenology leading to different yield components adequately.

References

Bouman, B., Kropf, M. J., Tuong, T. P., & Wopereis, M. C. (2001). *ORYZA2000: Modeling lowland rice* (Vol. 1). Los Baños: IRRI.

Guhathakurta, P., & Rajeevan, M. (2008). Trends in the rainfall pattern over India. *International Journal of Climatology, 28*(11), 1453–1470.

Naidu, C. V., Durgalakshmi, K., Muni Krishna, K., RamalingeswaraRao, S., Satyanarayana, G. C., Lakshminarayana, P., & Malleswara Rao, L. (2009). Is summer monsoon rainfall decreasing over India in the global warming era? *Journal of Geophysical Research: Atmospheres, 114*(D24).

Raychaudhuri, A., & Das, T. K. (2005). *WB economy: Some contemporary issues*. Delhi: Allied Publishers.

Sadhukhan, I., Lohar, D., & Pal, D. K. (2000). Pre monsoon season rainfall variability over Gangetic WB and its neighborhood, India. *International Journal of Climatology, 20*(12), 1485–1493.

Chapter 5
Mapping Agriculture Dynamics and Associated Flood Impacts in Bihar using Time-series Satellite Data

C. Jeganathan and P. Kumar

Abstract Agriculture is the prime requirement for sustaining human life on earth, and agriculture sustainability depends on soil health and suitable climatic variations. Human have adopted many local-weather-dependent crop types and its cultivation patterns based on knowledge about long term climatic and environmental conditions. Any anomaly in these factors would result in unforeseen reduction in the food production and associated socio-economic chaos at local/regional to global scale. Due to anthropogenic activities like expansion of urban area, industrialization, deforestation etc. have increased the greenhouse gases (GHGs) level and hence the mean earth surface temperature has increased by 0.74 °C during 1900 to 2000 AD and it is anticipated to rise by 1.4–5.8 °C during 2000 to 2100 AD with notable local differences which would result in increase in the frequency of drought, flood, sea level rise etc. and will drastically affect the crop production. Bihar is one of the fertile regions in India, gifted with numerous water resources like Ganga, Gandak and Kosi and many more rivers. But these rivers are both boon and bane to Bihar because most of the rivers flood during monsoon season. Hence it would be interesting to know the Agriculture cropping pattern over a decade, its changing scenario and the impact of flood on agriculture area in Bihar. In this regard, current study attempted to use time-series remote sensing data from 2001 to 2012 in deriving spatio-temporal, seasonal and annual cropping pattern, and as well as flood scenario purely based on space based observation.

Keywords MODIS EVI time series · Cropping system · Flood impact · Climate change · Bihar · India

---

C. Jeganathan (✉) · P. Kumar
Department of Remote Sensing, Birla Institute of Technology, Mesra, Ranchi, India
e-mail: jeganathanc@bitmesra.ac.in

S. Sheraz Mahdi (ed.), *Climate Change and Agriculture in India: Impact and Adaptation*, https://doi.org/10.1007/978-3-319-90086-5_5

5.1 Introduction

Agriculture is one of the primary activity for maintaining economic and social well-being of Indian population. Although the contribution of agriculture in the employment and Gross Domestic Product (GDP) has declined over time. The contribution of agriculture in GDP has declined from 39% in 1983 to 24% in 2000–01 and its contribution in total employment from 63% to 57% during the same period (Mall et al. 2006). Indian cropping system is widely reliant on monsoon from the ancient times and hence it will be severely affected in case of any drastic anomaly in the monsoon trend. Climate change is influencing the rainfall and temperatures conditions across India. The number of rainy days may decrease but the intensity is anticipated to increase, and the availability of gross per capita water will decline from 1820 m^3/ year in 2001 to as low as 1140 m^3/year in 2050 in India (Mahato 2014). Increasing concentrations of CO2 and temperature may affect the crop yield both positively and negatively at different regions, however the rain-fed/un-irrigated crops cultivated over 60% of cropland of India is more venerable to climate change. As per Kalra et al. (2003) the rise in winter temperature by 0.5 °C is likely to reduce rain-fed wheat yield by 0.45 tonnes per hectare in India. The economy of Bihar is mainly based on agriculture with 88.70% of its population living in villages (Census of India 2011) and 81% of the population is associated with agro based work and contributing 42% of GDP of the state (2004–05) (including forestry and fishing) (GOB 2012). Bihar's population in 1951, 1971 and 1991 were 29.08, 42.12 and 64.53 Million, respectively, and has crossed 104 Million in 2011 (Census of India 2011).

To analyse and understand climate change and its impact on agriculture has been recognised as an important first step to strategize adaptation (Turner II et al. 2003; Parry et al. 2007; UNDP. 2010; USAID. 2013). The Space-borne remote sensing technology has been extensively used to map cropping scenario, intensity and change (Xiao et al. 2003; Galford et al. 2008; Wardlow and Egbert 2008). In this study, Moderate Resolution Imaging Spectro radiometer (MODIS) Terra sensor derived vegetation index data (1 km spatial resolution and 16 days temporal resolution) has been used to study the cropping scenario and flood impact in Bihar during 2001 to 2012. Advantage of having high temporal resolution and coarse spatial resolution covering large area using MODIS data helps in monitoring agriculture situation in a continuous manner which otherwise would not be possible using high spatial resolution data due to coarse revisit period and cloud coverage issues.

5.2 Materials and Methods

5.2.1 Study Area

Bihar, the land of Buddha, has been a cradle of knowledge in Indian history. Patna is the capital of the state, which is located on the bank of the holy river Ganga. The river divides the state into two unequal halves from west to east. The northern part of the state is mostly affected by flooding and the southern part usually by drought. The state falls in the sub-tropical climatic zone. The state's average annual rainfall is about 1200 mm (Chowdary et al. 2008). The spatial extent of the state is between 83° 19.83′ to 88° 17.67′ E longitude and 24° 20.17′ to 27° 31.25' N latitude. It comprises of 38 districts and total area of the state is 94,163 km^2 (GOB 2014).

5.2.2 *Satellite Data*

Moderate Resolution Imaging Spectro radiometer Enhanced Vegetation Index (MODIS EVI) (MOD13A2, 1 km, 16 days composite) Terra's time series data over 12 years (2001 to 2012) were used to analyse cropping pattern of the state, and detailed results were given in this paper for the years 2001, 2006 and 2011. The satellite data were downloaded from USGS website (source: https://lpdaac.usgs.gov/). The flood map from IWMI were used in this study. These maps were derived using MODIS Data (MOD09A1, 500 m, 8 day composites) with validation based on Landsat TM and ALOS AVINIR / PALSAR data (Amarnath et al. 2012).

5.2.3 *Research Methods*

The methodology used to analyse the spatio-temporal change in cropping pattern and flood impact assessment on agriculture growing areas of Bihar is provided in the Fig. 5.1.

Smoothing based on Fourier technique and phenology detection algorithms were used to identify the starting of greening phase in annual EVI data. Based on these annual data horizontal expansion as well as vertical intensification (net sown area, Rabi, Zaid and Kharif Area) of the cropping pattern using were quantified spatially and temporally in this region. The detailed procedure can be seen in Kumar and Jeganathan (2016). The flood impact has been assessd only for the Kharif crop (monsoon crop). Since the actual loss of crop cannot be accurately quantified, Potential Kharif crop loss map for different years have been prepared based on Kharif frequency information along with Annual Net Sown area and flood maps (see Fig. 5.1 for details).

5.3 Results and Discussion

Seasonal cropping system (Rabi, Zaid & Kharif) based on MODIS EVI (MOD13A2) time series data seems well distributed all over the state (Fig. 5.2). The Tables 5.1 to 5.3 provides district wise percentage area under different cropping pattern of the state for the yeras 2001, 2006 and 2011 respectively. It can be seen that most of the

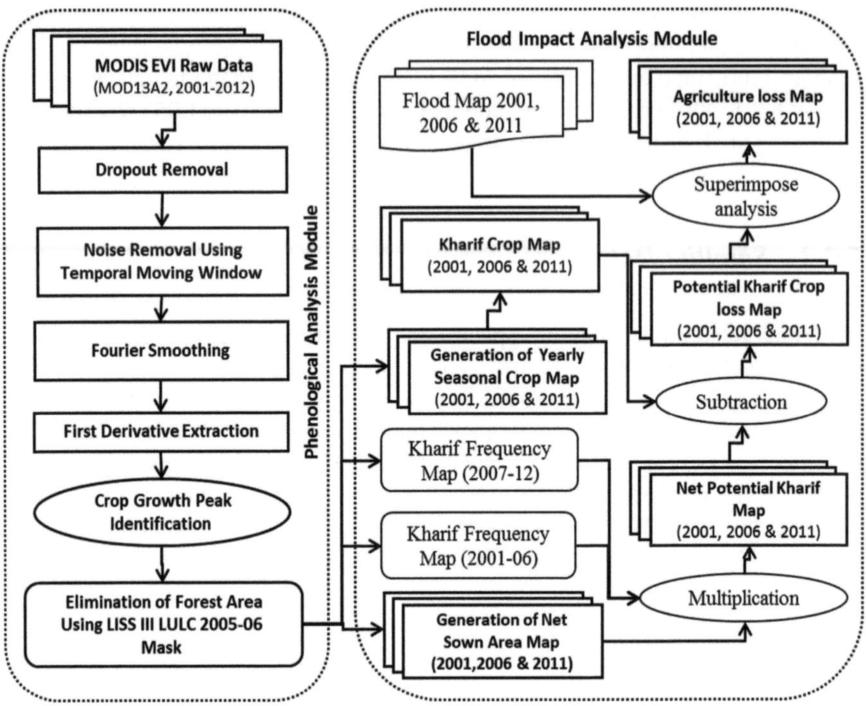

Fig. 5.1 Methodology of Research

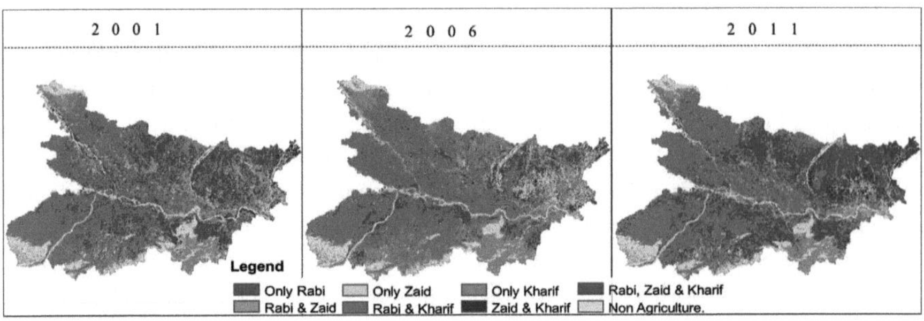

Fig. 5.2 MODIS EVI based cropping system dynamics of Bihar (2001, 2006 & 2011)

area in Bihar has double cropping pattern (*Rabi and Kharif* - dark green colour in Fig. 5.2) covering 43.66, 48.56 and 44.85% of the state during 2001, 2006 and 2011 respectively. The districts Araria and Supaul had more than 50% of area having triple cropping system in 2011. The year 2011 found to be having more area under triple cropping pattern (crop in all the 3 seasons - *Rabi, Zaid and Kharif*) (13.47%) than 2001 and 2006 in the state and this may be because of good rainfall throughout the year (DSV 2014). In the year 2001, Rabi and Zaid cropping area was more (11.77%) in comparison to 2006 and 2011. The second maximum area seems in the

Table 5.1 District-wise agriculture area (%) distribution of seasonal crops in the Bihar during 2001

Districts	Non Agriculture	Only Rabi	Only Zaid	Only Kharif	Rabi & Zaid	Rabi & Kharif	Zaid & Kharif	Rabi, Zaid & Kharif
W. Champaran	20.63	1.54	5.65	27.70	2.87	38.58	2.44	0.59
E. Champaran	2.34	4.76	2.38	7.47	4.76	75.95	1.22	1.12
Sitamarhi	0.25	4.97	0.71	2.09	7.27	72.40	0.29	12.03
Madhubani	0.65	5.16	2.83	19.93	8.66	35.37	2.71	24.70
Gopalganj	3.25	4.18	5.53	5.49	12.08	64.60	1.94	2.92
Sheohar	1.04	4.63	2.69	3.73	11.49	63.88	1.94	10.60
Araria	0.00	7.18	4.81	2.80	37.92	21.58	1.40	24.32
Supaul	11.04	3.63	2.26	4.21	23.44	11.65	2.55	41.23
Kishanganj	4.81	1.33	13.09	17.21	4.12	5.97	32.66	20.82
Darbhanga	0.15	5.79	4.75	5.34	24.22	36.43	2.17	21.15
Muzaffarpur	2.12	5.22	4.89	11.54	16.59	47.70	1.43	10.51
Siwan	2.25	2.68	2.33	5.01	6.06	80.72	0.08	0.86
Saran	8.61	9.09	3.33	3.14	11.03	62.99	0.23	1.59
Purnia	1.66	6.04	6.89	4.75	30.03	24.73	4.22	21.67
Madhepura	0.47	4.38	4.71	2.98	25.96	18.27	5.50	37.74
Samastipur	2.19	7.71	3.78	2.87	26.74	49.59	0.35	6.77
Saharsa	5.53	3.55	9.13	2.89	26.59	20.14	7.26	24.91
Vaishali	9.00	7.17	4.52	7.00	7.81	62.24	0.13	2.13
Katihar	7.98	4.44	18.39	8.96	27.48	16.04	9.15	7.56
Begusarai	7.88	5.80	3.50	4.12	19.74	58.34	0.13	0.49
Buxar	2.39	1.62	0.30	5.69	1.02	86.19	0.00	2.79
Bhojpur	3.51	3.83	0.61	6.54	1.05	80.01	0.04	4.41
Patna	9.28	8.71	1.98	3.48	7.06	64.82	0.00	4.68
Khagaria	4.37	7.25	6.38	1.32	42.90	29.50	2.01	6.27
Bhagalpur	8.59	5.59	7.73	10.82	26.91	17.15	10.26	12.95
Munger	34.98	5.76	8.39	6.70	10.64	7.13	3.94	22.47
Nalanda	2.31	6.60	0.59	1.50	11.14	60.35	0.07	17.44
Kaimur	33.46	0.31	0.10	13.08	0.05	52.89	0.03	0.08
Rohtas	23.40	0.18	0.33	8.29	0.42	66.05	0.04	1.28
Lakhisarai	14.34	16.01	5.89	7.17	7.81	29.45	2.18	17.16
Jehanabad	0.51	6.57	4.65	5.97	17.19	42.77	0.20	22.14
Sheikhpura	0.72	7.16	2.87	8.60	13.75	61.03	0.14	5.73
Jamui	20.79	0.86	11.74	55.15	1.50	2.96	3.35	3.65
Aurangabad	9.61	1.72	0.55	26.83	0.76	59.06	0.08	1.38
Banka	10.74	0.25	5.34	53.17	0.59	7.46	7.46	14.98
Nawada	20.18	3.25	2.06	19.73	3.25	44.81	0.38	6.34
Gaya	16.89	3.27	2.43	33.68	3.48	37.68	0.14	2.43
Arwal	6.31	1.69	0.56	6.99	1.92	79.71	0.11	2.71
Bihar	9.46	4.34	4.42	13.62	11.77	43.66	2.73	10.00

Table 5.2 District-wise agriculture area (%) distribution of seasonal crops in the Bihar during 2006

Districts	Non Agriculture	Only Rabi	Only Zaid	Only Kharif	Rabi & Zaid	Rabi & Kharif	Zaid & Kharif	Rabi, Zaid & Kharif
W. Champaran	19.36	0.16	1.79	43.55	0.18	34.42	0.29	0.25
E. Champaran	0.88	0.54	0.39	15.32	0.24	81.98	0.30	0.36
Sitamarhi	0.08	0.92	0.33	33.74	1.17	63.34	0.25	0.17
Madhubani	0.55	3.23	4.07	43.36	6.83	39.09	0.97	1.91
Gopalganj	2.03	0.34	1.27	17.74	0.38	77.95	0.21	0.08
Sheohar	0.45	0.30	0.60	20.90	1.04	75.97	0.60	0.15
Araria	0.00	3.47	8.40	18.78	21.70	25.47	8.40	13.79
Supaul	10.21	4.24	9.63	10.68	21.96	17.07	5.64	20.56
Kishanganj	1.93	0.64	14.59	47.81	1.63	4.76	22.66	5.97
Darbhanga	0.04	4.19	4.97	22.38	14.31	53.03	0.52	0.56
Muzaffarpur	0.86	4.46	1.16	21.35	3.07	67.65	0.51	0.94
Siwan	1.48	0.16	0.12	8.32	0.08	89.51	0.31	0.04
Saran	5.89	2.59	2.07	8.31	3.62	77.42	0.06	0.03
Purnia	0.61	1.27	22.80	42.54	11.30	11.96	7.65	1.87
Madhepura	0.47	2.19	21.02	18.69	6.62	34.39	9.04	7.60
Samastipur	1.95	6.09	1.63	20.33	7.74	61.94	0.24	0.09
Saharsa	4.72	1.47	15.07	28.97	1.62	32.57	12.18	3.40
Vaishali	6.70	1.45	1.71	12.93	3.07	73.72	0.26	0.17
Katihar	5.66	0.92	21.99	38.95	13.98	7.20	8.93	2.37
Begusarai	7.53	1.81	0.80	12.04	2.08	75.70	0.04	0.00
Buxar	1.88	2.34	0.46	5.38	3.10	84.87	0.05	1.93
Bhojpur	2.31	2.93	1.45	6.62	4.05	75.81	0.33	6.51
Patna	7.46	9.01	1.71	5.16	2.57	69.98	0.24	3.88
Khagaria	2.42	5.75	1.50	6.04	5.69	75.96	2.59	0.06
Bhagalpur	7.23	5.69	4.43	22.78	18.41	37.70	1.30	2.46
Munger	32.60	4.38	1.88	20.15	3.82	31.85	0.44	4.88
Nalanda	3.00	4.36	0.22	9.53	0.37	81.35	0.00	1.17
Kaimur	33.44	0.47	0.16	15.60	0.42	49.51	0.05	0.36
Rohtas	22.98	0.46	0.58	9.75	0.55	63.61	0.04	2.01
Lakhisarai	13.12	8.64	0.77	29.96	1.15	45.01	0.38	0.96
Jehanabad	0.40	1.01	0.30	26.29	0.20	70.88	0.10	0.81
Sheikhpura	1.43	2.29	0.14	35.10	0.00	60.74	0.00	0.29
Jamui	20.38	0.08	0.14	73.39	0.00	4.57	0.69	0.75
Aurangabad	8.96	0.94	2.17	45.30	0.57	37.02	1.52	3.53
Banka	10.23	0.08	0.62	60.58	0.00	20.30	2.94	5.26
Nawada	20.14	0.10	0.00	29.12	0.00	50.33	0.03	0.27
Gaya	16.46	0.34	0.22	52.98	0.05	29.68	0.02	0.24
Arwal	5.41	0.56	2.59	6.88	1.01	76.44	0.34	6.76
Bihar	8.57	2.22	4.07	27.32	4.44	48.56	2.24	2.58

Table 5.3 District-wise agriculture area (%) distribution of seasonal crops in the Bihar during 2011

Districts	Non Agriculture	Only Rabi	Only Zaid	Only Kharif	Rabi & Zaid	Rabi & Kharif	Zaid & Kharif	Rabi, Zaid & Kharif
W. Champaran	19.86	1.72	4.31	25.57	0.88	46.76	0.48	0.43
E. Champaran	1.54	1.59	0.64	4.14	1.14	89.70	0.15	1.09
Sitamarhi	0.08	2.38	2.59	7.52	5.14	48.60	2.05	31.65
Madhubani	0.22	2.33	1.74	10.03	5.14	46.89	4.39	29.26
Gopalganj	4.06	2.24	1.18	3.46	1.14	87.03	0.34	0.55
Sheohar	0.00	2.84	1.64	5.37	3.28	68.51	1.49	16.87
Araria	0.00	0.85	2.62	3.20	7.91	12.17	12.02	61.23
Supaul	10.50	2.77	2.91	2.52	12.47	8.81	5.46	54.57
Kishanganj	2.79	0.17	4.98	9.83	0.09	3.56	62.40	16.18
Darbhanga	0.15	4.67	2.95	4.11	14.46	39.95	2.80	30.90
Muzaffarpur	1.24	7.34	2.20	5.19	9.90	58.32	1.13	14.68
Siwan	1.59	2.25	0.58	0.82	1.83	92.15	0.19	0.58
Saran	6.57	7.41	1.39	1.49	10.81	69.07	0.26	3.01
Purnia	0.77	2.16	12.01	14.23	12.93	7.50	31.91	18.50
Madhepura	0.61	6.24	4.85	5.17	9.74	20.13	19.48	33.78
Samastipur	1.74	9.90	2.98	12.65	17.70	47.84	0.56	6.62
Saharsa	5.43	4.26	4.16	8.78	5.02	29.22	20.95	22.17
Vaishali	7.21	5.42	2.52	5.59	19.37	59.09	0.09	0.73
Katihar	6.47	3.60	12.30	12.56	23.44	9.26	22.24	10.13
Begusarai	8.63	9.61	1.11	8.54	15.32	56.71	0.04	0.04
Buxar	2.03	3.20	0.30	1.68	5.03	80.76	0.20	6.80
Bhojpur	2.42	10.20	0.69	2.21	8.46	65.76	0.18	10.09
Patna	8.85	9.65	2.81	3.37	14.54	47.53	0.27	12.99
Khagaria	2.47	23.06	1.38	1.55	18.80	46.58	3.39	2.76
Bhagalpur	8.59	11.29	4.90	9.16	31.37	18.25	3.23	13.22
Munger	34.61	15.64	1.25	10.89	6.38	16.71	1.31	13.20
Nalanda	2.53	4.91	0.55	3.22	3.70	68.52	0.37	16.20
Kaimur	33.44	1.59	0.13	10.40	1.98	50.81	0.00	1.66
Rohtas	22.87	0.84	0.73	6.53	0.91	66.25	0.00	1.88
Lakhisarai	13.25	12.10	2.75	9.22	11.72	32.71	1.66	16.58
Jehanabad	0.71	2.83	1.21	16.89	0.40	72.90	0.30	4.75
Sheikhpura	0.00	3.30	0.57	6.16	4.58	48.71	1.29	35.39
Jamui	20.40	0.72	2.13	57.86	0.17	9.44	2.38	6.89
Aurangabad	8.28	2.77	3.68	15.78	3.24	57.13	3.13	5.98
Banka	9.98	0.14	3.27	49.40	0.14	14.03	6.51	16.52
Nawada	20.01	2.91	3.05	9.22	4.56	38.40	1.10	20.76
Gaya	16.56	4.64	2.64	25.21	4.32	40.78	0.63	5.21
Arwal	5.41	3.27	1.01	3.16	2.82	67.87	0.68	15.78
Bihar	8.87	4.54	2.96	12.11	7.73	44.85	5.48	13.47

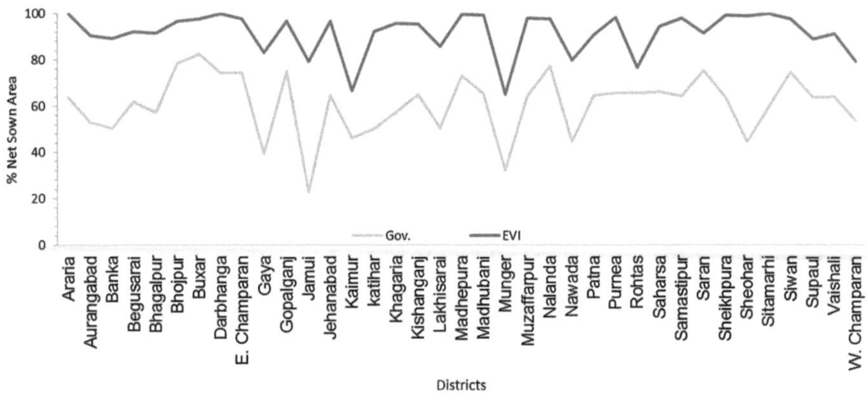

Fig. 5.3 Percentage net sown area comparison between Bihar Government report 2001–02 and MODIS EVI 2001

only Kharif cropping system 13.62, 27.32 and 12.11% respectively 2001, 2006 and 2011 (Table 5.1 to 5.3). These results would help in understanding cropping system and its dependency on monsoon season. It can be seen that few districts like West Champaran in the North -West and Jamui and Banka districts of South -East of the state, have *only kharif* crop (see orange colour in Fig. 5.2) which could be because of gaps in irrigation facility and they depend only on monsoon rain.

The net sown area of the state for the years 2001, 2006 and 2011 is 90.54, 91.42 and 91.13% respectively. Fig. 5.3 provides comparison of net sown area extracted using Government of Bihar report (2001–02) and MODIS EVI (2001). The Bihar Government report 2001–02 (DES 2016) and the district wise result of net sown area based on MODIS EVI 2001 found some difference but the pattern seems same. This trend reflects strong inter-relation of these statistics (Fig. 5.3). The major draw-back of MODIS data is its coarse spatial resolution however the trend is found to be similar. It is also noticed that there were contradictions about net-sown area amongst the government departments.

For example, Government of Bihar report (GOB 2007) revealed the net sown area in Gopalganj, Saran and Siwan as 98%, in Seohar, Muzaffarpur, Vaishali and Sitamarhi districts as 95%, in Darbhanga, Samastipur and Madhubani as 87%, in Jehanabad, Aurangabad and Gaya as 86.7%. These government statistics are closely related to MODIS EVI results.

The Fig. 5.4 shows flood impact on Kharif crop during 2001, 2006 and 2011 in the state. The flood has affected crops mainly in the northern districts from Ganges closer to the rivers like Gandak, Burhi, Bagmati, Gandak, Kosi and Mahananda. In the southern part it has affected mainly in the Sheikhpura and Lakhisarai districts (Fig. 5.4). In the year 2001, the kharif crop area lost due to flood is 8.22%. The least impact of flood was observed in the year 2006 (only 3.56%) in comparison to 2001 and 2011. In the year 2011, flood impact (5.87% loss) was mainly seen along the banks of Ganges and lower part of Kosi River.

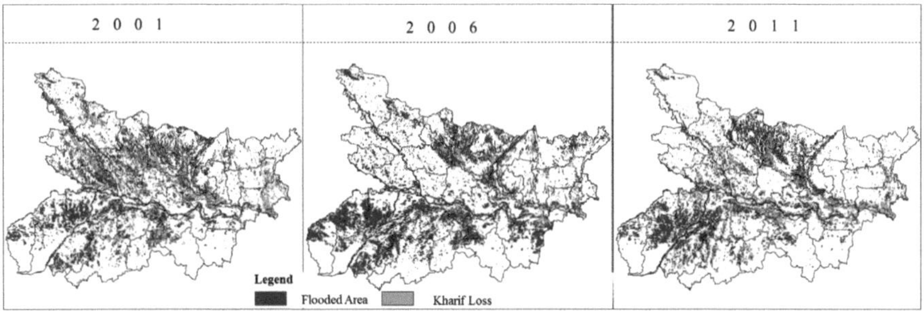

Fig. 5.4 Flood impact assessment on Kharif crop in Bihar (2001, 2006 & 2011)

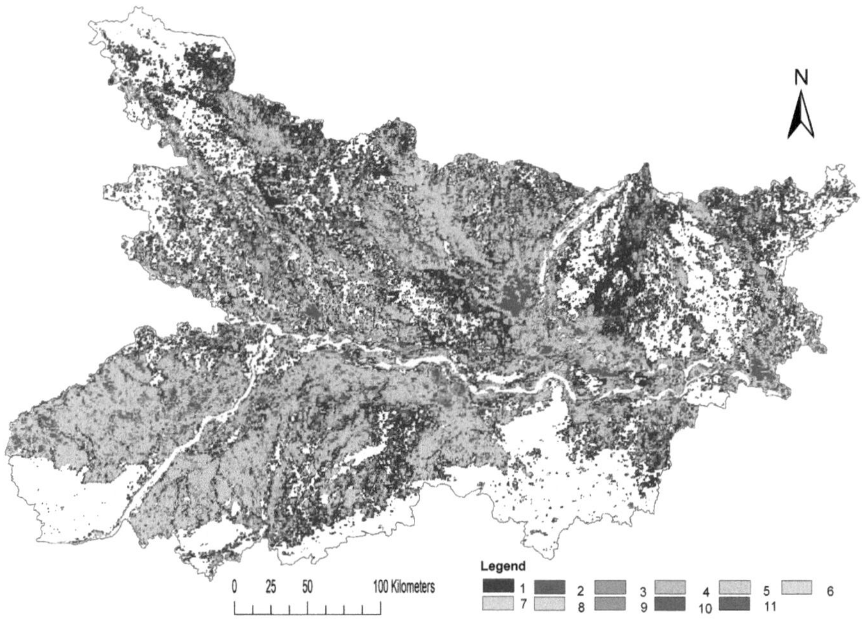

Fig. 5.5 Flood frequency of Bihar during 2001–11

In the Fig. 5.5 has shown flood frequency in Bihar during 2001 to 2011. The maximum concentration of flood frequency is seen at the junction of Burhi Gandak, Kamla and Balan near the embankment of Kosi River. It has also revealed concentrations of flood freqancy at the junction of Mahananda and Ganga rivers.

5.4 Conclusion

The study has successfully used the MODIS EVI time series data for seasonal cropping system assessment as well as the flood impact on agriculture in Bihar. The study has revealed spatio-temporal variations in the cropping pattern during 2001,

2006 and 2011. There are contradictions between results from our study and the government reports which requires further scrutiny. Due to varying rainfall scenario in different years the agriculture pattern in Bihar is highly dynamic and only Satellite based observation can provide a reliable estimates. Any loss estimate based on previous year cropping situation should not be used as the current study has revealed high inter-annual variation in the cropping pattern in this State. The changing climatic condition and its impact on agriculture pattern need to be further quantified with more authentic ground information.

References

Amarnath, G., Ameer, M., Aggarwal, P., & Smakhtin, V. (2012). Detecting spatio-temporal changes in the extent of seasonal and annual flooding in South Asia using multi-resolution satellite data. *International Society for Optics and Photonics*, 853–818. https://doi.org/10.1117/12.974653.

Census of India (2011). "Bihar Population Census." Ministry of Home Affairs, Office of the registrar general & census commission, Government of India. http.//www.censusindia.gov.in/2011-prov-results/prov_data_products_bihar.html.

Chowdary, V. M., Chandran, R. V., Neeti, N., Bothale, R. V., Srivastava, Y. K., Ingle, P., Ramakrishnan, D., Dutta, D., Jeyaram, A., Sharma, J. R., & Singh, R. (2008). Assessment of surface and sub-surface waterlogged areas in irrigation command areas of Bihar state using remote sensing and GIS. *Agricultural Water Management, 95*, 754–766.

DES (2016). Bihar Statistical Hand Book 2006. Directorate of Economics and Statistics, Department of Planning & Development, Patna, Government of Bihar. http.//dse.bih.nic.in/.

DSV (2014). Monthly rainfall Statistics during 1989 to 2011. Directorate of Statistics Evaluation, Department of Planning and Development, Patna, Government of Bihar.

Galford, G. L., Mustard, J. F., Melillo, J., Gendrin, A., Cerri, C. C., & Cerri, C. E. P. (2008). Wavelet analysis of MODIS time series to detect expansion and intensification of row-crop agriculture in Brazil. *Remote Sensing of Environment, 112*, 576–587.

GOB (2007). State of Environment Report." Bihar State Pollution Control Board and Department of Environment and Forest, Government of Bihar.

GOB (2012). *State Action Plan on Climate Change*. Supported by United Nations Development Programme India.

GOB (2014). *Five year plan 2012–17 and annual plan 2012–13*. Planning and development Department, Government of Bihar. http://planning.bih.nic.in/

Kalra, N., Aggarwal, P. K., Chander, S., Pathak, H., Choudhary, R., Choudhary, A., Mukesh, S., Rai, H. K., Soni, U. A., Anil, S., Jolly, M., Singh, U. K., Owrs, A., & Hussain, M. Z. (2003). In P. R. Shukla, S. K. Sharma, N. H. Ravindranath, A. Garg, & S. Bhattacharya (Eds.), *Impacts of climate change on agriculture. Climate change and India: Vulnerability assessment and adaptation* (pp. 193–226). Hyderabad: Orient Longman Private Ltd..

Kumar, P., & Jeganathan, C. (2016). Monitoring horizontal and vertical cropping pattern and dynamics in Bihar over a decade (2001–2012) Based on Time-Series Satellite Data. *Journal of Indian Society of Remote Sensing*. https://doi.org/10.1007/s12524-016-0614-1.

Mahato, A. (2014). Climate change and its impact on agriculture. *International Journal of Scientific and Research Publications, 4*(4), 1–6.

Mall, R. K., Singh, R., Gupta, A., Srinivasan, G., & Rathore, L. S. (2006). Impact of climate change on Indian agriculture: A review. *Climatic Change, 78*(2–4), 445–478.

Parry, M. L., Canziani, O. F., Palutikof, J. P., van der Linden, P. J., & Hanson, C. E. (2007). In M. L. Parry, O. F. Canziani, J. P. Palutikof, P. J. van der Linden, & C. E. Hanson (Eds.), *Climate change 2007: Impacts, adaptation and vulnerability. Contribution of working group II to the fourth assessment report of the intergovernmental panel on climate change* (pp. 23–78).

Turner, B. L., II, Kasperson, R. E., Maston, P. A., McCarthy, J. J., Corell, R. W., Christensen, L., Eckley, N., Kasperson, J. X., Luers, A., Martello, M. L., Polsky, C., Pulsipher, A., & Schiller, A. (2003). A framework for vulnerability analysis in sustainability science. *PNAS, 100*(14), 8074–8079.

UNDP (2010). Mapping climate change vulnerability and impact scenarios: A guidebook for sub-national planners.1–83.

USAID (2013). Uganda climate change vulnerability assessment report African and Latin American Resilience to Climate Change (ARCC) report.

Wardlow, B. D., & Egbert, S. L. (2008). Large-area crop mapping using time-series MODIS 250 m NDVI data: an assessment for the US Central Great Plains. *Remote Sensing of Environment, 112*, 1096–1116.

Xiao, X. M., Liu, J. Y., Zhuang, D. F., Frolking, S., Boles, S., Xu, B., Liu, M. L., Salas, W., Moore, B., & Li, C. S. (2003). Uncertainties in estimates of cropland area in China: A comparison between an AVHRR-derived dataset and a Landsat TM-derived dataset. *Global and Planetary Change, 37*, 297–306.

Chapter 6
Impact Assessment of Bio Priming Mediated Nutrient Use Efficiency for Climate Resilient Agriculture

Amitava Rakshit

Abstract Since environmental stress negatively affects crop growth and productivity throughout the world and the energy crisis threatens the sustainability of both irrigated and rain fed system, it is becoming increasingly evident that priming techniques can enhance and improve the performance of crops without deteriorating the natural resource base. Among the available options, on-farm seed priming is a simple, proven technology that has been an age old practice, tested, and refined in laboratories, in experimental plots, and by farmers themselves in their fields. It's easy to use with a wide range of crops in many different farming conditions. Farmers in the indo-gangetic plains of Uttar Pradesh, India prime rice, wheat, maize and pulse seed before sowing. This simple method is now spreading to other parts of the country as well. Although priming with water or tiny amounts of phosphorus, boron and zinc is common but use of microbes can make a huge difference. Biopriming is becoming a potentially prominent technique to induce profound changes in plant characteristics and to encourage desired attributes in plants growth associated with fungi and bacteria coatings. Biological factors such as fungi and bacteria are used in biopriming which includes: fungi and antagonist bacteria and the most important of all are *Trichoderma, Pesodomonas, Glomus, Bacillus, Agrobacterium* and *Gliocladium.* Therefore, seed priming in combination with low dosage of bio control agents has been used to improve the plant performance, stabilize the efficacy of biological agents in the present set up of agriculture and reducing dependency on chemical inputs.

Keywords Bio priming · N use efficiency · Crop yield and climate resilient agriculture

---

A. Rakshit (✉)
Department of Soil Science and Agricultural Chemistry, Institute of Agricultural Science, Banaras Hindu University, Banaras, UP, India
e-mail: amitavar@bhu.ac.in

S. Sheraz Mahdi (ed.), *Climate Change and Agriculture in India: Impact and Adaptation*, https://doi.org/10.1007/978-3-319-90086-5_6

Table 6.1 Input statistics in agriculture

Input	Usage (10^6 tonnes)		Subsidy (Rs. billion)		Size of the industry (Rs. billion)		Energy Involvement (MJ kg^{-1})
	India	Global	India	Global	India	Global	
Fertilizer	24.5	170	750	–	30	5000	78.2(N);17.5(P);13.8(K)
Pesticide	0.85	2.6	–	–	180	2500	215(Harbicide), 238(Insecticide) and 92(Fungicide)
Bio pesticide	0.25	25	–	–	2	200	–
Bio fertilizer	0.28	200	–	–	4	180	0.01(liquid); 0.3(solid)

Source: Rakshit et al. (2015)

6.1 Huge Energy Requirement in Agriculture

Energy has always been essential for the production of food. Prior to the industrial revolution, the primary energy input for agriculture was the sun; photosynthesis enabled plants to grow, and plants served as food for livestock, which provided fertilizer (manure) and muscle power for farming. However, as a result of the industrialization and consolidation of agriculture, food production has become increasingly dependent on energy derived from fossil fuels (Table 6.1). Today, industrial agriculture consumes fossil fuels for several purposes i.e., fertilizer production, water consumption, farm equipment and processing, packaging and transportation.

6.2 Rising Energy Prices Affect Agricultural Sector

There are three ways by which agriculture sector is influenced by rising energy prices. Farmers have to pay increased prices for energy related inputs such as fuel and fertilizers. This requires more expenditure which, in turn, is reflected in increased cost of production or lower net farm income. Basically rising prices of crude oil or energy add to the general inflation in the country which in turn is fed into the agricultural sector through an increase in prices of all other agricultural inputs. In essence, rising energy prices lead to increased cost of cultivation. The impact of rising prices on agriculture will primarily depend upon the time taken for adjustments. Two time runs may be considered either short term or medium to long term. For farmers, a short term period can be up to one crop growing season or a year. Here, the adjustment time to react to rising energy prices is so short that farmers hardly change their production plans, search for new energy conservation measures, or substitute less energy intensive and relatively cheaper inputs for costlier energy related inputs. In most of the cases, farmers cut back on use of fertilizers to a certain extent. In the medium to long term situation, which could be more than a year or several years, farmers can react in many ways. They can change their production plans, grow less energy intensive crops, adopt energy saving technologies such as low input sustainable agricultural practices, and so on.

6.2.1 Nutrient Balance Sheet: Another Concern

An assessment of nutrient additions, removals, and balances in the agricultural production system generates useful, practical information on whether the nutrient status of a soil is being maintained, built up, or depleted. Nutrient balance sheets in most soils of India have been deficient and continue to be so. This is primarily because nutrient removals by crops far exceed the nutrient additions through manures and fertilizers. For the past 50 years the gap between removals and additions has been estimated at 8 to 10 M t N + P2O5 + K2O per year which is being manifested to the multiplicity of nutrient deficiency problems across the last few decades presented in Fig. 6.1.

This has been the case in the past, at present, and this will likely continue into the future. To this extent, the soils are becoming depleted – the situation is akin to mining soils of their nutrient capital, leading to a steady reduction in soil nutrient supplying capacity (Table 6.2). On top of this deficit are the nutrient losses through

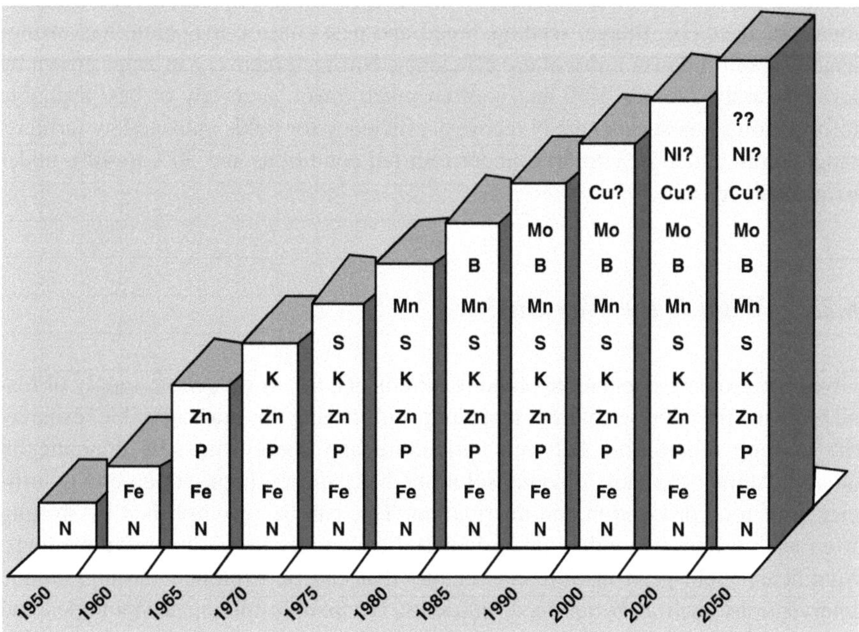

Fig. 6.1 Nutrient deficiency across the last few decades

Table 6.2 Nutrient use efficiency of major nutrients

Nutrient	Efficiency
Nitrogen	30–50
Phosphorus	15–20
Potassium	70–80
Sulphur	<5
Zinc	2–5
Iron	1–2
Copper	1–2
Boron	2–3
Molybdenum	2–5

various other means. For example, nutrient losses through soil erosion are alarmingly large, but are rarely taken into account.

Again awareness of and interest in improved nutrient use efficiency has never been greater. Driven by a growing public belief that crop nutrients are excessive in the environment and farmer concerns about rising fertilizer prices and stagnant crop prices, the fertilizer industry is under increasing pressure to improve nutrient use efficiency.

Data on N use efficiency for cereal crops from experimental plots reported that single-year fertilizer N recovery efficiencies averaged 65% for corn, 57% for wheat, and 46% for rice. However, experimental plots do not accurately reflect the efficiencies obtainable on-farm. Differences in the scale of farming operations and management practices (i.e. tillage, seeding, weed and pest control, irrigation, harvesting) usually result in lower nutrient use efficiency. Nitrogen recovery in crops grown by farmers rarely exceeds 50% and is often much lower. A review of best available information suggests average N recovery efficiency for fields managed by farmer's ranges from about 20% to 30% under rain fed conditions and 30% to 40% under irrigated conditions.

6.2.2 *What Is the Way Out?*

Given the growing population's food requirements, the world's finite supply of fossil fuels and the adverse environmental impacts of using this nonrenewable resource, the existing relationship between agriculture and energy must be dramatically altered. Among the most obvious solutions is to simply improve the energy efficiency of food production and distribution. This can be accomplished by shifting from energy-intensive industrial agricultural techniques to less intensive methods. Significant yield gaps in different regions that can be exploited through simple interventions such as better seed, nutrients, compatible mixing technologies and water management. However, it is generally necessary to move towards more sophisticated, more knowledge-intensive forms of agriculture – and provide the technologies and incentives that make it viable for farmers to adopt and adapt them. In crop production, agro-ecological intensification primarily implies to implement good agronomic management principles in a local context, including:

- Profitable and sustainable crop rotations
- Choosing quality seed of a well-adapted high-yielding variety or hybrid that also meets market demands
- Planting at the right time to maximize the attainable yield by capturing light, water and nutrients
- Maximize the capture and efficient utilization of available water
- Integrated soil and nutrient management, including conservation agriculture, balanced and more efficient use of fertilizers, as well as utilization of available biological and organic sources

With this back ground priming can be a viable option (Rakshit et al. 2013) which is simple, inexpensive and improves plant acclimatizability under both biotic and abiotic stress.

6.3 Bio-Priming Can Be the Answer

Bio-priming is a process of biological seed treatment that refers combination of seed hydration (physiological aspect of disease control) and inoculation (biological aspect of disease control) of seed with beneficial organism to protect seed (Rakshit et al. 2014). It is an ecological approach using selected fungal antagonists against the soil and seed-borne pathogens. Biological seed treatments are becoming a safe aid for biotic and abiotic stress management.

6.3.1 Simple Tool

The methods of introducing biological agents is popular and user friendly (Bisen et al. 2015; Fig. 6.2).Seeds have to be pre-soaked the in water for 12 hours the formulated product of bioagent (*Trichoderma harzianum* and/or *Pseudomonas fluorescens*) have to be mixed with the pre-soaked seeds at the rate of 10 g per kg seed. Biological factors such as fungi and bacteria are used in biopriming which includes: fungi and antagonist.

Bacteria and the most important of all are *Trichoderma*, *Pesodomonas, Bacillus, Rhizobacteria, Gliocladium* and *Agrobacterium.*Treated seeds are placed as a heap. The heap is to be covered with a moist jute sack to maintain high humidity and incubate the seeds under high humidity for about 48 h at approx. 25 to 32 °C. Bio agent adhered to the seed grows on the seed surface under moist condition to form a protective layer all around the seed coat. The seeds have to be sown in nursery bed. The seeds thus bio primed with the bio agent provide protection against seed and

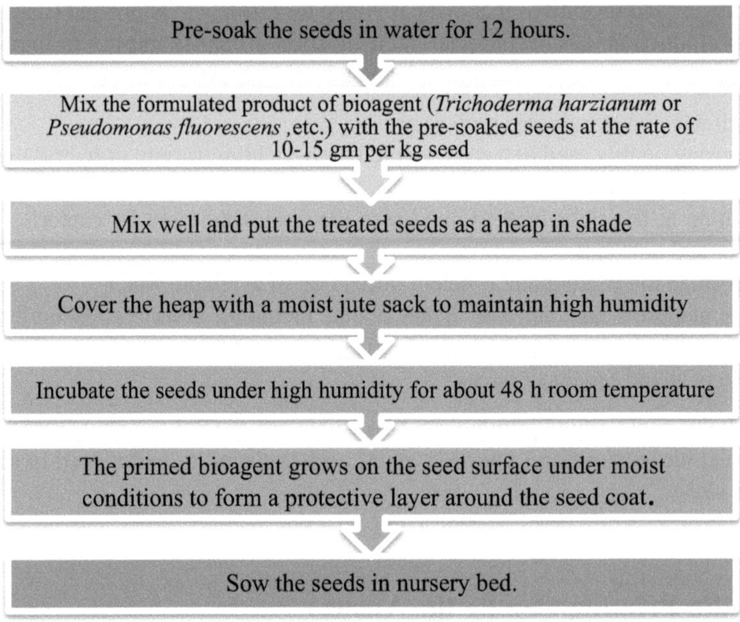

Fig. 6.2 General Procedure of seed biopriming

soil borne plant pathogens, improved germination and seedling growth. For seedling treatment suspension have to be prepared by mixing 1 kg bio-fertilizer culture in 10–15 litres of water. Seedlings required for one acre need to be tied in small bundles of seedlings. Root portion of these seedlings is dipped the in this suspension for 15–30 min. and transplant immediately. Generally, the ratio of inoculant and water is 1:10 For vegetables like chilly, tomato, cabbage, cauliflower, 250 g of bio-fertilizer is sufficient for 0.1 ha of land.

Many small-scale farmers rely on basic tools, minimal technology to cultivate their crops, which exacerbates their vulnerability to changing energy scenario, weather patterns and other stressors that affect food security.

6.3.2 Tremendous Opportunities

Bio-priming provides a simple inexpensive method for improving plant nutrition as well as improving yield of crops. Bio-priming methods have practical advantages of minimal waste material produced compared to others.

- Decrease time between sowing and emergence and to improve seedling vigour.
- Increase nutrient use efficiency

- Improves WUE of several crop species
- Increases plant growth and nutrient uptake
- Competence with weeds
- Uniform growth of plants
- Eliminates seed borne pathogens
- Increases rate of germination
- Overcome or alleviate phytochrome-induced dormancy in plants
- Extend the temperature range at which a seed can germinate
- Application of microbial agents in agriculture has four beneficial effects for plants. First, it can colonize plant root and its rhizosphere, second, control plant pathogens thought parasitism, and antibiosis production, and promote systemic resistance. Third, it improve plant health through increasing plant growth. Ultimately, these microorganisms stimulate root growth and improve plant growth. Studies showed that these group of microorganisms treatment increased fresh weight of root and leaf area index, lateral roots and the nutrient use efficiency in a number of crop species (Table 6.3).

6.3.3 Biopriming Mediated Nitrogen Use Efficiency

Improving NUE is an important goal to harvest better crop yield on sustained basis. Studies have shown that the growth-promoting ability of microbes may be highly specific to certain plant species, cultivar and genotype (Table 6.4).

6.3.4 Where Do We Stand Now?

Bi-priming technology has been distributed through various channels to improve the livelihoods of resource-poor farmers in marginal environments. In addition detailed advice, research protocols and supporting materials concerning seed priming, have been provided in response to enquiries from individuals and organisations across the length and breadth of the sate who have expressed an interest. The technology has been featured on the radio and on TV on several occasions. Thousands of farmers, researchers and extensionists through RKVY training, farmers training have been exposed to this technology and enough time has elapsed to allow us to follow up and learn from their experiences. In addition, studies on bio-priming technology have been funded directly by ICAR under seed platform. However, more fundamental research will certainly be necessary to understand the nature of the relation between seed priming and integrated resource management.

Table 6.3 Effect of bio priming on different crop species

Biopriming agent	Crops	Response	Reference
Mycorrhiza	Common bean (*Phaseolus vulgaris),* Kinnow, Rice and Mung bean	Plant nutrition	Tajini and Drevon (2012), Shamshiri et al. (2012), Li et al. (2009)
	Wheat and black gram	Resource distribution in plant communities	Shukla et al. (2012)
	Wheat	Buffering against host plant stress	Gamal et al. (2012)
	Maize, green gram, sorghum, millet, mash bean and mung bean	Benefits in the context of agro ecosystems sustainability and role in the stability of ecosystem with climate change	Mobasser and Moradgholi (2012), Ananthi et al. (2011), Sharif et al. (2011), Amanullah et al. (2011)
Trichoderma	Soybean, Cucumber, Tomato, Tea, Tomato, Rice	Enhancement of nutrient(macro and micro) use efficiency	Entesari et al. (2013), de Santiago et al. (2012), Molla et al. (2012), Moharam and Negim (2012), Thomas et al. (2010), Azarmi et al. (2011), Shaharoona et al. (2008), Yedidia et al. (2001)
	Maize, Mustard, Bitter gourd, Betelvine, Muskmelon	Overall performance (above ground and below ground) with reference to yield and buffering against biotic stress	Karthika and Vanangamudi (2013), P Lalitha et al. (2012), Lavania et al. (2006), Kaveh et al. (2011)
Pseudomonas	Arabidopsis, Sunflower (*Helianthus annuus* L.), Sweet Corn	Increased growth attributes, nutrient use efficiency	Ryu et al. (2005), Moeinzadeh et al. (2010), Callan et al. (1991), Shaharoona et al. (2008)
	Broad beans	Phytoremidiation potential	Radwan et al. (2005)
	Strawberry, Radish	Buffering against biotic stress	Pirlak and Kose (2009), Kaymak et al. (2008)
	Wheat, Pea	Buffering against abiotic stress	Egamberdieva (2008), Zahir et al. (2008)
Bacillus	Rice, Pea,Mungben	Growth promotion under abiotic stress	Palmqvist (2011), Zahir et al. (2008), Ahmad et al. (2011)
	Cotton	Buffering against biotic stress	Flavio et al. (2011)
	Mangroves rhizosphere	Bioremidiation potential	Syakti et al. (2013)

(continued)

Table 6.3 (continued)

Biopriming agent	Crops	Response	Reference
Agrobacterium	Radish	Buffering against abiotic stress	Kaymak et al. (2009)
Gliocladium	**Greenhouse cucumber, pepper & tomato**	Buffering against biotic stress	Sabaratnam (2012)
PGPR Consortia	Radish (Raphanus sativus L.)	Buffering against biotic stress	Çağlar et al. (2009)
	Mung bean, Maize, Chickpea, Rajma, Barley	Increased growth attributes	Shaharoona et al. (2006), Karthika and Vanangamudi (2013), Yadav et al. (2013), Mirshekari et al. (2012)
	Cucumber, Arabidopsis(*Arabidopsis thaliana*)	Buffering against biotic stress	Ryu et al. 2004, Ryu et al. (2005)

Table 6.4 Biopriming mediated nitrogen use efficiency in different crop cultivars

Crop	Bio agent	Nitrogen use efficiency	Reference
Rice(*Oryza sativa*)	A. amazonense	N (3.5–18.5%)	Rodrigues et al. (2008)
Maize (Zea mays)	*T. harzianum*	8.8–9.76% N in root; 3.5% N in shoot	Ageeb et al. (2012)
Soybean (*Glycine max*)	*Trichoderma harzianum AS 19–2*	N (15.8%)	Entesari et al. (2013)
Broccoli (*Brassica oleracea*)	AM fungi	N- 102.08%	Tanwar et al. (2013)
Cucumber (*Cucumis sativus*)	*T. harzianum 4*	N (13%); P (12%); K (11.7%)	Moharam and Negim (2012)
Melon (*Cucumis melo*)	*T. harzianum*	N (27.03%)	Martínez-Medina et al. (2011)
Tea (*Camellia sinensis*)	*Trichoderma harzianum*	N (44%)	Thomas et al. (2010)

6.4 Conclusions

Rising concern among scientists and general public regarding serious health hazards on human health associated with use of chemical in food supplies have propelled research for ecofriendly alternative approaches for integrated crop management for overall growth promotion and performance. Seed biopriming using biological control agents and growth promoter may be an appropriate alternate choice as biopriming with different beneficial microbes may not only enhance seed quality but also boost seedling vigor and ability to withstand abiotic and biotic stresses, and thus offer an innovative crop protection tool for the sustainable improvement of crop yield.

References

Ageeb, A. S., & Abbas, S. M. (2012). Application of Trichoderma harziunum T22 as a biofertilizer supporting maize growth. *African Journal of Biotechnology, 11*(35), 8672–8683.

Ahmad, M., Zahir, Z. A., Asghar, H. N., & Asghar, M. (2011). Inducing salt tolerance in mung bean through coinoculation with rhizobia and plant-growth-promoting rhizobacteria containing1 aminocyclopropane-1-carboxylate deaminase. *Canadian Journal of Microbiology, 57*(7), 578–589.

Amanullah, M. M., Ananthi, T., Subramanian, K. S., & Muthukrishnan, P. (2011). Influence of mycorrhiza, nitrogen and phosphorus on growth, yield and economics of hybrid maize. *The Madras Agricultural Journal, 98*, 62–66.

Ananthi, T., Amanullah, M. M., & Subramanian, K. S. (2011). Influence of fertilizer levels and mycorrhiza on root colonization, root attributes and yield of hybrid maize. *The Madras Agricultural Journal, 98*, 56–61.

Azarmi, R., Hajieghrari, B., & Gigloul, A. (2011). Effect of Trichoderma isolates on tomato seedling growth response and nutrient uptake. *African Journal of Biotechnology, 10*(31), 5850–5855.

Bisen, K., Keswani, C., Mishra, S., Saxena, A., Rakshit, A., & Singh, H. B. (2015). Unrealized potential of seed bio priming for versatile agriculture. In A. Rakshit et al. (Eds.), *Nutrient Use Efficiency: From Basics to Advances*. https://doi.org/10.1007/978-81-322-2169-2_12.

Çağlar, H. K., Guvenc, I., Yarali, F., & Donmez, M. F. (2009). The effects of bio-priming with PGPR on germination of radish (*Raphanus sativus* L.) seeds under saline conditions. *Turkish Journal of Agriculture and Forestry, 33*, 173–179.

Callan, W., Mathre, D. N. E., & Miller, J. B. (1991). Field performance of sweet corn seed bio-primed and coated with *Pseudomonas fluorescens* AB254. *Hortscience, 26*(9), 1163–1165.

de Santiago, A., García-López, A. M., Quintero, J. M., Avilés, M., & Delgado, A. (2012). Effect of *Trichoderma asperellum* strain T34 and glucose addition on iron nutrition in cucumber grown on calcareous soils. *Soil Biology & Biochemistry, 57*, 598–605.

Egamberdieva, D. (2008). Plant growth promoting properties of rhizobacteria isolated from wheat and pea grown in loamy sand soil. *Turkish Journal of Biology, 32*(1), 9–15.

Entesari, M., Sharifzadeh, F., Ahmadzadeh, M., & Farhangfar, M. (2013). Seed biopriming with *Trichoderma* species and *Pseudomonas fluorescent* on growth parameters, enzymes activity and nutritional status of soybean. *International Journal of Agronomy and Plant Production, 4*(4), 610–619.

Flavio, H. V., Ricardo, M. S., Fernanda, C., Medeiros, L., Huiming, Z., Terry, W., Paxton, P., Henrique, M. F., & Paul, W. P. (2011). Transcriptional profiling in cotton associated with *Bacillus subtilis* (UFLA285) induced biotic-stress tolerance. *Plant and Soil, 347*(1–2), 327–337.

Gamal, E., Salah, A. A., & Mutlaq, O. (2012). Nutritional quality of biscuit supplemented with wheat bran and date palm fruits (*Phoenix dactylifera* L.). *Food and Nutrition Sciences, 3*, 322–328.

Karthika, C., & Vanangamudi, K. (2013). Biopriming of maize hybrid COH (M) 5 seed with liquid biofertilizers for enhanced germination and vigour. *African Journal of Agricultural Research, 8*(25), 3310–3317.

Kaveh, H., Jartoodeh, S. V., Aruee, H., & Mazhabi, M. (2011). Would Trichoderma affect seed germination and seedling quality of two muskmelon cultivars, khatooni and qasri and increase their transplanting success? *Journal of Biological and Environmental Sciences, 5*(15), 169–175.

Kaymak, H. C., Yarali, F., Guvenc, I., & Donmez, M. F. (2008). The effect of inoculation with plant growth rhizobacteria (PGPR) on root formation of mint (*Mentha piperita* L.) cuttings. *African Journal of Biotechnology, 7*, 4479–4483.

Kaymak, Ç. H., Guvenec, İ., Yarali, F., & Donmez, M. F. (2009). The effects of bio-priming with PGPR on germination of radish (*Raphanus sativus* L.) seeds under saline conditions. *Turkish Journal of Agriculture and Forestry, 33*, 173–179.

Lalitha, P., Srujana, & Arunalakshmi, K. (2012). Effect of *Trichoderma viride* on germination of mustard and survival of mustard seedlings. *International Journal of Life Sciences Bio technology and Pharma Research, 1*(1), 137–140.

Lavania, M., Chauhan, P. S., Chauhan, S. V. S., Singh, H. B., & Nautiyal, C. S. (2006). Induction of plant defense enzymes and phenolics by treatment with plant growth-promoting rhizobacteria *Serratia marcescens* NBRI1213. *Current Microbiology, 52*(5), 363–368.

Li, Y., Ran, W., Zhang, R., Sun, S., & Xu, G. (2009). Facilitated legume nodulation, phosphate uptake and nitrogen transfer by arbuscular inoculation in an upland rice and mung bean intercropping system. *Plant and Soil, 315*, 285–296.

Martínez-Medina, A., Roldán, A., Albacete, A., & Pascual, J. A. (2011). The interaction with arbuscular mycorrhizal fungi or Trichoderma harzianum alters the shoot hormonal profile in melon plants. *Phytochemistry, 72*, 223–229. https://doi.org/10.1016/j.phytochem.2010.11.008.

Mirshekari, B., Hokmalipour, S., Sharifi, R. S., Farahvash, F., & Gadim, A. E. K. (2012). Effect of seed biopriming with Plant Growth Promoting Rhizobacteria (PGPR) on yield and dry matter accumulation of spring barley (*Hordeum vulgare* L.) at various levels of nitrogen and phosphorus fertilizers. *Journal of Food, Agriculture & Environment, 10*(3&4), 314–320.

Mobasser, H. R., & Moradgholi, A. (2012). Mycorrhizal bio-fertilizer applications on yield seed corn varieties in Iran. *Annals of Biological Research, 3*, 1109–1116.

Moeinzadeh, A., Sharif-Zadeh, F., Ahmadzadeh, M., & Tajabadi, F. H. (2010). Biopriming of sunflower (*Helianthus annuus* L.) seed with *Pseudomonas fluorescens* for improvement of seed invigoration and seedling growth. *AJCS, 4*(7), 564–570.

Moharam, M. H. A., & Negim, O. O. (2012). Bio control of fusarium wilt disease in cucumber with improvement of growth and mineral uptake using some antagonistic formulations. *Communications in Agricultural and Applied Biological Sciences Ghent University, 77*(3), 53–64.

Molla, A. H., Haque, M. M., Haque, M. A., & Ilias, G. N. M. (2012). *Trichoderma*-enriched biofertilizer enhances production and nutritional quality of tomato (*Lycopersicon esculentum* mill.) and minimizes NPK fertilizer use. *Journal of Agricultural Research, 1*(3), 265–272.

Palmqvist, M. (2011). *Biotic priming of growth and stress tolerance in cereals*. M.Sc. Thesis submitted to Department of Plant Biology and Forest Genetics, Swedish University of Agricultural Science, Uppsala.

Pirlak, L., & Kose, M. (2009). Effects of plant growth promoting rhizobacteria on yield and some fruit properties of strawberry. *Journal of Plant Nutrition, 32*(7), 1173–1184.

Radwan, S. S., Dashti, N., & El-Nemr, I. M. (2005). Enhancing the growth of Vicia faba plants by microbial inoculation to improve their phytoremediation potential for oily desert areas. *International Journal of Phytoremediation, 7*(1), 19–32.

Rakshit, A., Pal, S., Rai, S., Rai, A., Bhowmick, M. K., & Singh, H. B. (2013). Micronutrient seed priming: A potential tool in integrated nutrient management. *Satsa Mukhapatra (Annual Technical Issue), 17*, 77–89.

Rakshit, A., Pal, S., Meena, S., Manjhee, B., Preetipriya, Rai, S., Rai, A., Bhowmik, M. K., & Singh, H. B. (2014). Bio-priming: A potential tool in the integrated resource management. *Satsa Mukkhapatra (Annual Technical Issue), 18*, 94–103.

Rakshit, A., Sunita, K., Pal, S., Singh, A., & Singh, H. B. (2015). Bio-priming mediated nutrient use efficiency of crop species. In A. Rakshit et al. (Eds.), *Nutrient Use Efficiency: from Basics to Advances*. https://doi.org/10.1007/978-81-322-2169-2_12.

Rodrigues, E. P., Rodrigues, L. S., de Oliveira, A. L. M., Baldani, V. L. D., Teixeira, K. R. D., Urquiaga, S., & Reis, V. M. (2008). *Azospirillum amazonense* inoculation: Effects on growth, yield and N_2 fixation of rice (*Oryza sativa* L.). *Plant and Soil, 302*(1–2), 249–261.

Ryu, C. M., Murphy, J. F., Mysore, K. S., & Kloepper, J. W. (2004). Plant growth-promoting rhizobacteria systemically protect *Arabidopsis thaliana* against cucumber mosaic virus by a salicylic acid and NPR1-independent and jasmonic acid-dependent signaling pathway. *The Plant Journal, 39*(3), 381–392.

Ryu, C. M., Hu, C. H., Locy, R. D., & Kloepper, J. W. (2005). Study of mechanisms for plant growth promotion elicited by rhizobacteria in *Arabidopsis thaliana*. *Plant and Soil, 268*(1), 285–292.

Sabaratnam, S. (2012). *Pythium diseases of greenhouse vegetable crops*. Abbotsford Agriculture Centre British Columbia Ministry of Agriculture, Canada.

Shaharoona, B., Bibi, R., Arshad, M., Zahir, Z. A., & Zia, U. H. (2006). 1-Aminocylopropane-1-carboxylate (ACC) deaminase rhizobacteria extenuates ACC-induced classical triple response in etiolated pea seedlings. *Pakistan Journal of Botany, 38*(5), 1491–1499.

Shaharoona, B., Naveed, M., Arshad, M., & Zahir, Z. A. (2008). Fertilizer-dependent efficiency of *Pseudomonas* for improving growth, yield, and nutrient use efficiency of wheat (*Triticum aestivum* L.). *Applied Microbiology and Biotechnology, 79*(1), 147–155.

Shamshiri, M. H., Usha, K., & Singh, B. (2012). Growth and nutrient uptake responses of kinnow to vesicular arbuscular mycorrhizae. *ISRN Agronomy*, 1–7.

Sharif, M., Ahmad, E., Sarir, M. S., Muhammad, D., Shafi, M., & Bakht, J. (2011). Response of different crops to arbuscular mycorrhiza fungal inoculation in phosphorus-deficient soil. *Communications in Soil Science and Plant Analysis, 42*, 2299–2309.

Shukla, A., Kumar, A., Jha, A., Ajit, & Rao, D. V. K. N. (2012). Phosphorus threshold for arbuscular mycorrhizal colonization of crops and tree seedlings. *Biology and Fertility of Soils, 48*, 109–116.

Syakti, A. D., Yani, M., Hidayati, N. V., Siregar, A. S., Doumenq, P. I. M., & Sudiana, M. (2013). The bioremediation potential of hydrocarbonoclastic bacteria isolated from a mangrove contaminated by petroleum hydrocarbons on the cilacap coast, Indonesia. *Bioremediation Journal, 17*(1), 11–20.

Tajini, F., & Drevon, J. J. (2012). Phosphorus use efficiency in common bean (*Phaseolus vulgaris* L.) as related to compatibility of association among arbuscular mycorrhizal fungi and rhizobia. *African Journal of Biotechnology, 11*(58), 12173–12182.

Tanwar, A., Aggarwal, A., & Panwar, V. (2013). Arbuscular mycorrhizal fungi and Trichoderma viride mediated Fusarium wilt control in tomato. *Biocontrol Science and Technology, 23*(5), 485–498.

Thomas, J., Ajay, D., Raj Kumar, R., & Mandal, A. K. A. (2010). Influence of beneficial microorganisms during in vivo acclimatization of in vitro-derived tea (*Camellia sinensis*) plants. *Plant Cell, Tissue and Organ Culture, 101*, 365–370.

Yadav, S. K., Dave, A., Sarkar, A., & Singh, H. B. (2013). Co-inoculated Biopriming with *Trichoderma, Pseudomonas* and *Rhizobium* Improves Crop Growth in *Cicer arietinum* and *Phaseolus vulgaris*. *International Journal of Agriculture, Environment & Biotechnology, 6*(2), 255–259.

Yedidia, I., Srivastva, A. K., Kapulnik, Y., & Chet, I. (2001). Effect of Trichoderma harzianum on microelement concentrations and increased growth of cucumber plants. *Plant and Soil, 235*, 235–242.

Zahir, Z. A., Munir, A., Asghar, H. N., Shaharoona, B., & Arshad, M. (2008). Effectiveness of rhizobacteria containing ACC deaminase for growth promotion of peas (*Pisum sativum*) under drought conditions. *Journal of Microbiology and Biotechnology, 18*(5), 958–963.

Chapter 7
Greenhouse Gas Emissions from Selected Cropping Patterns in Bangladesh

M. M. Haque, J. C. Biswas, M. Maniruzzaman, A. K. Choudhury, U. A. Naher, B. Hossain, S. Akhter, F. Ahmed, and N. Kalra

Abstract There are many cropping systems followed in Bangladesh for enhancing cropping intensity and increasing crop production, but greenhouse gas (GHG) emission from agricultural fields are mostly reported on country basis. In order to estimate of GHG emission from agriculture fields, Cool Farm Tool Beta-3 was used to determine total GHG from selected cropping systems. It was found that non-rice based cropping system had lower global warming potential (GWP) than rice based cropping systems. Among the rice based cropping systems, Onion-Jute-Fallow, Jute-Rice-Fallow, Wheat-Mungbean-Rice and Maize-Fallow-Rice systems are relatively more suitable for reducing GHG emission and subsequent GWP. There are spatial variations in CH_4 emissions and the higher amounts were found in Mymensingh and Dinajpur districts in Bangladesh. In 2013–14, about 1.56 Tg year^{-1} CH_4 emissions took place from paddy field in Bangladesh. Further study is required for validation and suggesting suitable mitigation strategies to check the GHG emission in Bangladesh.

7.1 Introduction

The demand for food is increasing in Bangladesh due to rapid population growth. Farmers are growing different crops in a year to increase food production by following different cultural management options including use of variable amounts of fertilizers. Most of the farmers use excessive urea fertilizer (Biswas et al. 2004) and try to keep paddy field continuously flooded. These practices, not only increase cost of production, but also enhance additional GHG emission from crop fields. Annual total GHG emissions from agriculture are estimated to be 1.4–1.6 Gt CO_2-C equivalent (CO_2-Ce) yr^{-1}, corresponding to the attributed 10–12% of the human-induced

M. M. Haque · J. C. Biswas (✉) · M. Maniruzzaman · A. K. Choudhury
U. A. Naher · B. Hossain · S. Akhter · F. Ahmed
Bangladesh Rice Research Institute, Gazipur, Bangladesh

N. Kalra
Division of Agricultural Physics, IARI, New Delhi, India

S. Sheraz Mahdi (ed.), *Climate Change and Agriculture in India: Impact and Adaptation*, https://doi.org/10.1007/978-3-319-90086-5_7

warming effect (IPCC 2014). Major GHGs in general are emitted from agriculture field (Rice, barley, wheat and cereal crop) are CH_4, CO_2 and N_2O (Haque et al. 2015a, b). Rice crop covers about 85% of agricultural land in Bangladesh and contribute to global warming potential (GWP) (Solomon et al. 2007; Lee 2010). Globally 81% of agricultural emissions come from nitrogenous fertilizer production and its use (Iserman 1994). These imply that climate smart agricultural practices need to be followed for reducing GHG emission from crop fields; but pattern based emission data are lacking in Bangladesh. So, GWP for selected major cropping patterns and CH_4 emission from paddy fields were computed for subsequent ecosystem modification and adaptation in crop production.

7.2 Materials and Methods

Field experimental data were collected from Hand Book of Agricultural Technology, Proceedings Research Review and Planning Workshop of Soils Program of NARS institutes (2015) and different research organization in Bangladesh. Crop area data of 2013–14 were collected from Year Book of Agricultural Statistics-2014. Cool Farm Tool Beta-3 (CFT) was used to determine total GHG gas emission under different cropping systems and expressed as GWP. Many major cropping systems are followed in different districts. Among them Jute-T. Aman (rainfed lowland)-Fallow, Onion-Jute-Fallow, Boro (dry season irrigated rice)-Fallow-T. Aman, Mustard-Boro-T. Aman, Mustard-Boro-Fallow, Wheat-T. Aus (pre-monsoon)-T. Aman, Potato-Boro-T. Aman, Maize-Fallow-T. Aman, Potato-Maize-T. Aman, Wheat-Mungbean-T. Aman and Grass pea-T. Aus-T. Aman cropping systems were undertaken for estimation of total GHG and GWP. Emission factors, input variables and outputs of CFT are as follows.

Emission factor	Input variables	CFC output
Fertilizer induced N_2O	Fertilizer types/application rate ha^{-1}/ management practices ha^{-1}	kg CO_2eq/ha, kg CO_2eq/ kg product
Fertilizer production	Fertilizer type/ application rate, production technology	kg CO_2eq/ha, Kg CO_2eq/ kg product
Pesticide production	Number of applications	kg CO_2eq/ha, Kg CO_2eq/ kg product
Diesel use	Liters used	kg CO_2eq/ha, Kg CO_2eq/ kg product
Electricity use	Kwh	Kg CO_2eq/ha, Kg CO_2eq/ kg product
Crop residue management	kg/management practice	kg CO_2eq/ha, Kg CO_2eq/ kg product
Water management	Liters/management practice	kg CO_2eq/ha, Kg CO_2eq/ kg product

7.2.1 *Correction Factor Determine of GHG*

Using static close chamber method (Haque et al. 2013, 2015a and Haque et al. 2016a, b, c) and CFT (Hiller et al. Hillier et al. 2011), correction factor was determined for actual GHG and GWP estimate under major cropping systems in Bangladesh.

7.2.2 *Statistical Analysis*

Statistical analyses were carried out using SAS software (SAS Institute 1995). Fisher's protected Least Significant Difference (LSD) was computed at 0.05 probability level for making treatment means comparison.

7.3 Results

7.3.1 *Cropping System Based GHG*

Total CH_4 emission was about 48 kg ha^{-1} under Jute-T. Aman-Fallow, Maize-Fallow-T. Aman, Potato-Maize-T. Aman and Wheat-Mungbean-T. Aman. However, rice based cropping system like Jute-T. Aman-Fallow showed significantly lower amounts of GHG than others systems (Table 7.1). Rice-Rice based cropping systems showed significantly higher amounts of CH_4 emission (97–295 kg ha^{-1}), but CO_2 and N_2O emissions were not significant. Rice-Rice-Fallow cropping systems

Table 7.1 Greenhouse gas emission from major cropping systems in Bangladesh

Cropping system	CO_2 (kg ha^{-1})	CH_4 (kg ha^{-1})	CH_4 (CrF) (kg ha^{-1})	N_2O (kg ha^{-1})
Jute-Fallow-Onion	836.6i	0f	0f	4.3bcd
Jute-T. Aman-Fallow	668.4 k	48.4e	40.17e	4.2 cd
Boro (IF)-T. Aman-Fallow[a]	1114.2ef	196.6b	163.17b	3.9d
Boro (CF)-T. Aman-Fallow[a]	1141de	295.4a	245.18a	3.9d
Mustard-Boro-T. Aman	1516.1c	196.6b	163.17b	4.8abc
Mustard-Boro-Fallow	1082.6gh	148.2c	123.0c	3.8d
Wheat-T. Aus-T. Aman	1109.1 fg	96.8d	80.34d	2.5f
Potato-Boro-T. Aman	1871.3b	196.6b	163.17b	3.5de
Maize-Fallow-T. Aman	1167.5d	48.4e	40.17e	5.5a
Potato-Maize-T. Aman	1924.9a	48.4e	40.17e	5.1ab
Wheat-Mungbean-T. Aman	1080.9 h	48.4e	40.17e	3.5de
Grass pea-T. Aus-T. Aman	799.2j	96.8d	80.34d	2.8ef

Small letters in a column compare means at 5% level of probability by LSD
[a]IF = Intermittent flooding, CF = Continuous flooding, CrF = After imposing correction factor

Table 7.2 Global warming potential from selected cropping pattern under standard chemical fertilization

Cropping system	GWP (CO_2eq kg ha^{-1})	GWP (CO_2eq kg ha^{-1}) (CrF)[a]
Onion-Jute-Fallow	2125 k	2125 k
Jute-T. Aman-Fallow	3129j	2923j
Boro (IF)-T. Aman-Fallow[a]	7191d	6355d
Boro (CF)-T. Aman-Fallow[a]	9688a	8432a
Mustard-Boro-T. Aman	7854b	7018b
Mustard-Boro-Fallow	6376e	5746e
Wheat-T. Aus-T. Aman	4592f	4180f
Potato-Boro-T. Aman	7811c	6975c
Maize-Fallow-T. Aman	3988 h	3782 h
Potato-Maize-T. Aman	4618f	4412f
Wheat-Mungbean-T. Aman	3315i	3109i
Grass pea-T. Aus-T. Aman	4055 g	3643 g

Small letters in a column compare means at 5% level of probability by LSD
[a]*IF* = Intermittent flooding, *CF* = Continuous flooding, *CrF* = Correcting factor

increased about 102–515% CH_4 and reduced 8–41% N_2O than Jute-T. Aman-Fallow, Maize-Fallow-T. Aman, Potato-Maize-T. Aman and Wheat-Mungbean-T. Aman based systems in Bangladesh (Table 7.1). Carbon dioxide emission significantly increased under Potato-Maize-T. Aman cropping system but it was significantly the lowest under Jute-T. Aman-Fallow system. Non rice cropping systems showed the lowest GHG emission.

7.3.2 *Cropping System Based GWP*

The GWP among different cropping systems varied significantly. Computed GWP was significantly the lowest (2125 CO_2eq kg ha^{-1}) in Jute-Fallow-Onion and the highest (9688 and 7854 CO2eq kg ha^{-1}) in Boro (CF)-T. Aman-Fallow and Mustard-Boro-T. Aman cropping systems (Table 7.2). Major cereal cropping systems viz. Jute-T. Aman-Fallow (3129 CO_2eq kg ha^{-1}), Maize-Fallow-T. Aman (3988 CO_2eq kg ha^{-1}), Potato-Maize-T. Aman (4618 CO_2eq kg ha^{-1}), Grass pea-T. Aus-T. Aman (4055 CO_2eq kg ha^{-1}), Wheat-Mungbean-T. Aman (3315 CO_2eq kg ha^{-1}) systems showed significantly the lowest GWP than other cropping systems.

7.3.3 *Management and GHG Emission*

Water management significantly influenced CH_4 emission but not $CO_2$and N_2O emissions (Table 7.3). About 40% CH_4 emission was reduced because of intermittent drainage. The choice of variety also influences GHG emissions. For example,

Table 7.3 GHG and GWP as influenced by water management

	GHG (kg ha^{-1})			
Water management	CH_4	CO_2	N_2O	GWP
Intermittent drainage	148b	543a	1.1a	4585b
Continuous flooding	247a	570a	1.1a	70821a

Small letters in a column compare means at 5% level of probability by LSD

Table 7.4 GHG and GWP as influenced by varietal differences

	Greenhouse gas emission (kg ha^{-1})			
Variety (wet season)	CH_4	CO_2	N_2O	GWP
HYV Rice	48.4b	406a	0.9a	1875b
Local Rice	195a	199b	0.6b	5255a

Small letters in a column compare means at 5% level of probability by LSD

high yielding rice varieties (HYV) showed significantly higher emission of CO_2 and N_2O than local rice varieties but significantly lower amounts of CH_4 emission than local rice varieties (Table 7.4).

7.3.4 Methane Production Area in Bangladesh

Our result indicated that Mymensingh and Dinajpur had significantly higher amounts of CH_4 emission than other districts of Bangladesh (Fig. 7.1). Among the 64 district, the lowest CH_4 was found in Ramgati and Bandarban. In terms of CH4 emission rate, it varied from 89 to 148 kg ha^{-1} $year^{-1}$ depending on locations of the country and types of rice culture and variety used (Fig. 7.2). In total, computed CH_4 emission was about 1.56 Tg $year^{-1}$ in Bangladesh (Fig. 7.3).

7.4 Discussion

In Bangladesh, very suitable (VS), suitable (S) and moderately suitable (MS) areas for T. Aus (Pre-monsoon), T. Aman (Monsoon) and Boro (Dry season irrigated rice) rice covers about 2.01, 2.01 and 2.43 million ha (Mha) of cultivable land, respectively (Hossain et al. 2012). In future, such suitable areas will be affected because of increase in temperature. Boro rice based cropping system gave higher GWP than T. Aus and T. Aman rice based cropping systems because of variations in growth duration, fertilizer and water requirements than other rice varieties. Haque et al. (2016a, c) found that fertilizer and irrigated water increases total GHG emission and subsequent GWP. Efficient water and efficient fertilizer management practices need to be followed for reducing GWP from rice based cropping systems.

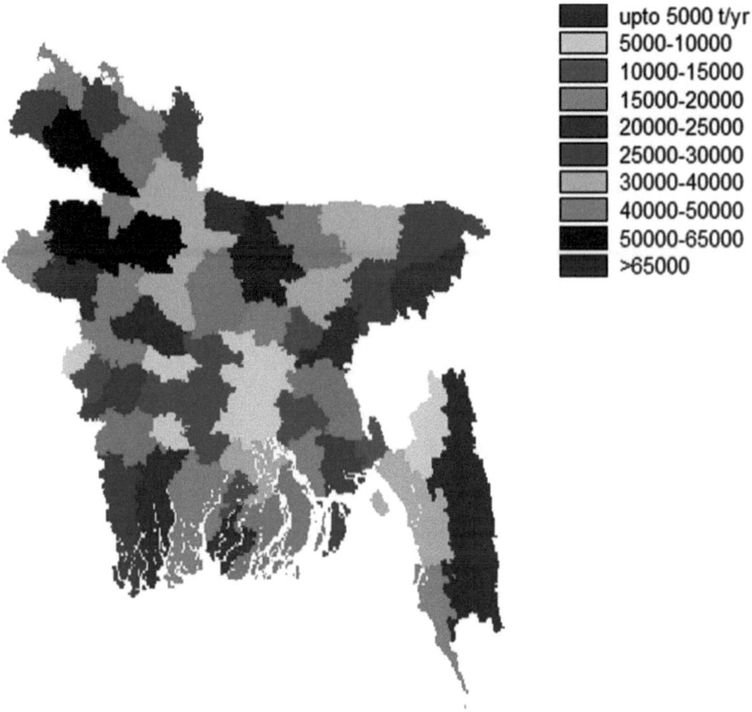

Fig. 7.1 Annual methane emission from paddy fields in different regions of Bangladesh

Jute-Rice-Fallow (3129 CO_2eq kg ha^{-1}), Wheat-Mungbean-Rice (3315 CO_2eq kg ha^{-1}) and Maize-Fallow-Rice (3988 CO_2eq kg ha^{-1}) systems could also be alternate options for mitigation of GHG emission in Bangladesh. However, Wheat-Rice-Rice (4592 CO_2eq kg ha^{-1}) and Potato-Maize-Rice (4618 CO_2eq kg ha^{-1}) systems need to be practised considering food security of the country (Table 7.2). Earlier findings also mentioned that rice based cropping system gave higher CH_4 and GWP than other cropping systems (Haque et al. 2015a). However, it is clear that adopting more effective and efficient cropping systems play a key role in increasing crop yields while mitigating emission of GHG in agriculture. Integrated soil-crop management practices are advocated to address the key constraints to yield improvement and alleviate environmental impacts, specifically reducing GHG emission (Fan et al. 2011; Zhang et al. 2012).

Amongst different cereal crops grown worldwide, rice emits the highest GHG, especially when grown under irrigated conditions (Table 7.1). This is because CH_4 emission is partly mediated by rice plants. Methane emission varies across different regions of the country because of rice culture types, varieties and water management conditions. In low lying areas of Bangladesh, local rice cultivars are gown in flooded lands and remain water stagnant almost up to maturity of the crop. This flooding condition favors greater CH_4 emission from paddy fields. Similar findings were also reported by Gupta et al. 2009 and Alberto et al. 2014. Intermittent drainage at critical stages of

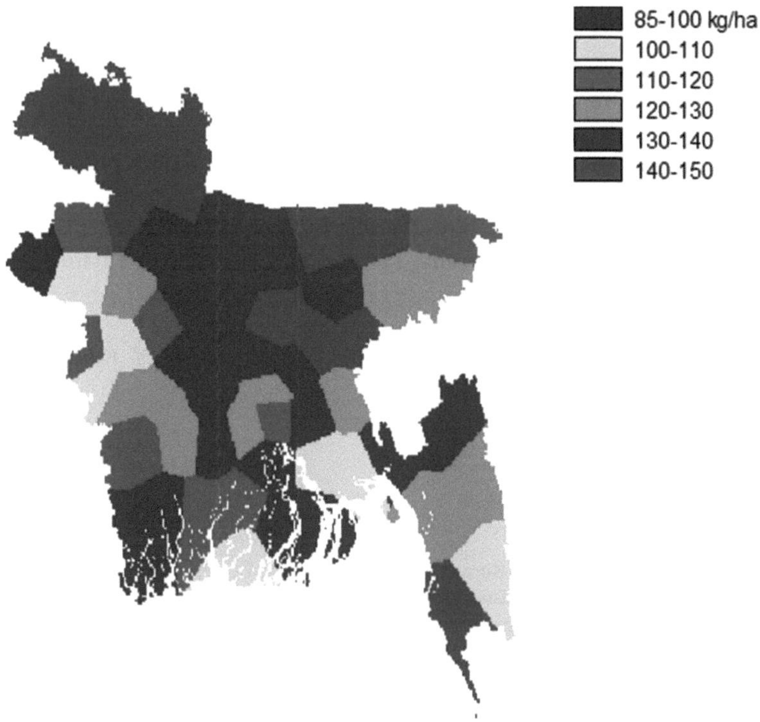

Fig. 7.2 Methane emission rates from paddy field in different regions of Bangladesh

Fig. 7.3 Total methane emission (Tg year^{-1}) from paddy fields in Bangladesh

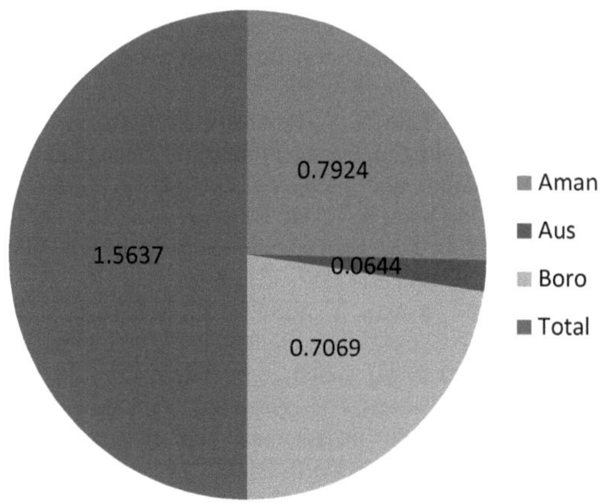

crop growth is one of the significant options to reduce carbon footprint of paddy cultivation. Livestock also plays a key role in CH_4 emission and also needs to be accounted. There is a need to validate the model results with specific test studies. However, the results as obtained from model run in the present investigation are in agreement with the reports for other similar production environments in South Asia.

7.5 Conclusion

There are spatial variations in the GHG emissions over Bangladesh, primarily because of cropping systems differences and inputs and/or management practices followed. Jute-Rice-Fallow and Wheat-Mungbean-Rice cropping systems are suitable to reduce GHG emission and subsequent GWP than other cropping systems. There is a need to validate models results with location specific studies. The options needed to mitigate GHG emission for various productions environments of Bangladesh are to be delineated.

Acknowledgement We greatly acknowledge the financial support of Krishi Gobeshona Foundation (KGF) through CRP-II project.

References

Alberto, M. C. R., Wassmann, R., Buresh, R. J., Quality, J. R., Correa, T. Q., Jr., Sandor, J. M., & Centeno, C. A. R. (2014). Measuring methane flux from irrigated rice fields by Eddy covariance method using open-path gas analyzer. *Field Crops Research, 160*, 12–21.

Biswas, J. C., Islam, M. R., Biswas, S. R., & Islam, M. J. (2004). Crop productivity at farmers fields: Options for soil test based fertilizer use and cropping patterns. *Bangladesh Agronomy Journal, 10*, 31–41.

Fan, M. S., Shen, J. B., Yuan, L. X., et al. (2011). Improving crop productivity and resource use efficiency to ensure food security and environmental quality in China. *Journal of Experimental Botany, 63*, 13–24.

Gupta, P. K., Gupta, V., Sharma, C., Das, S. N., Purkait, N., Adhya, T. K., Pathak, H., Ramesh, R., Baruah, K. K., Venkataraman, L., Singh, G., & Iyer, C. S. P. (2009). Development of methane emission factors for Indian paddy fields and estimation of national methane budget. *Chemophore, 74*, 590–598.

Haque, M. M., Kim, S. Y., Pramanik, P., Kim, G. Y., & Kim, P. J. (2013). Optimum application level of winter cover crop biomass as green manure under considering methane emission and rice productivity in paddy soil. *Biology and Fertility of Soils, 49*, 487–493.

Haque, M. M., Kim, S. Y., Ali, M. A., & Kim, P. J. (2015a). Contribution of greenhouse gas emissions during cropping and fallow seasons on total global warming potential in mono-rice paddy soils. *Plant and Soil, 387*, 251–264.

Haque, M. M., Kim, S. Y., Kim, G. W., & Kim, P. J. (2015b). Optimization of removal and recycling ratio of cover crop biomass using carbon balance to sustain soil organic carbon stocks in a mono-rice paddy system. *Agriculture, Ecosystems and Environment, 207*, 119–125.

Haque, M. M., Biswas, J. C., Kim, S. Y., & Kim, P. J. (2016a). Intermittent drainage in paddy soil: Ecosystem carbon budget and global warming potential. *Paddy and Water Environment*. https://doi.org/10.1007/s10333-016-0558-7.

Haque, M. M., Biswas, J. C., Waghmode, T. R., & Kim, P. J. (2016b). Global warming as affected by incorporation of variably aged biomass of hairy vetch for rice cultivation. *Soil Research*. https://doi.org/10.1071/SR15061.

Haque, M. M., Biswas, J. C., Kim, S. Y., & Kim, P. J. (2016c). Suppressing methane emission and global warming potential from rice fields through intermittent drainage and green biomass amendment. *Soil Use and Management*. https://doi.org/10.1111/sum.12229.

Hillier, J., Walter, C., Malin, D., Garcia-Suarez, T., Mila-i-Canals, L., & Smith, P. (2011). A farm-focused calculator for emissions from crop and livestock production. *Environmental Modelling & Software, 26*, 1070–1078.

Hossain Sk, G., Chowdhury, M. K. A., & Chowdhury, M. A. H. (2012). *Land suitability assessment and crop zoning of Bangladesh* (pp. 1215–1110). Dhaka: Bangladesh Agricultural Research Council.

IPCC. (2014). Climate change 2014: Mitigation of climate change. In O. Edenhofer, R. Pichs-Madruga, Y. Sokona, E. Farahani, S. Kadner, K. Seyboth, A. Adler, I. Baum, S. Brunner, P. Eickemeier, B. Kriemann, J. Savolainen, S. Schomer, C. von Stechow, T. Zwickel, & J. C. Minx (Eds.), *Contribution of working group III to the fifth assessment report of the intergovernmental panel on climate change*. Cambridge and New York: Cambridge University Press.

Iserman, K. (1994). Agriculture's share in the emission of trace gases affecting the climate and some cause-oriented proposals for sufficiently reducing this share. *Environmental Pollution, 83*, 95–111.

Lee, Y. H. (2010). Evaluation of no-tillage rice cover crop cropping system for organic farming. *Korean Journal of Soil Science and Fertilizer, 43*, 200–208.

SAS Institute. (1995). *System for windows release* (Vol. 6, p. 11). Cary: SAS Institute.

Solomon, S., Qin, D., Manning, M., Chen, Z., & Marquis, M. (2007). *Contribution of working group I to the fourth assessment report of the intergovernmental panel on climate change*. Cambridge: Cambridge University Press.

Xu, X., Zhang, B., Liu, Y., Yanni, Y., & BDi, X. (2013). Carbon footprints of rice production in five typical rice districts in China. *Acta Ecologica Sinica, 33*, 227–232.

Zhang F S, Cui Z L, Chen X P et al (2012). Integrated nutrient management for food security and environmental quality in China. In: Sparks, D L (ed) Advances in Agronomy 116, 1–40.

Chapter 8
Global Climate Change and Inland Open Water Fisheries in India: Impact and Adaptations

B. K. Das, U. K. Sarkar, and K. Roy

Abstract India has crossed the fisheries production of 10 million tonnes in 2015 and presently on its way to achieve the second blue revolution. Among all the major factors impeding sustainability of fisheries, factor of climate change is the recent addition. Climate change trends along major river basins of India have revealed a warming trend (0.2–0.5 °C), declining rainfall (257–580 mm) and shifting seasonality of rainfall occurrence. Rising sea levels (1.06–1.75 mm/year), receding Himalayan glaciers and frequent occurrence of extreme weather events are also a matter as per IPCC AR5. The present article discusses the contributions made by ICAR-CIFRI since 2004 on climate change vulnerability assessment framework, changes in breeding phenology of fishes, models on fish reproduction and diversity, thermal tolerance of fishes, carbon sequestration potential of wetlands and indigenous climate smart fisheries adaptation strategies. In addition, understanding the response and adaptation capacity of fishing and fishers to the physical and biological changes have also been discussed in the chapter.

Keywords Climate change · Open water fisheries · Impact and mitigation

8.1 Introduction

A comprehensive and exclusive database of climate change impact on fisheries production is non-existent in India. It is quite tedious to quantify and precisely focus on climate change effects on fish production statistics. In last 10 years (2005–2015) the inland fisheries production have increased from 3.52 MT to 6.92 MT, nearly a two fold increase with an average annual growth rate of 6.06 percent. Aquaculture and culture based fisheries alone contribute 70–80% of fish production from Inland waters. Rest is contributed by capture fisheries of wild fish stocks. The contribution of fisheries sector to the GDP has gone up from 0.46 per cent in 1950–51 to

B. K. Das (✉) · U. K. Sarkar · K. Roy
ICAR-Central Inland Fisheries Research Institute, Barrackpore/Kolkata, West Bengal, India

S. Sheraz Mahdi (ed.), *Climate Change and Agriculture in India: Impact and Adaptation*, https://doi.org/10.1007/978-3-319-90086-5_8

0.83percent in 2013–14. The share of fisheries in agricultural GDP has impressively increased during this period from a mere 0.84 per cent to 4.75 per cent.

Climate change is expected to alter the water temperatures, water levels and stream flow in the inland water bodies and have significant impacts. This may have an effect on inland fish species and their associated fisheries, and consequently will impact recreational and subsistence fishers, human communities, and economies. Climate change is predicted to affect aquatic ecosystems in diverse ways with implications for management of inland fishes and fisheries (Carlson and Lederman 2016). For example, the frequency of weather events that alter the availability and movement of water (e.g., droughts, heavy precipitation, heat waves) is predicted to increase with climate change (Saha et al. 2006). Rising sea levels are predicted to cause saltwater intrusion (i.e., replacement of freshwater by saltwater) in coastal aquifers (Iyalomhe et al. 2015), which may alter habitat suitability for freshwater and marine fishes. Climate change is altering the physical, chemical, and biological characteristics of freshwater habitats (Hartmann et al. 2013), with concomitant effects on freshwater and diadromous fishes. Warmer air temperatures resulting from climate change are expected to increase water temperatures, with effects on growth, reproduction, and survival of fishes and their prey (Hershkovitz et al. 2015, Kanno et al. 2015). Moreover, climate change is predicted to alter species interactions, the timing of important life history events (e.g., migration, spawning), and the spatial distribution of fish populations (Lynch et al. 2016). On a physiological level, effects of climate change on individual fish include reduced immune function, decreased cardiovascular performance, and changes in reproductive investment (Whitney et al. 2016). Chemical characteristics of water bodies, such as dissolved oxygen (Ito and Momii 2015), salinity (Bonte and Zwolsmen 2010), and nutrient concentrations (Moss et al. 2011), are directly influenced by these climate induced changes in thermal and hydrologic regimes. In recent years climate variability has threatened the sustainability of inland fisheries and dependent fishers in India. Some of the changes in the hydrology having potential repercussions on inland fisheries are: flood magnitude and frequency could increase owing to more intense precipitation events; water temperature will increase; low flows would be more severe owing to increased evaporation; peak stream flow would move from spring to winter owing to earlier thaw.

In nutshell, the fisheries sector is booming and contributing increasingly to the economic growth of the nation while climate change impact has become a concern that needs to be addressed. The present communication deals with climate change scenario in India, its probable impact on inland open water fisheries and suggest adaptation strategies for increasing their resilience to climate change.

8.2 Climate Change Trends

The latest climate change forecast based on IPCC (2014) AR5 predictions for regions of South-East Asia, of which India is a part, includes the following:

- Intense summer
- Increased occurrence of thunderstorms
- More non-seasonal rains causing gradual shifting of monsoon proper
- Accelerated hydro-cycle resulting in frequent occurrence of high-intensity short-duration dry season followed by low-intensity long-duration wet seasons in a loop
- Shortening of winter
- High Places (high altitude) will become warmer. Wet places will be wetter and dry places will be drier
- Rate of sea level rise and momentum of global warming showing no sign of recession and will continue to progress as predicted in IPCC (2014)

8.2.1 Climate Change Trends Along Major River Basins

Although high spatial differences occur in terms of changing climate along river Ganga but in general it can be mentioned that a warming temperature (+0.20 to +0.47 °C) and a decreasing total annual rainfall (−257 to - 580 mm) have occurred over the last 3 decades. On the contrary, along a southern peninsular river (River Cauvery), an extension of monsoon season (+14 days) and an increasing trend of annual rainfall (+80 mm) have occurred in addition to the warming temperature scenario. This depicts the high regional variability of both climate change and its anticipated impacts on fisheries making it hard to conclude a unanimous trend.

8.3 Impacts of Climate Change on Inland Fisheries

It has been found that the impact of climate change on inland fisheries is quite wide. It ranges from changes in range distribution, breeding and spawning behavior, growth rates, thermal tolerance, stress physiology, invasion of exotics to impact on aquatic primary productivity, habitat quality through sedimentation, water stress, aquatic weed proliferation and salt water intrusion. The major contributions made by ICAR-CIFRI under NICRA Project are summarized.

8.3.1 Impact of Enhanced Thermal Regime

8.3.1.1 Changes in Geographical Distribution and Diversity of Fishes

Expansion in the range of non-native warm water fishes (*Glossogobius giuris, Puntius ticto, Xenentodon cancila* and *Mystus vittatus*) from middle stretch of river Ganga up to Haridwar have been observed due to an increase in mean air temperature by +0.99 °C.

8.3.1.2 Effects on Growth, Stress and Reproductive Physiology

It is evident from the recent ongoing farming practices in the Uttarakhand hills (1200–1600 msl) that Indian major carps, particularly *Labeo rohita* are thriving well in the pond conditions at Pati, Champawat district, where it did not survived in earlier trials made during past 10–15 years back, due to low temperature. This could be due to decrease in frost duration in the region and resultant increase of water temperature. This indicates that the increased water temperature might support culture of Indian major carps in the upland regions in coming years. But at the same time this would be an alarming signal for existence of valuable trout fishery.

Increasing temperatures could bring the advantages of faster growth rates and longer growing seasons. Specific growth rates in IMCs remain comparatively high between water temperatures of 29–34 °C. However at temperatures near or beyond thermal tolerance limits, impairment of homeostasis occurs. The changes evident are hypercholesterolemia indicating impaired sterol mechanism, hyperglycemia and decreased blood sugar regulatory mechanism. Pituitary activation as evidenced by interrenal ascorbic acid depletion and cortisol elevation is pronounced. Oxygen consumption in both the fishes increased as judged by increased haemoglobin. Serum glucose, protein, triglycerides, T3 and T4 were increased in both in fast and slower rate of temperature increment. These indicate that the homeostatic mechanism of the fish is stressed which makes them highly susceptible to diseases.

All the stages of reproduction in fish viz., gametogenesis and gamete maturation, ovulation/spermiation, spawning and early development stages are affected by temperature. Temperature change modulates the hormone action at all levels of reproductive endocrine cascade. If stress is maintained then the effects start manifesting by the inhibition of reproductive function, cessation of ovulation, depression of reproductive hormones in blood and ovarian failure.

8.3.1.3 Changes in Breeding Behaviour and Recruitment

Fish farmers in aquaculture hatcheries of major fish breeding states are witnessing an extended breeding period of Indian major carps (*C. catla, L. rohita and C. mrigala)*. In recent decades the phenomenon of IMC maturing and spawning as early as March is observed due to increase in both mean air temperature and rainfall during pre-monsoon (Fig. 8.1).

An extended breeding period in golden mahseer *Tor putitora* and snow trout *Schizothorax richardsonii* primarily due to enhanced thermal regime and prolonged erratic monsoon was observed in recent years when compared with historical records.

The anadromous Indian river shad *Tenualosa ilisha* commonly known as 'Hilsa' forms a very important commercial fishery along the lower stretch of River Ganga especially in West Bengal. Gonadal maturation and spawning of Hilsa in river Ganga is influenced by water temperature. To be precise, water temperatures between 29–32 °C are necessary for attainment of gonadal ripeness and subsequent

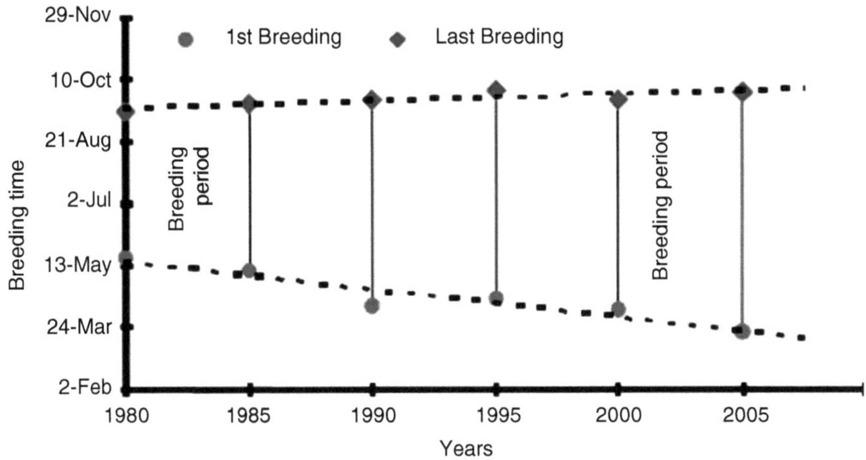

Fig. 8.1 Graph showing advancement and extension of breeding in Indian Major Carps

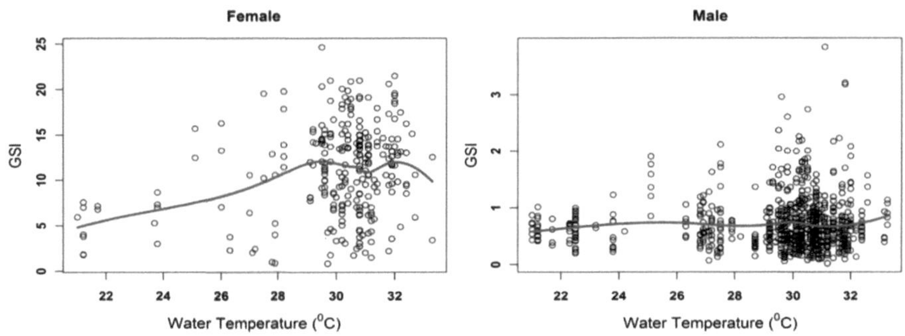

Fig. 8.2 Optimum water temperature for breeding of Hilsa in Ganga lower stretch

spawning in this species. Earlier studies had reported optimal water temperature range between 26–30 °C was ideal for spawning of Hilsa in River Ganga but recent studies of CIFRI have updated this range to 29–32 °C stating that if the projected water temperature in the region due to global warming lies within this range then spawning will not be disturbed given that the precipitation remains normal (Fig. 8.2).

8.3.1.4 Thermal Tolerance Limits of Fish

A total of around 15 commercially important freshwater inland fish species were screened for their *ex-situ* upper thermal tolerance levels. It was found that for most of the fish species the tolerance limits ranged around 39–41 °C among which cat-fishes, snakeheads and gobies were the most thermally tolerant fish species while the carps were weaker in terms of their upper thermal tolerance. Bottom dwelling fish species are generally more tolerant to water and thermal stress followed by pelagic and surface inhabiting fishes. Fishes with accessory air breathing organs adapted better under warm and low water conditions. Fingerlings show better acclimatization potential to warmer waters in comparison with adults.

8.3.1.5 Changes in Water Quality

Primary productivity in inland waters might increase under enhanced thermal regime. Although global warming manifested into warmer water temperatures round the year may initially result in increased fish food organisms (algal population > zooplankton) in rivers but in the long run this effect might not remain beneficial especially during summer. As surface water temperatures increase with predicted climate change, the solubility of dissolved oxygen (DO) in those waters will decrease. A tendency of progressive eutrophication also exists in the semi closed or closed water bodies owing to the enhanced availability of nutrients under high temperature-low water conditions.

8.3.2 Impact of Changing Precipitation Pattern

8.3.2.1 Changes in Fish Assemblage Pattern

Under a continuously warming climate scenario, highest congregation or assemblage of fishes in specific stretches of river is expected to coincide with the periods of maximum rainfall. Models have been developed for predicting fish assemblage patterns in specific stretches of river Ganges under variable climatic conditions (Fig. 8.3).

Analysis of catch composition from fish landings at specific stretches of River Ganga and Cauvery indicated a reduction in the share of commercially important indigenous fishes like IMCs in the Ganga catch while the contribution of exotics especially Common carp, Tilapia and other exotic fishes like African catfish, Suckermouth catfish have increased significantly to the extent of 25–50%. Simultaneously a warming trend, decreasing monsoonal rainfall and increasing rainfall during pre and post monsoon prevailed in Allahabad. Similarly in a southern peninsular river, Cauvery, comparison with historical records of fish composition revealed a decline of various earlier abundant fish species like *Hemibagrus punctatus, Puntius carnaticus, Gonoproktopterus dubius, Tenualosa ilisha, Cirrhinus cirrhosa* alongside an unusual increase in abundance of exotic species *Oreochromis mossabicus, O. niloticus* and transplanted *Catla catla.* The climatic trend during the same period showed a decrease in the number of rainy days during south west monsoon and an increase during the north east monsoon.

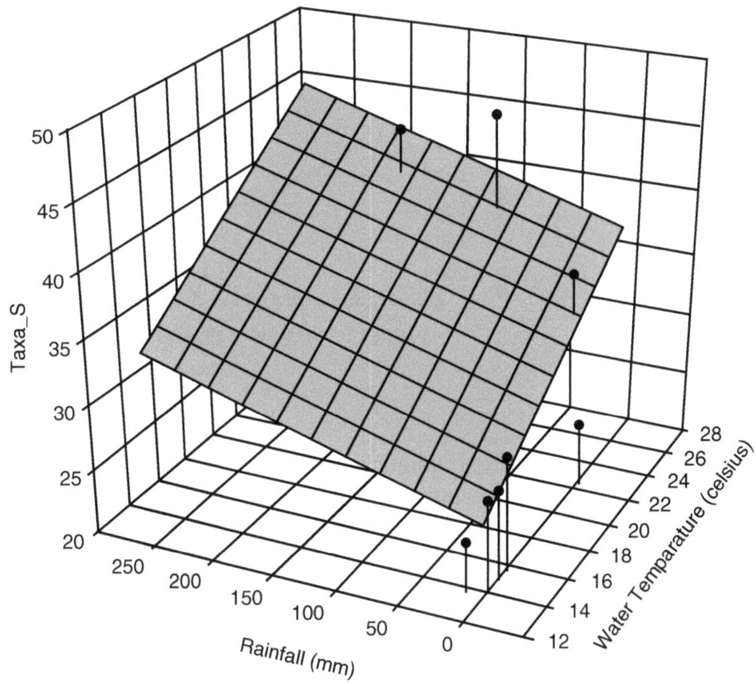

Taxa_S = 12.400 + (0.965 * Water Temparature) + (0.0271 * Rainfall)

R = 0.838

Fig. 8.3 A typical multi-parameter 3D scatter-mesh regression model for predicting fish assemblage

8.3.2.2 Changes in Recruitment and Riverine Fish Seed Availability

The fish spawn availability index of IMCs in river Ganga declined from an average of 1529 ml during (1965–69) to an average of 568 ml in recent years (2005–2009) due to decreasing trend of rainfall exclusively during breeding months (May–August) (Fig. 8.4).

8.3.2.3 Changes in Breeding Periodicity and Adaptation of Reproductive Phenology

Recent studies with some commercially important target fish species in River Ganga and associated wetlands has hinted that region-specific changes in breeding periodicity of some riverine catfishes and floodplain inhabiting snakeheads may likely occur through extension, shortening or shift in breeding season primarily due to changes in rainfall pattern followed by rise in mean air temperature manifested into

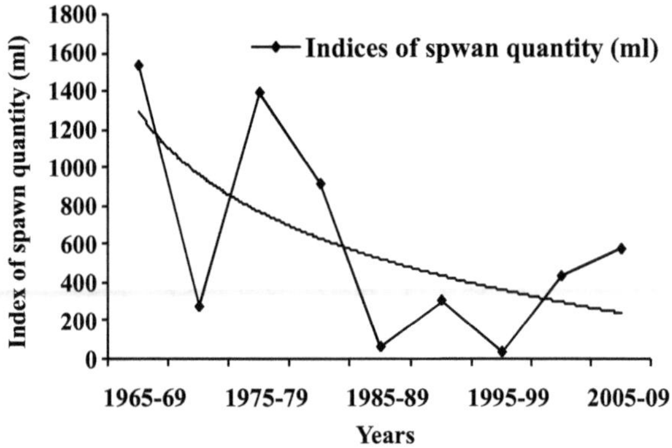

Fig. 8.4 Graph showing decreasing riverine fish seed availability of IMC (Sharma et al. 2015)

water temperature. Among the two climatic variables *i.e.* - Temperature and Rainfall, the latter seems to impart higher influence on breeding behaviour in natural open waters in a warm tropical climatic setup like that of India. Region-specific adaptation in reproductive phenology based on local trends of changing climate along River Ganga has also been suspected for some fishes.

8.3.2.4 Changes in Environmental Flows and Habitat Availability

Sub-optimum environmental flows in rivers and water stress in wetlands are the primary threats for inland fisheries. On the other hand, changing rainfall pattern during pre-monsoon and/or post-monsoon may either result in earlier flow pulse or delayed flow pulse in rivers having serious consequences on larval survivability and reproductive success of seasonally breeding fishes. Studies have been initiated on this line to map the possible impacts species-wise.

Floodplain wetlands are perceived to be the most impacted due to changing precipitation pattern (Sarkar et al. 2016). Studies on some floodplain wetlands revealed that problems like *Water stress, Wetland accretion/sedimentation, Aquatic weed proliferation, Loss of wetland connectivity with Parent River* are expected to aggravate in future climate scenario which may likely cause decline in fish production from these waters due to habitat unavailability.

8.3.3 Impact of Inter-Annual Climatic Variability and Extreme Climatic Events

8.3.3.1 Impact on Culture Fisheries Production from Inland Waters

It was found that inter-annual climatic variability have much lesser impact on fish production than management decisions like stocking and feeding. If proper fish culture management protocols are consistently maintained then fish production might increase in future without much detrimental impact caused due to local trends of changing climate.

8.3.3.2 Vulnerability of Low Lying Coastal Areas

Inland coastal lowlands are most vulnerable to extreme weather events like storm surges and sea level rise. Studies in District South 24 Parganas, West Bengal after occurrence of a tropical cyclone *Aila* had revealed massive salt water intrusion into coastal low lands due to ingression of Sea water from the Bay of Bengal. Saline waters intrusion from the Bay of Bengal into the inland water areas of Coastal South 24 Parganas, West Bengal increased water salinity from 8 to 23 ppt. These fertile lands were rendered unfit for agriculture which later turned out to be good sites for brackishwater aquaculture of highly priced seabass and mullets as an alternative livelihood. It was observed that during cyclones causing sea level may rise from 1 to 2 meters causing 3 to 11% of the land area of the district to be submerged.

8.3.3.3 Impact of Storm Surges and Floods

Storm surges and subsequent flooding are most likely to cause inundation and flooding of culture waters, unanticipated salinity changes, introduction of pathogens/diseases, introduction of exotics and predators, escape of cultured fish/prawn stocks, dispersal of preferable microscopic fish food organisms, damage to culture facilities and increased insurance costs.

8.3.3.4 Impact of Water Stress and Drought

Droughts are often associated with reduced/limited water level, cessation of fish hatchery functioning, decreased dissolved oxygen concentration, increased salinity, algal bloom, ammonia toxicity, increased occurrence of fish diseases and mortalities. Moreover there also occurs an increased social conflict on water use.

8.3.3.5 Vulnerability of Fisher Folk

The greatest limitation towards developing an effective adaptation strategy for climate change in India is the lack of vulnerability assessment framework and vulnerability mapping of various climate sensitive sectors of Indian economy. In order to address that issues ICAR-CIFRI under Project NICRA had innovated such a tool capable of comprehensively assess the vulnerability of inland fisheries at spatio-temporal scale from the regional trends of changing climate and vulnerability index and vulnerability mapping of the inland aquatic resources and fisheries was conducted in 14 districts of West Bengal. Low adaptive capacity of the fishers limited their capacity to cope up with the extensive loss to fish production and infrastructural facility associated with extreme events of climate change (Fig. 8.5).

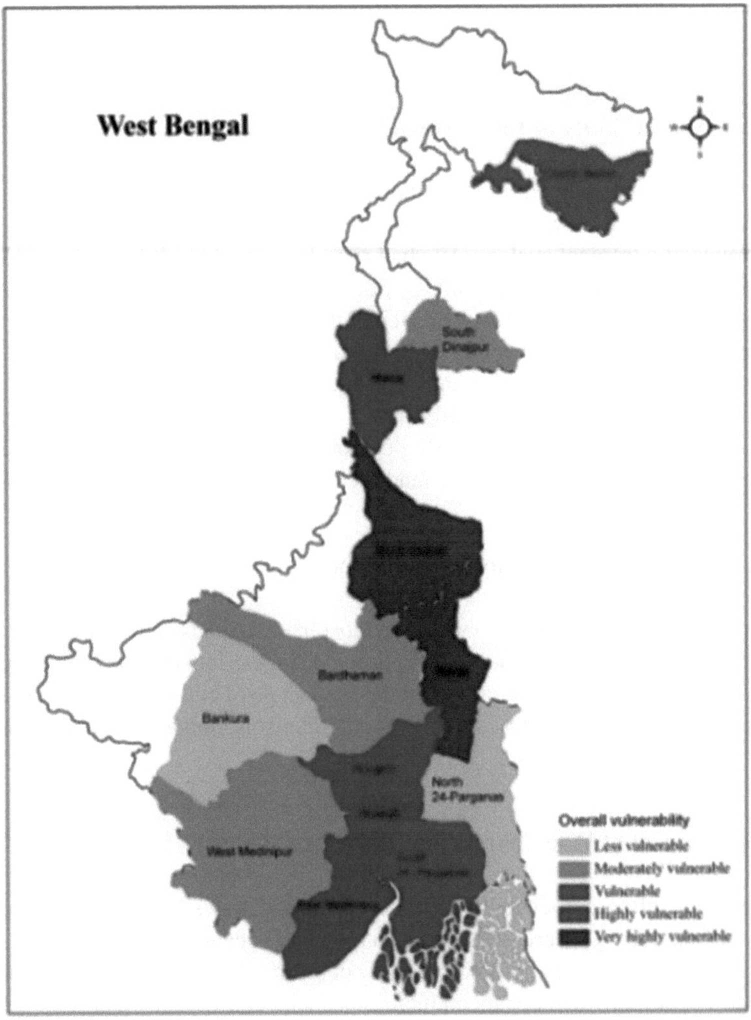

Fig. 8.5 Graph showing vulnerability map of West Bengal's fisher folk to climate change (Sharma et al. 2015)

8.4 Climate Change Adaptation in Inland Fisheries

Approaching climate change typically involves actions that either reduce the amount of carbon dioxide and other greenhouse gases (GHGs) in the atmosphere or prepare the fishing community for the impacts associated with climate change via adaptation. Even though it is hard to generalize the impacts of climate change on fisheries, climate change is very likely going to lead to fluctuations in fish stocks which in turn will have major economic consequences for many vulnerable communities and national economies that heavily depend on fisheries (Brander 2010).

The important adaptive measures that can rejuvenate the fisheries sector are as follows:

- Rebuilding stocks and improving fisheries governance
- Managing declining incomes if fish catches fall, and efforts aimed at diversification and fostering alternative livelihood activities
- Disaster preparedness and response
- Aquaculture development and ecosystem based adaptation

For inland fisheries, adaptation involves adjusting fishing pressure to sustainable levels. Setting catch limits based on changes in recruitment, growth, survival and reproductive success can be done via adaptive management, monitoring and precautionary principles. This may also require changes in craft or gear types.

8.4.1 General Adaptation Measures

Some climatic situation specific general adaptation measures as recommended by the CIFRI over the years are tabulated below (Table 8.1).

Additionally, some climate change impact specific adaptation measures for inland fisheries as recommended by FAO (De Silva and Soto 2009) are listed in Table 8.2.

8.4.2 Climate Smart Adaptation Strategies for Wetland Fisheries

Consequences of climate change like extreme precipitation events (heavy rain storm, cloud bursts etc.) and prolonged dry spells had occurred in various states. In order to mitigate the adverse effect of climate change in wetland fisheries, potential advisories were recommended to the fishers/ stakeholders of different wetlands of West Bengal and Assam. The details are given below.

Creation of deep pools in shallow wetlands have been adopted in several wetlands of Assam and West Bengal for providing summer time refuge to fishes and also maintain base stocks for recruitment in the next season.

Holding fish in net pen enclosures have been advised in 30 wetlands of Assam and 3 wetlands in West Bengal for safeguarding fish stocks against escape during monsoon and also to sustain the livelihood of dependent fishers through a provision of temporary pre-summer enclosure during water deficient months.

Creation of weed refuge, branch pile refuge in weed/algae infested wetlands also needs to be popularized as these will effectively make use of the proliferating

Table 8.1 General adaptation measures for inland fisheries under changing climate

Situation	Adaptation measures
Enhanced temperature	*Making changes in feed formulations and feeding regimes of fishes*
	Exploring substitution by alternate species of fish
	Providing monetary input to the changes in operational costs in ponds and hatcheries
Pre-flood	*Harvesting fish at smaller size*
	Giving importance to fish species that require short culture period and minimum expense in terms of input
	Increasing infrastructure sophistication of hatcheries for assured seed production of 34,000 million carp fry, 8000 and10000 million scampi and shrimp PL respectively
Post-flood	*Continuous supply of fish seed from hatcheries or raising of fish seed in hatcheries is required.*
	Cage culture in large water logged bodies for raising seed from fry to fingerlings
Storm surge	*Early detections systems of extreme weather events*
	Communication of early warning system
	Accept certain degree of loss
	Development and implementation of alternative strategies to overcome these culture periods
	Maximizing production and profits during successful harvest
	Suitable site selection and risk assessment work through GIS modelling
	Increasing infrastructure sophistication of hatcheries for assured seed production of 34,000 million carp fry, 8000 and10000 million scampi and shrimp PL respectively
	The ingressed saline water inundated coastal low lands which become unfit for agriculture provided temporary opportunities for converting these areas into ponds for fish culture with saline tolerant fish species viz., Mugil parsia, M. tade and Lates calcarifer
Pre-drought	*Hatcheries affected by drought condition may divert from rearing Indian major carps to other fish species which favourably adapt to water stress, oxygen deficiency and high temperature conditions.*
Post-drought	*Smaller ponds that retain water for 2–4 months can be used for fish production with appropriate fish species (catfish, tilapia etc.) and management practices.*
Water stress	*Multiple use, reuse and integration of aquaculture with other farming systems*
	Intensification of aquaculture practices in resources of wastewater and degraded water such as ground saline water
	Smaller ponds (100-200 m2) of seasonal nature (1–4) months can be used for rearing /culture of appropriate species of fish/prawn

aquatic weeds/algae under climate change scenario, in three ways i.e. serve as fish refuge, support fisheries and promote carbon sequestration in wetlands.

Interventions made by ICAR-CIFRI on climate smart wetland fisheries planning: The indigenous fisheries strategies practiced by the fisher folk in the wetlands are not given much importance by the fisher folk themselves as they are constantly in

Table 8.2 Impact specific adaptation measures for inland fisheries under changing climate (adopted from De Silva and Soto 2009)

Impact	Adaptation measure
Reduced yields	Access higher-value markets
	Increase fishing effort (risks overexploitation) moving/planning siting of cage aquaculture facilities
	Migration as fish distribution changes (risks overexploitation)
	Research and investments into predicting where fish populations will move to.
Increased yield variability	Diversify livelihood portfolio (e.g. algae cultivation for biofuels or engage in non- fishery economic activity such as ecotourism)
	Precautionary management
	Ecosystem approach to fisheries/aquaculture and adaptive management
	Shift to culture-based fisheries
Reduced profitability	Diversify livelihoods, markets and/or products
	Reduce costs to increase efficiency
	Shift to culture-based fisheries
Increased risk	Weather warning systems
	Improved communication networks
	Workshops to train data gathering and interpretation
	Monitoring of harmful algal blooms
	Improved vessel stability/safety
	Compensation for impacts
Increased vulnerability for those living near rivers	Early warning systems and education
	Rehabilitation and disaster response
	Post-disaster recovery
	Encourage native aquaculture species to reduce impacts if fish escape damaged facility

search of better production systems. But after CIFRI sensitized the climate smart nature of these ignored traditional fishery strategies through several awareness-cum-training programmes, a new attitude among 70–90% fisher folk community towards these pre-existing traditional methods had emanated recently.

8.4.3 *Climate Change Resilient Adaptation Strategies for Riverine Capture Fisheries*

Particularly for riverine fisheries, adaptation should involve adjustment of fishing pressure to sustainable levels. This may be accomplished through setting catch limits (based on changes in recruitment, growth, survival and reproductive success), adaptive management, stock monitoring and precautionary conservation initiatives. This may also require changes in craft or gear types. Rebuilding of depleting stocks should be given highest priority followed by aquaculture development in the associated floodplains along the river basin to diversify the income of fishers.

8.4.4 Wetland Fishery and Conservation as a Tool to Adapt and Mitigate Climate Change

Wetland ecosystems are perhaps among the largest carbon (C) sinks in the world. Soils of wetlands play an important role in global C-cycle. According to Ramsar secretariat about 1/3rd of the world terrestrial carbon is trapped and stored in wetlands, double of that of forests. As per the estimations, C sequestration potential of restored wetlands (over 50 year period) comes out to be about 0.4 tonnes C/ha/year (IPCC 2001). From the studies of ICAR-CIFRI, the rate of C accumulation in the floodplain wetlands in Eastern Indian conditions was estimated to be >0.15 Mg/ha/year which is almost double of the global estimates given for large lakes and inland seas, and also greater than terrestrial upland soils. Primary producers (phytoplankton, aquatic weeds) are the major pathway of carbon capture. In terms of carbon harvested through fish flesh (*i.e.* blue carbon), detritivorous and bentho-pelagic fish species have highest carbon uptake from wetland system compared to surface inhabiting planktivores.

8.4.5 Policy Document on Climate Change and Inland Fisheries

ICAR-CIFRI, in 2015, had published a comprehensive policy paper under Project NICRA encompassing various issues and forecasted situations of climate change with its implications on inland fisheries of India. The policy paper titled "Inland Fisheries and Climate Change: Vulnerability and Adaptation Options" stands out to be the very first of its kind and is expected to augment knowledge of researchers, experts and aid policymakers in devising climate smart policies for the sustainable development of inland fisheries in the country. The document not only provides research based information of climate change impacts on various aspects like aquatic ecology, fish biology, fisheries socio-economics and vulnerability but also recommends DO's and DON'Ts for specific climatic scenarios or weather events. The recommendations of this document have been well consulted and appreciated by states like Assam and Orissa through incorporation in their state action plans for fisheries development.

8.5 Recommendations to Offset Negative Impact of Climate Change

- Following some principle of responsible fisheries in rivers, reservoirs and wetlands *viz.* closed season, closed area, mesh size regulations, by-catch reduction devices and catch limit (CPUE) regulations. This is to avoid recruitment overfishing and growth overfishing of already stressed fish stocks in natural ecosystems under changing climate.
- Proper pre-damming environmental impact assessment to be done for ascertaining environmental flow requirements of the river systems in order to sustain ecosystem services and avoid habitat fragmentation. It is required for minimizing anthropogenic impacts on the already impacted river systems due to climate change.
- Creation of temporary pre-summer enclosures (pen), creation or conservation of deep pools in deepest parts of the wetlands and rivers. This will serve as summer refuge to fishes or sustain fisheries under low water conditions.
- Creation of weed refuge, branch pile refuge in weed/algae infested wetlands. This will effectively make use of the proliferating aquatic weeds/algae under climate change scenario, in three ways *i.e.* serve as fish refuge, support fisheries and promote carbon sequestration in wetlands.
- Immediate implementation of dredging/de-silting programmes in the wetlands using a combination of mechanical and manual methods.
- Promote culture fisheries in reservoir with planktivorous or herbivorous fishes low in the food chain. Such fishes have highest carbon assimilation efficiency. Short duration crop cycles in seasonal waters and long duration crop cycles in perennial waters is recommended.
- Promote wetland fishery from carbon economics point of view, *i.e. – to convert invaluable inorganic carbon into high value organic carbon of fish flesh.*
- Identification of indigenous fisheries and aquaculture strategies those are capable of adapting to present state of climate change. Such practices can be optimized with scientific intervention for promoting climate change resilient fisheries.
- Carry out wet lab experiments with brood stock, fish larvae of commercially important inland fish species in heat chambers to assess the chronic changes in physiology at neuro-endocrinal levels when fishes are subjected to realistic elevated temperatures and water stress.
- Research on carbon sequestration in wetlands with more emphasis of Green House Gas (GHG) emissions from different types of wetlands and assessment of carbon sequestration potential. Assessment of carbon footprint in enclosure based fish culture like cages /pens and promotion of wetland fisheries from carbon economics point of view need to be done.

8.6 Conclusion

The impact of climate change on inland fish production is mixed in nature ranging from being detrimental to beneficial in various cases. The true impact of climate change on inland fisheries production should be reflected from the revenue generated and environmental costs incurred. This is due to the fact that localized loss of certain fish stocks due to habitat degradation, recruitment failures, changes in reproductive phenology, larval survivability, disease outbreak, range contraction, competitive displacement by non-native alien species is often offset by a new replacement to restore ecological integrity under natural conditions. Climate change often acts in conjunction with anthropogenic impacts. The already climate stressed aquatic ecosystems and fishes living therein should not be stressed further through adding more anthropogenic impacts. Environmental impact assessment and code of conduct for responsible fisheries should be given higher priority than before in order to minimize or offset climate change impacts on inland fisheries. Public awareness and capacity building of stakeholders on climate change will aid in effective implementation of the policies and to respond effectively to the threats or opportunities posed by climate change.

References

Bonte, M., & Zwolsmen, J. J. G. (2010). Climate change induced salinization of artificial lakes in The Netherlands and consequences for drinking water production. *Water Research, 44*, 4411–4424.

Brander, K. (2010). Impacts of climate change on fisheries. *Journal of Marine Systems, 79*(3–4), 389–402.

Carlson, A. K., & Lederman, N. J. (2016). Climate change and fisheries education. *Fisheries, 41*(7), 411–412. https://doi.org/10.1080/03632415.2016.1182510.

De Silva S, Soto D (2009). Climate change and aquaculture: Potential impacts, adaptation and mitigation. In K Cochrane, C De Young, D Soto & T Bahri (eds.), (n.d.). *Climate change implications for fisheries and aquaculture: Overview of current scientific knowledge*. FAO fisheries and aquaculture technical paper no. 530. Rome, FAO. pp 212. Also available at www.fao.org/docrep/012/i0994e/i0994e00.htm.

Hartmann, D. L., Klein Tank, A. M. G., Rusticucci, M., Alexander, L. V., Brönnimann, S., Charabi, Y., Dentener, F. J., Dlugokencky, E. J., Easterling, D. R., Kaplan, A., Soden, B. J., Thorne, P. W., Wild, M., & Zhai, P. M. (2013). Observations: Atmosphere and surface. In T. F. Stocker, D. Qin, G. K. Plattner, M. Tignor, S. K. Allen, J. Boschung, A. Nauels, Y. Xia, V. Bex, & P. M. Midgley (Eds.), *Climate change 2013: The physical science basis. Contributions of working group I to the fifth assessment report of the intergovernmental panel on climate change* (pp. 159–254). Cambridge/New York: Cambridge University Press.

Hershkovitz, Y., Dahm, V., Lorenz, A. W., & Hering, D. (2015). A multi-trait approach for the identification and protection of European freshwater species that are potentially vulnerable to the impacts of climate change. *Ecological Indicators, 50*(2015), 150–160.

IPCC. (2001). *Climate change: The scientific basis. Intergovernmental panel on climate change: Working group I*. Cambridge: Cambridge University Press.

IPCC. (2014). Technical summary. In C. B. Field, V. R. Barros, D. J. Dokken, K. J. Mach, M. D. Mastrandrea, T. E. Bilir, M. Chatterjee, K. L. Ebi, Y. O. Estrada, R. C. Genova, B. Girma, E. S. Kissel, A. N. Levy, S. MacCracken, P. R. Mastrandrea, & L. L. White (Eds.), *Climate change 2014, Impacts, Adaptation, and Vulnerability. Part A: Global and Sectoral Aspects. Contribution of Working Group II to the Fifth Assessment Report of the Intergovernmental Panel on Climate Change* (pp. 35–39). Cambridge/NY: Cambridge University Press.

Ito, Y., & Momii, K. (2015). Impacts of regional warming on longterm hypolimnetic anoxia and dissolved oxygen concentration in a deep lake. *Hydrological Processes, 29*, 2232–2242.

Iyalomhe, F., Rizzi, J., Pasini, S., Torresan, S., Critto, A., & Marcomini, A. (2015). Regional risk assessment for climate change impacts on coastal aquifers. *Science of the Total Environment, 537*(2015), 100–114.

Kanno, Y., Pregler, K. C., Hitt, N. P., Letcher, B. H., Hocking, D. J., & Wofford, J. E. B. (2015). Seasonal temperature and precipitation regulate brook trout young-of-the-year abundance and population dynamics. *Freshwater Biology*. https://doi.org/10.1111/fwb.12682.

Lynch, A. J., Myers, B. J. E., Chu, C., Eby, L. A., Falke, J. A., Kovach, R. P., Krabbenhoft, T. J., Kwak, T. J., Lyons, J., Paukert, C. P., & Whitney, J. E. (2016). Climate change effects on North American inland fish populations and assemblages. *Fisheries, 41*(7), 346–361. https://doi.org/10.1080/03632415.2016.1186016.

Moss, B., Kosten, S., Meerhoff, M., Battarbee, R. W., Jeppesen, E., Mazzeo, N., Havens, K., Lacerot, G., Liu, Z., De Meester, L., Paerl, H., & Scheffer, M. (2011). Allied attack: Climate change and eutrophication. *Inland Waters, 1*, 101–105.

Saha, S. K., Rinke, A., & Dethloff, K. (2006). Future winter extreme temperature and precipitation events in the Arctic. *Geophysical Research Letters, 33*(15), L15818.

Sarkar UK, Roy K, Karnatak G, Nandy SK. (2016). *Climate change resilient fisheries in the floodplain wetlands of West Bengal: Adaptation strategies*. Proceedings of national academy of biological sciences (*communicated*).

Sharma AP, Joshi KD, Naskar M, Das MK. (2015). Inland fisheries and climate change: Vulnerability and adaptation options, ICAR-CIFRI Special Publication, Policy paper No. 5. ISSN 0970-616X.

Whitney, J. E., Chokhachy, R. A., Bunnell, D. B., Caldwell, C. A., Cooke, S. J., Eliason, E. J., Rogers, M., Lynch, A. J., & Paukert, C. P. (2016). Physiological basis of climate change impacts on North American inland fishes. *Fisheries, 41*, 332–345.

Chapter 9
Looking at Climate Change and its Socio-Economic and Ecological Implications through BGC (Bio-Geo-Chemical Cycle)-Lens: An ADAM (Accretion of Data and Modulation) and EVE (Environmentally Viable Engineering Estimates) Analysis

J. S. Pandey

Abstract Viewing climate change merely as a change in temperature or humidity levels is a very narrow way of looking at the problem, which in essence has a much wider implication and ramification. Transmission and distribution of impacts under interactive and integrated influence of climate change and environmental pollution is an important area of research, which helps in quantifying human and ecosystem health risks. This also helps ultimately in converting them into the real economic terms - financial benefits accrued or costs incurred. Moreover, alterations and aberrations in temperature and humidity are very closely linked with significant perturbations in bio-geo-chemical cycles (BGCs). In essence, this means that climate change problem should ideally be viewed as a significant perturbation in BGCs. It is now gradually getting understood and established that alterations in land use and cover changes also affect (very significantly) the transmission and distribution dynamics of various diseases. The need of the hour, therefore, is to study the directions of these alterations and quantifications of their implications and impacts through ADAM (Accretion of Data and Modulation) and EVE (Environmentally Viable Engineering Estimates) Analysis. While discussing above-mentioned issues in the present paper, some case studies are also presented.

Keywords Carbon foot print · Ecological footprint · Life style solutions · Climate change

---

J. S. Pandey (✉)
Chief Scientist & Head, Climate Sustainability, CSIR-National Environmental Engineering Research Institute (NEERI), Nagpur, India
e-mail: js_pandey@neeri.res.in

S. Sheraz Mahdi (ed.), *Climate Change and Agriculture in India: Impact and Adaptation*, https://doi.org/10.1007/978-3-319-90086-5_9

9.1 Introduction

Whether scientific research should follow a policy driven approach or a curiosity driven approach? This question assumes its significance and importance especially in view of the financial crunch and continuously depleting global natural resources. Transmission and distribution of impacts under interactive and integrated influence of climate change and environmental pollution is an important area of research, which helps in quantifying human and ecosystem health risks. This also helps ultimately in converting them into the real economic terms – financial benefits accrued or costs incurred. The real scientific challenge today is to find out the ways and means of quantifying the following scientific facts:

- Intricate interaction between climate change and environmental pollution, and their impacts in terms of human and ecosystem health
- Alterations in land cover, biodiversity, air, water and soil pollution, and their implications on climate change impacts and on availability of food-items and nutrients
- Impacts of above-mentioned alterations on disease transmission and dynamics

In this context, it is important to reiterate that forests have one very important ecological role i.e. they regulate all the natural bio-geo-chemical cycles – for instance, carbon cycle, water cycle, nitrogen cycle and phosphorus cycle etc. One very pertinent fact which should not be overlooked is the fact that alterations and aberrations in temperature and humidity are also very closely linked with significant perturbations in bio-geo-chemical cycles (BGCs).

In essence, this means that climate change problem should ideally be viewed as a significant perturbation in BGCs. Viewing climate change merely as a change in temperature or humidity levels is a very narrow way of looking at the problem, which in essence has a much wider implication and ramification. Ideally, there is no such anthropogenic activity (be it residential, commercial or industrial), which does not involve alterations in surrounding ecosystems and bio-geo-chemical cycling. Moreover, pollution generation can simply be construed as a land-use impact. Therefore, a given shift in land use pattern would essentially result in a consequential shift in pollution levels (air, water and soil) and generation of various green house gases. It is now gradually getting understood and established that alterations in land use and cover changes also affect (very significantly) the transmission and distribution dynamics of various diseases. The need of the hour, therefore, is to study the directions of these alterations and quantifications of their implications and impacts through ADAM (Accretion of Data and Modulation) and EVE (Environmentally Viable Engineering Estimates) Analysis. While discussing above-mentioned issues in the present paper, some case studies are also presented.

9.2 Climate Change

Long-term changes in the average weather conditions, which are unique combinations of factors like temperature, precipitation and wind, denote Climate Change. According to the United Nations Intergovernmental Panel on Climate Change (IPCC), our climate is undergoing significant changes as a result of day by day increasing greenhouse gas (GHG) emissions from several anthropogenic activities. GHG's are those gases in the atmosphere that act together like a glass roof around the earth. This glass roof (imaginary), traps in heat that would have otherwise escaped into space. This heat-trapping-effect is known as the "greenhouse effect".

Amongst these GHGs, Carbon dioxide (CO_2) happens to be the most significant and prevalent GHG. It is mostly released due to the burning of fossil fuels like coal, oil and natural gas. Methane (CH_4), sulphur hexafluoride (SF_6), nitrous oxide (N_2O), nitrogen trifluoride, hydrofluorocarbons (HFCs) and perfluorocarbons (PFCs) were later on included in the Kyoto Protocol. Kyoto Protocol, which was signed on 11th December, 1997 and came into force on 16th February, 2005, is an international agreement for emission reduction. Chlorofluorocarbons (CFCs) are being phased out under the requirements of Montreal Protocol (http://www.carbonneutral.com/resource-hub/climate-change-summary).

9.3 Global Warming and Climate Change

In simple terms, global warming and climate change are results of continuously rising global average temperatures. The rise in global average temperature is caused mainly due to increase in many green house gases - carbon dioxide being the main culprit. Green house gases are mainly being emitted through various anthropogenic activities, which involve fossil fuel burning. Here it is also pertinent to note the difference between the times involved in the deposition of fossil fuels and that in their consumption. It would amount to several billion years. Temperature being one of the key determinants of weather and climate, a global warming automatically converts into unpredictable changes in local, regional and global weather-patterns. And, over a long period of time, the event automatically converts into climate change (http://www.globalissues.org/print/article/233).

9.4 Main Indicators of Climate Change

According to the National Oceanic and Atmospheric Administration (NOAA), there are 10 main indicators of climate change. 7 of these indicators (Tropospheric temperature, temperature over oceans, sea surface temperature, oceanic heat content, temperature over land, sea level and humidity) are expected to increase as they are already doing and 3 indicators (glaciers, snow cover and sea ice) should show a declining trend. These events are already being witnessed.

9.5 Green House Effect

Energy from the sun heats the earth's surface, Earth radiates some of this energy back to the space. Part of the energy is absorbed by the green house gases (water vapor, carbon dioxide, and other gases) present in the earth's atmosphere. They retain heat in the atmosphere same as the way glass panels of a greenhouse does. This is what drives the earth's weather and climate. That's the reason why these gases are known as greenhouse gases (GHG).

There are 3 main GHGs such as:

- Carbon dioxide (CO_2)
- Methane (CH_4) with global warming potential (GWP) almost 20 times higher than carbon dioxide
- Nitrous oxide (N_2O).

These are emitted through both natural and man-made sources. However, emissions through man-made sources have recently increased substantially and have ultimately become serious cause of concern. In addition, there are some GHGs which have only anthropogenic sources of emissions. They are:

- Hydrofluorocarbons (HFC_s)
- Perfluorocarbons (PFC_s) and
- Sulphur hexafluoride (SF_6).

Water vapor is also considered a greenhouse gas.

9.6 Climate Change and Bio-Geochemical Cycling

Bio-geochemical cycles (Carbon cycle, water cycle, nitogen cycle, phosphorus cycle and sulphur cycle etc.), green house effect, global warming and climate change are all interlinked. All the time, we are mining coal and extracting oil from the Earth's crust. Subsequently, we are burning these fossil fuels for various industrial, commercial and residential activities such as energy production, transportation, cooling, heating, cooking, and manufacturing. Ultimately the rate at which fossil fuels are being burnt, GHG_s are being emitted into the atmosphere is much higher than the rate at which GHGs are removed from the atmosphere. This leads to continuous GHG-build-up in the atmosphere.

On the other hand, forests, which are very effective filters or sinks for many GHG_s (mainly CO_2) are continuously being cleared for various agricultural, residential, commercial and industrial purposes. This results in doing away with one

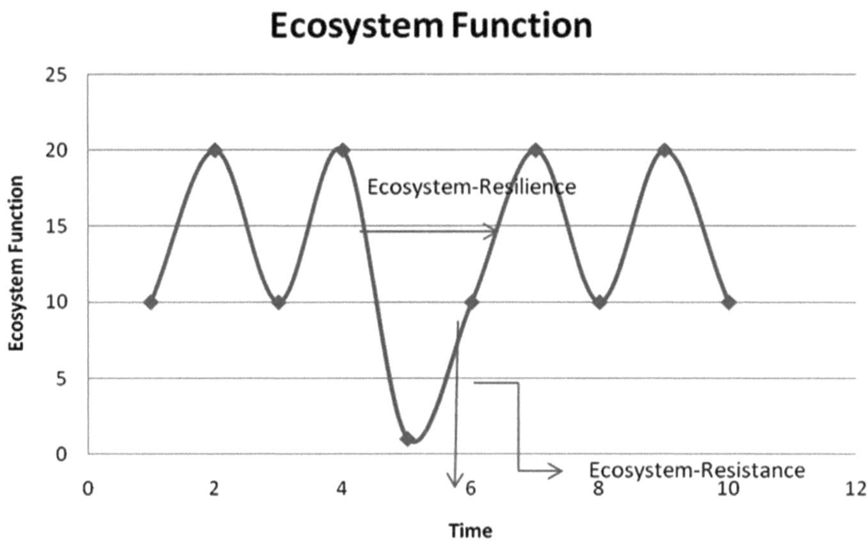

Fig. 9.1 Ecosystem resistance and resilience

of the most effective sinks of pollution and consequently enhancing GHG-build-up in the atmosphere. Mainly because of these reasons, GHG-concentrations today are much higher than what they were several decades and centuries back.

There could be a very simple analogy, which would make us understand the problem related to GHG-built-up in the atmosphere. GHGs' requirement and presence in the atmosphere can be construed as that of salt in our food. A given amount of salt is essential for providing good taste to our food and is also essential for our health. But if we add too much salt into our food, it not only spoils the taste of our food but would also spoil our health leading to diseases like hypertension etc.

The other problems with human induced land use changes and consequent GHG-emissions are that they have recently become so rapid that the surrounding ecosystems are not able to adjust with respect to them. In short, it affects ecosystem's resilience and resistance (Fig. 9.1).

9.7 Impact of Life Style Changes on Carbon Cycle and Climate Change

There are always some issues, queries and questions raised about the fact that since the climate has been always and continuously changing, why there should be so much fuss about it now. The main difference lies in the rapidity of weather related aberrations observed recently. And some of the significant indicators, which point towards the human footprint on climate change, can be delineated as follows:

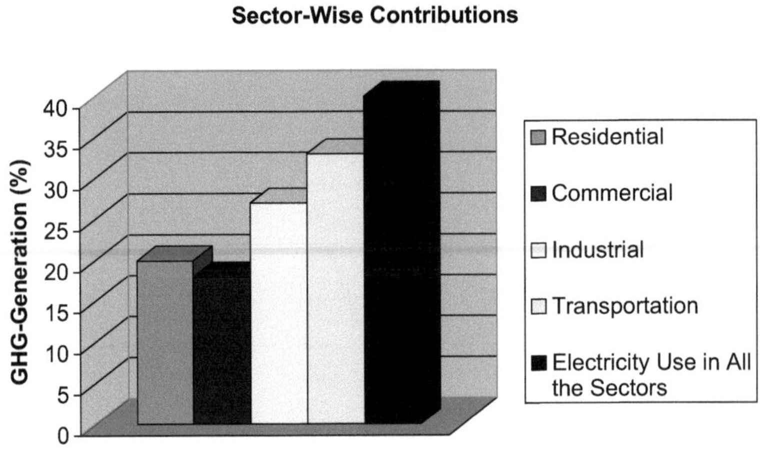

Fig. 9.2 Sector-wise CO_2-emissions

- Less oxygen in the air
- Shrinking thermosphere
- Elongation of tropopause
- More fossil fuel carbon in the air, and
- More fossil fuel carbon in the coral reef

Ultimate solutions to climate change problems (Adger et al. 2005) lie in regulating and controlling three key sectors: production, consumption and life-style. The third one automatically takes care of the first two, because our production and consumption patterns and trends depend directly on our life-styles. Food, clothing and shelter are our primary requirements and they all contribute significantly to increasing concentrations of Greenhouse Gases (GHGs) (USEPA 2005; Pandey et al. 1997) and ultimately to climate change problem. The present paper illustrates some model development exercises based on realistic and relevant parameters, which are easy to measure and monitor. Subsequently, the article also highlights research-needs, which need to be pursued in various educational and research Institutions so as to gradually make every citizen of the society environmentally aware and responsible. Figure 9.2 provides the percentage contribution from various sectors (industrial, commercial, residential and transportation) towards GHG-generation (http://www.eia.doe.gov/oiaf/1605/ggrpt/carbon.html#emissions).

9.8 Carbon Foot-Printing (CF) and Ecological Foot-Printing (EF)

Carbon Foot-printing (CF) and Ecological Foot-printing (EF) (http://www.ecologicalfootprint.com; Rees and Wackernagel 1996, 1999; Mishra et al. 2008) are some of the recent Environmental Impact Assessment tools. They (CF & EF) not only help in understanding and quantifying impacts due to various activities including solid waste disposal, waste water treatment, air pollution control etc., but also in evolving appropriate cost-effective environmental management plans. Moreover, there is a significant awareness in respect of greenhouse gases (GHGs), global warming, climate change and carbon footprints (CFs). Institutions like US Environmental Protection Agency (USEPA) and Water Utility Climate Alliance are already working vigorously in this direction. CO_2 (carbon dioxide), CH_4 (methane), N_2O (nitrous oxide) and fluorinated gases such as hydro fluorocarbons, per fluorocarbons and sulfur hexafluoride are the main GHGs, which contribute significantly to total CF. These GHGs have widely different global warming potentials according to the Intergovernmental Panel on Climate Change (IPCC 2006). Activities that lead to GHG-emissions are said to be ***carbon-positive***, while those which remove GHGs from the environment are known as ***carbon-negative*** or carbon-sinks. When GHG-emissions equal GHG-assimilation or absorption, the activities are known as ***carbon-neutral.***

9.9 CF (Carbon Footprint) – Calculators

CF-calculators (Padgett et al. 2008) are very frequently being used for estimating GHG (CO_2-e) emissions. However, not all CF-calculators give same or similar results. The differences amongst them could be as high as 5–6 million MT per year per individual. Generally, the kind of inputs, these CF-calculators require can be summarized as follows:

- Electricity / Energy / Oil / Natural Gas / Propane / Kerosene / Wood consumption and related emission factors
- Waste generation and related emission factors
- Number of individuals / institutions / activities (as the case may be)
- Distance covered in transportation (flight / rail / road) and related emission factors
- Number of vehicles and their emission factors
- Use of air conditioners and their emission factors

And, the variations in CF-results are normally attributed to the following factors:

- Methodologies
- Individual Behavioral Features
- Conversion and Emission Factors, and
- Lack of Transparency

The greatest uncertainty, however, is associated with emission (conversion) factors. Most of the calculators do not display or explicitly explain the methodologies behind these factors. As a result, they land up using significantly different emission factors. And since these calculators are supposed to influence and guide the citizens and policy makers for taking appropriate pollution (carbon) reduction measures and strategies, these uncertainties and variations cannot be ignored. However, even with these uncertainties and sources of errors, these calculators do generate awareness amongst common masses about environmental protection and conservation.

9.10 Ecological Footprint

The ecological footprint (EF) [http://www.ecologicalfootprint.com/] is a broad measure of resource use which highlights the areas where consumption is exceeding environmental limits. It mainly depends on the following parameters (http://steppingforward.org.uk/calc/):

1	Travelling (Transport)
2.	Living Space (Residential)
3.	Sharing of Apartment (Life Style)
4.	Heating / Cooling Bills (Energy/Electricity Consumption)
5.	Use of Electricity (Type)
6.	Energy Conservation Measures
7.	Food Habits
8.	Food Imports/Exports
9.	Waste Generation
10.	Type of Waste

9.11 ADAM and EVE Applications: Some Case Studies

In the following section, some results from ADAM and EVE applications to different areas like Paper Industry (Fig. 9.3), Waste Water Treatment System (Figs. 9.4 & 9.5), Himalayan Ecosystem (Fig. 9.6), Various Indian Coastal Zones (Fig. 9.7) and Mithi River Ecosystem (Fig. 9.8) are presented for illustration.

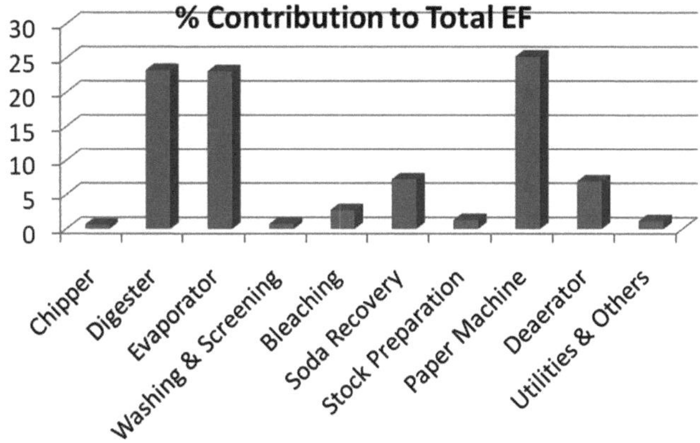

Fig. 9.3 Paper industry

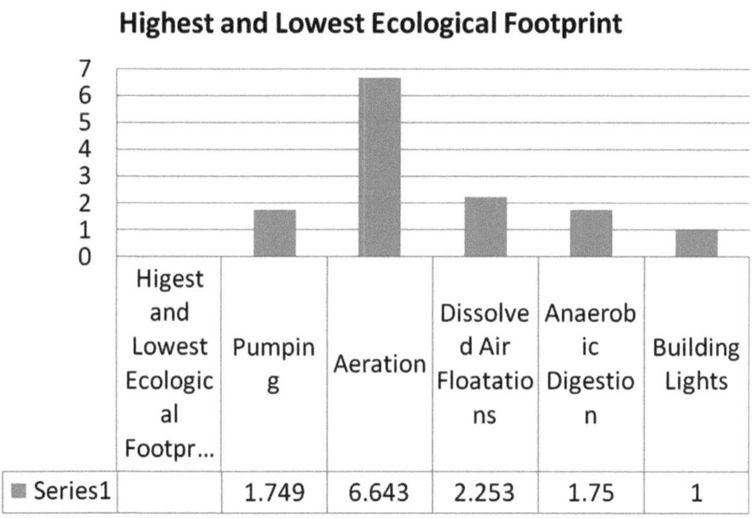

Fig. 9.4 Waste water treatment

9.12 Need for Urgency

The following areas need immediate attention:

- Emission reduction targets need to be given the highest priority.
- Nationally Appropriate Mitigation Actions (NAMA) should focus on enhancing energy efficiency as the most important means for reducing GHG-emissions.
- Investment in energy efficient programmes in developing countries should need continuous and consistent support from well developed countries.

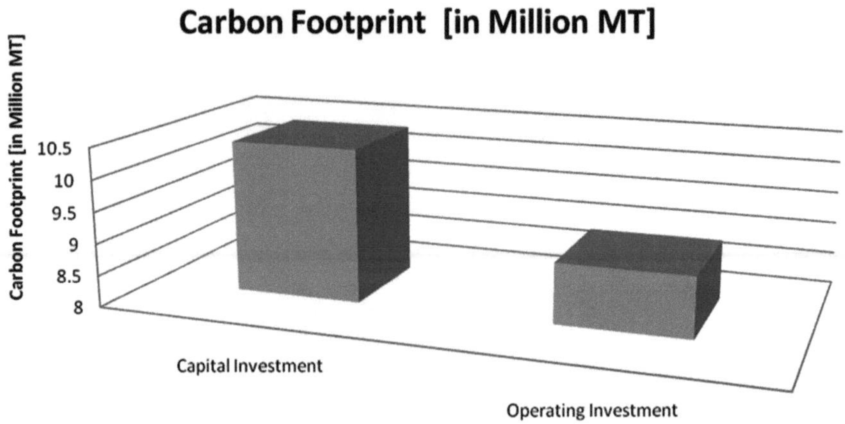

	Capital Investment	Operating Investment
Carbon Footprint [in Million MT]	10.417	8.951

Fig. 9.5 Investment in waste water treatment system

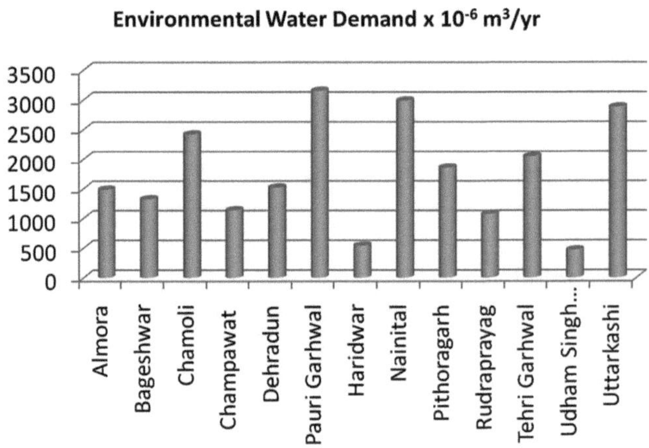

Fig. 9.6 Himalayan ecosystem

- Continuous monitoring and energy / environment audits are the most important segments of any well-intentioned and effective environmental management plan.

In short, the environmental audit process should be guided through the following diagram (Fig. 9.9).

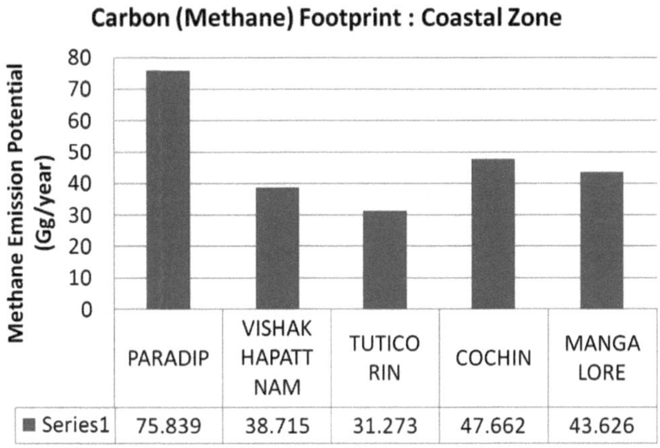

Fig. 9.7 Coastal zone

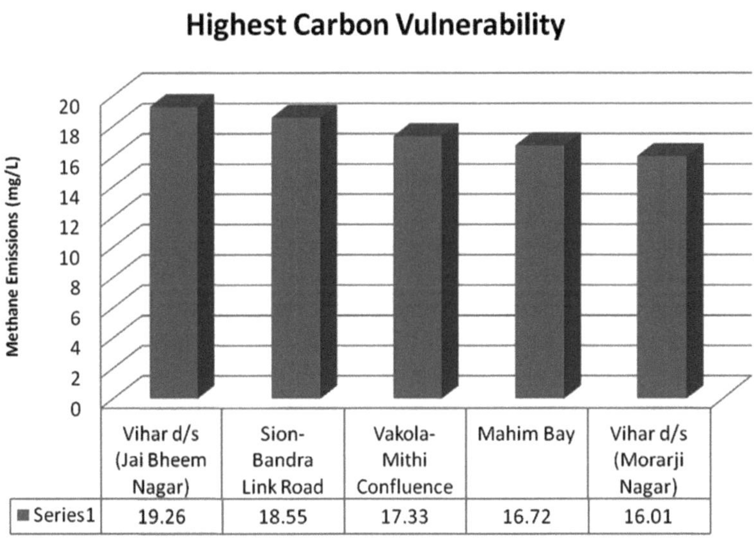

Fig. 9.8 River ecosystem

9.13 Conclusion

Ultimate solutions to climate change problems lie in regulating and controlling the three key sectors: production, consumption and life-style. In fact the third one automatically takes care of the first two, because our production and consumption patterns and trends depend directly on our life-styles. This boils down to the fact that key to climate change mitigation lies in our life styles. Food, clothing and shelter are our primary requirements and they all contribute significantly to

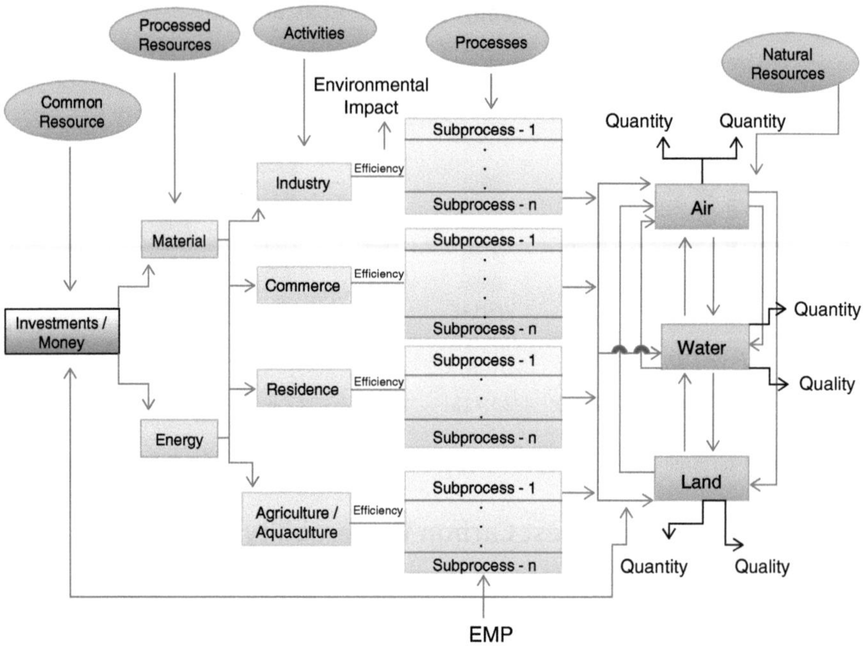

Fig. 9.9 Environmental and ecological auditing (Flow diagram)

increasing concentrations of Greenhouse Gases (GHGs) and ultimately to climate change problem.

The present paper illustrates some model development exercises based on realistic and relevant parameters, which are easy to measure and monitor. Subsequently, the article also shows what kind of researches need to be pursued in various educational and research Institutions so as to gradually make every citizen of the society environmentally aware and responsible. In short, the presentation discusses and recommends the kind of activities which need regular pursuance, refinement, modification and application in regard to evolving site-specific, region-specific and ecosystem-specific environmental management plans aimed at combating the climate regulated environmental crisis which is unfolding before us every day with a newer dimension.

Inter alia, the article deals with the kind of research, which is needed in the area of Climate Change. Side by side, these researches need to be extended and pursued further so as to strike a balance between ecology and economy. Future exercises are needed, which should aim at the dynamics of Ecological Footprints (Pandey et al. 2001a, b); analysis of Environmental Risks (Pandey and Joseph 2001) by way of developing models which deal with the issues like Temporal Risk Gradients (TRG) (Pandey et al. 2001b) and Ecological Economics of Natural Resources (Pandey et al. 2004). There is also a need for quantifying region-specific-emission factors for different GHGs (Pandey et al. 2007). On the basis of these emission factors, region-specific ecosystem health (Pandey and Khanna 1992) and human health risk assessment (Pandey et al. 1992, 1993, 1994, 2005) can be carried out.

Subsequently, appropriate region-specific environmental management plans can be developed. Ecology works very much on the concepts of species-specific, ecosystem-specific and process-specific bio-rhythms (Pandey and Khanna 1995). It has a perfect analogy with the way a musical concert or consortia works or in terms of Electronics Engineering, the way an integrated circuit (IC) works. All these features are embedded in Ecological Engineering.

Acknowledgement The present article, *inter alia*, is a synthesis of earlier works done by the author. He is grateful to all those sources of information including discussions, discourses, workshops and conferences, which have helped in shaping this article by way of a systematic data analysis and appropriate synthesis and conversion of the information into useful models (ADAM and EVE approach). The views expressed in the article are mainly his and his Institution may or may not share the same views.

References

Adger, W. N., Arnell, N. W., & Tompkins, E. T. (2005). Successful adaptation to climate change across scales. *Global Environmental Change, 15*, 77–86.

IPCC. (2006). IPCC. In *Guidelines for national greenhouse gas inventories*. Available at http://www.ipcc-nggip.iges.or.jp/public/2006gl/vol5.html

Mishra, A. P., Tembhare, M. W., Pandey, J. S., Kumar, R., & Wate, S. R. (2008, November 23–25). *Carbon footprint : Where India stands in global Scenario. Souvenir: International Conference on Recent trends in Environmental Impact Assessment (RTEIA – 2008)* (p. 34). Nagpur: NEERI.

Padgett, J. P., Steinmann, A. C., Clarke, J. H., & Vandenbergh, M. P. (2008). A comparison of carbon calculators. *Environmental Impact Assessment Review, 28*, 106–115.

Pandey, J. S., & Khanna, P. (1992). Speed-dependent modeling of ecosystem exposures from vehicles in the near-road environment. *Journal of Environmental Systems, 21*(2), 185–192.

Pandey, J. S., Pimparkar, S., & Khanna, P. (1992). Micro-environmental zones and occupancy factors in Jharia coal-field: PAH-health exposure assessment. *Journal of Environmental Systems, 21*(4), 349–356.

Pandey, J. S., Pimparkar, S., & Khanna, P. (1993). Health exposure assessment and policy analysis. *International Journal of Environmental Health Research, 3*, 161–170.

Pandey, J. S., Mude, S., & Khanna, P. (1994). Comparing indoor air pollution health risks in India and United States. *Journal of Environmental Systems, 23*(2), 179–194.

Pandey, J. S., & Khanna, P. (1995). Development of plant function types for studying impact of green house gases on terrestrial ecosystems. *Journal of Environmental Systems, 23*(1), 67–82.

Pandey, J. S., Deb, S. C., & Khanna, P. (1997). Issues related to greenhouse effect, productivity modelling and nutrient cycling: A case study of Indian wetlands. *Environmental Management, 21*(2), 219–224.

Pandey, J. S., & Joseph, V. (2001). A scavenging-dependent Air-Basin ecological risk assessment (SABERA) – model applied to acid rain impact around Delhi City, India. *Journal of Environmental Systems, 28*(3), 193–202.

Pandey, J. S., Khan, S., Joseph, V., & Singh, R. N. (2001a). Development of a dynamic and predictive model for ecological foot printing (EF). *Journal of Environmental Systems, 28*(4), 279–291.

Pandey, J. S., Khan, S., & Khanna, P. (2001b). Modeling and quantification of temporal risk gradients (TRG) for traffic zones of Delhi City in India. *Journal of Environmental Systems, 28*(1), 55–69.

Pandey, J. S., Joseph, V., & Kaul, S. N. (2004). A zone-wise ecological-economic analysis of Indian wetlands. *Environmental Monitoring and Assessment, 98*, 261–273.

Pandey, J. S., Kumar, R., & Devotta, S. (2005). Health risks of NO_2, SPM and SO_2 in Delhi (India). *Atmospheric Environment, 39*, 6868–6874.

Pandey, J. S., Wate, S. R., & Devotta, S. (2007, 27–28 September). Development of emission factors for GHGs and associated uncertainties. In *Proceedings: 2nd International Workshop on Uncertainty in Greenhouse Gas Inventories.* Laxenburg: International Institute for Applied Systems Analysis.

Rees, W. E., & Wackernagel, M. (1996). Urban ecological footprints : Why cities cannot be sustainable – And why they are a key to sustainability. *Environmental Impact Assessment Review, 16*, 223–248.

Rees, W. E., & Wackernagel, M. (1999). Monetary analysis: Turning a blind eye on sustainability. *Ecological Economics, 29*, 47–52.

USEPA. (2005). *Global anthropogenic non- CO2 greenhouse gas emission: 1990–2020.* Available at http://www.epa.gov/methane/pdfs/global_emissions.pdf

Chapter 10
Adaptation and Intervention in Crops for Managing Atmospheric Stresses

N. P. Singh, S. K. Bal, N. S. More, Yogeshwar Singh, and A. Gudge

Abstract Climate change is thevariation in the statistical distribution of weather patterns when that change lasts for an extended period of time.The relationship between climate change and agriculture is complex, both ecologically and politically. Climate variability is inherently linked to the productive capacity of agricultural production systems worldwide. In India population depends highly on agriculture and they created excessive pressure on natural resources with poor coping mechanisms.The significant negative impacts have been noticed with respect toclimate change, predicted to reduce yield by 4.5 to 9 per cent, roughly up to 1.5 per cent of GDP per year. India is more challenged withimpacts of looming climate change, and agricultural production in the country isbecoming increasingly vulnerable to climatevariability and change characterized by alteredfrequency, timing and magnitude of precipitation and temperature. Therefore, it is need of the hour to enhance resilience of agriculture to climate change through planned adaptation and mitigation strategies.

10.1 Introduction

Climate is the primary determinant of agricultural productivity. The crop productivity and produce qualityare greatly influencedby different climate changing events. This changing climate effectson physical processes in many parts of the world, leading to changes in temperature, rainfall patterns, wind direction and increased intensity and frequency of extreme events*viz.*, extremetemperatures, drought, flood, intense radiation, soil salinity and alkalinity, heavy metal toxicity, oxidative stress, etc.According to the various world estimates, an average of almost 50% yield losses in agricultural crops are caused by various abiotic factors (Oerke 2006); mostly shared by high temperature (20%), low temperature (7%), soil salinity (10%),

N. P. Singh (✉) · S. K. Bal · N. S. More · Y. Singh · A. Gudge
ICAR-National Institute of Abiotic Stress Management, Malegaon, Baramati, Pune, Maharashtra, India
e-mail: director.niasm@icar.gov.in

S. Sheraz Mahdi (ed.), *Climate Change and Agriculture in India: Impact and Adaptation*, https://doi.org/10.1007/978-3-319-90086-5_10

drought (9%) and miscellaneousstresses (4%). All these abiotic stresses are interconnected with each other (Mittler 2006) and affectsthe water relations ofplant at cellular as well as whole plant level causing specific as well as unspecific reactions (Beck et al. 2007). This leads to a series of morphological, physiological, biochemical and molecular changes that adversely affect plant growth and productivity.

Fifth assessment report of Inter-governmental Panel on Climate Change (IPCC 2013) revealed that the atmospheric concentrations of the greenhouse gases (GHGs) like carbon dioxide (CO_2), methane (CH_4), and nitrous oxide (N_2O) have increased since 1750 by about 40%, 150%, and 20%, respectively which can be attributed mainly to the anthropogenic developmental activities. Each of the last three decades has been successively warmer at the Earth's surface than any preceding decade since 1850. There are also indications that a warming climate would favour an increase in the intensity and frequency of extreme events and we are already witnessing some of these such as heat waves and precipitation extremes. India ismore vulnerablein view of the huge populationdependent on agriculture,excessive pressure on natural resources and poor copingmechanisms.Being a tropical country, India is more challenged withimpacts of looming climate change (IPCC 2013), and agricultural production in the country is becoming increasingly vulnerable to climate variability and change characterized by alteredfrequency, timing and magnitude of precipitation and temperature. Already, the productivity of Indian agriculture is limited by its highdependency on monsoon rainfall which is most often erratic andinadequate in its distribution. Due to the climate change effect, country is experiencing declining trend of agriculturalproductivity due to fluctuating temperatures (Aggarwal 2008), frequentlyoccurring droughts and floods (Samra 2003), problem soils, andincreased outbreaks of insect-pests and diseases (Dhawanet al.Dhawan et al. 2007). These problems arelikely to be aggravated further by changing climate which put forth major challenge to attain a goal of food security.

Planned adaptation is essential to increase the resilience of agricultural production to climate change. Several improved agricultural practices evolved over me for diverse agro-ecological regions in India have potential to enhance climate change adaptations, if deployed prudently. Management practices that increase agricultural production under adverse climatic conditions also tend to support climate change adaptation because they increase resilience and reduce yield variability under variable climate and extreme events. Some practices that help adapt to climate change in Indian agriculture are soil organic carbon build up through carbon sequestration, in-situ moisture conservation, residue incorporation instead of burning, water harvesting and recycling for supplemental irrigation, growing drought and flood tolerant varieties, water saving technologies, location specific agronomic and nutrient management, rearing of local breeds moretolerant to stresses, improved livestock feed and feeding methods.

Thus, there has been a long felt need to understand the nature and intensity of variousextreme weather events and their impacts on agriculture. This topic summarizes the impact of climate change on various atmosphere stresses and its adaptation and management interventions for climate resilient agriculture.

10.2 Atmospheric Stresses, their Impact on Crop Growth and Management

10.2.1 Temperature Extremes

Plants feel stress under high as well as low temperature exposure. Responses to temperature differ among crop species throughout their lifecycle and are primarily the phonological responses, i.e., stages of plant development. For each species, a defined range of maximum and minimum temperatures form the boundaries of observable growth. In high temperature exposure, stress tolerance can be induced by exposure to shorttermelevated temperature and this is known as acquired thermo-tolerance (Kotak et al. 2007), while at lower temperatures, stress tolerance can be induced by exposure to reduced temperatureand is known as chilling tolerance and/or cold acclimation.

10.2.2 Heat Stress or High Temperature Stress

Extreme temperature events may have short-term durations of a few days with temperature increases of over 5 °C above the normal temperatures. Extreme events occurring during the summer period would have the most dramatic impact on plant productivity. For example, apple fruit exposed to direct sunlight can reach in excess of 40 °C during the late season (leading up to harvest) can enhance susceptibility of apples to superficial scald. Sun scald of grapes, soft nose of mango, sun burn in papaya and citrus are mostly related to high temperature. Fruit cracking in pomegranate and litchi is associated with temperature aberrations coupled with fluctuation in moisture. It has been suggested that higher temperatures reduce net carbon gain by increasing plant respiration more than photosynthesis. In fact, the light-saturated photosynthesis rate of C_3 crops such as wheat and rice is at a maximum for temperature from about 20–32 °C, whereas total croprespiration shows a steep non-linear increase for temperatures from 15 to 40 °C, followed by a rapid and nearly linear decline (Porter and Semenov 2005). Increased temperature could potentially reducephotosynthetic capacity due to the heat ability of Rubisco on the one hand and the limitation of electron transport in thechloroplast on the other hand (Sage et al. 2008). In addition, by increasing the air evaporative demand, higher temperaturesare often implicated with stomatal closure, which further decreases photosynthesis due to smaller CO_2 flux into leaves. The predicted increase in average global temperature will accelerate crop development rates, and the negative effect tend to be larger for grain yield than for total biomass. These constraints associated with other on-goingclimatic changes such as more frequent and severe droughts andmore intense precipitation events will probably offset gains in carbonassimilation associated with elevated CO_2 (Sage et al. 2008).

At warmer temperatures, the yield of wheat may decline up to 10% per 1 °C rise in mean seasonal temperature. The timing of wheat floweringand grain maturity may be considerably earlier at warmertemperatures, thus shortening the time for carbon fixation andbiomass accumulation before seed set. Hot temperatures (>32–36 °C) can alsogreatly reduce seed set in many annual crops if elevated temperaturescoincide with a brief critical period of only 1–3 days around the time of flowering (Craufurd and Wheeler 2009). In groundnut, for example, Vara Prasad et al. (2000) noted that, between 32 and 36 °C and up to 42 °C, the percentage fruit set fell from 50% of flowers tozero and the decline in rate was linear, illustrating the sharpness of response of crop plants to temperatures during the flowering and fruiting periods. Similar patterns havebeen identified for other food crops; for example, in maize, pollen viability is reduced at temperatures above 36 °C, while grain sterilityis brought on by temperatures in the mid-30 °C in rice. In fact, the reproductive limits for mostcrops are narrow, with temperatures in the mid-30 °C representing the threshold for successful grain set (Porter and Semenov 2005). In Indian Punjab, decrease in productivity of rice-wheat system is expected because of the overpowering yield decreasing effect of increased temperature over yield-enhancing effect of rising CO_2 (Jalota et al. 2013). In soybean, peanut (*Arachis hypogaea* L.), and cotton (*Gossypium hirsutum* L.) anthesis occurs over several weeks and avoid a single occurrence of an extreme event affecting all of the pollening flowers. For peanut (and potentially other legumes) the sensitivity to elevated temperature for a given flower, extends from 6 days prior to opening (pollen cell division and formation) up through the day of anthesis (Prasad *et al.* 2001).

A heat-wave is a run of unusually hot days (Three or more consecutive days of unusually high maximum temperatures) but what constitutes a heat wave will differ from region to region given the normal temperatures for that area. Inland areas can expect to be normally hotter than areas closer to the coast. Heat-wave intensity is classified as low, severe or extreme based on the actual temperatures and the previous 30 days' temperatures.Fruit is less likely to set if heat waves occur during blossoming. There are no precise threshold temperatures or periods for reduced pollination because it is a complex interaction between bee activity and the availability of viable pollen and receptive stigmas.Sunburned fruit is caused by combined high air temperature and solar radiation. Sunburn typically occurs in the afternoon on the western side of trees.An early sign of fruit sunburn is white, tan or yellow skin. Severe sunburn results in dark brown to black sunken patches on the fruit surface. Browning associated with sunburn is likely when the fruit surface reaches 46–49 °C. Skin necrosis occurs at 52 °C. In grapes, hot and dry summers typically result in early harvests of mature fruits associate with lower-yield harvests.

10.2.3 Management Strategies

Producing an economically significant yield under heat stress conditions depends on several plant physiological parameters and mechanisms whichcontribute to heat tolerance in the field, such points are listed below,

- High temperature stress could be avoided by agronomic/crop management practices. The selection of type of tillage and planting methods play an important role in emergence and growth of a crop.
- The presence of crop residues/mulch on the soil surface protects the crop from high temperature during its initial growth period and keeps soil temperature lower than ambient during the day and higher at night and also conserves soil moisture.
- The selection of cultivar with respect to date of sowing and expected temperature rise during the crop growth period is important to get better yield under high temperature stress conditions.
- Appropriate irrigation methods based on soil type and amount and quality of irrigation water and irrigation scheduling according to growth stages and weather may help in mitigating the effects of heat stress on crop.
- Phyto hormones/ bioregulators like salicylic acid also plays significant role in the regulation of plant growth and development under drought and water logging stress environment.
- Proper nutrient management including beneficial nutrients like Silicon. The nutritional status of plants greatly affects their ability to adapt to adverse climatic conditions. N-adequate plants are able to tolerate excess light by maintaining photosynthesis at high rates and developing protective mechanisms.Potassium (K) plays a crucial role in survival of crop plants under environmental stress conditions. K is essential for many physiological processes, such as photosynthesis, translocation of photosynthates into sink organs, maintenance of turgidity and activation of enzymes under stress conditions (Mengel and Kirkby 2001). Calcium and magnesium plays a vital role in regulating a number of physiological processes in plants at tissue, cellular and molecular levels that influence both growth and responses to environmental stresses (Waraich et al. 2011). Similary Silicon helps in providing resistance against various biotic and abiotic stresses.
- Accumulation of osmo-protectants is an important adaptive mechanism in plants subjected to extreme temperatures, as primary metabolites participate directly in the osmotic adjustment (Sakamoto and Murata 2000). For instance, accumulation of proline, glycine betaine, and soluble sugars is necessary to regulate osmotic activities and protect cellular structures from increased temperatures by maintaining the cell water balance, membrane stability, and by buffering the cellular redox potential (Farooq et al. 2008).
- Plants are capable of adapting to a wide range of temperatures by reprogramming their transcriptome, proteome, and metabolome through biotechnological tools.

- Late embryogenesis abundant (LEA) proteins, ubiquitin, and dehydrins have been found to play important roles in protection from heat and drought stress.
- Agroforestry as an integral component of conservation agriculture has a major impact on crop performance as trees can buffer climate extremes that affect crop growth.It also helps in carbon di oxide removal from the atmosphere and increasing carbon in soil. Trees on farms bring favourable changes in microclimatic conditions by influencing radiation flux (radiative and advective processes), air temperature, wind speed, saturation deficit of understory crops all of which will have a significant impact on modifying the rate and duration of photosynthesis, transpiration etc. (Monteith et al. 1991).

10.2.4 Cold and Frost Stress

Cold and frost stresses, are comprised of chilling injury (0 to 20 °C) and frost injury (<0 °C) which has an impact on plant development, net photosynthesis and yield (Lang et al. 2005) and respiration rate (Palta et al. 1980) of agricultural and horticultural crops. The damage to fruit plantations is so severe that many a times farmers up root their well-established orchards. Crop yields were lower by 10 to 40 per cent in wheat, 25 per cent in gram, 50 to 70 per cent in mustard and 60 to 95 per cent in aonla compared to a normal year. The orchards of mango, litchi, guava, ber and kinnow in Punjab were hit resulting in reduced size or poor quality of the fruit. The plant mortality in mango varieties was highest in Dashehari, followed by Amrapali and least in Langra with 18 to 48 per cent flowering damage in parts of Jammu, Punjab, Haryana, Chandigarh and Himachal Pradesh. The mortality rate in papaya ranged from 40 to 83 per cent in the Shiwalik region, plains of Uttar Pradesh, Bihar and the northeast. The damage to fruit trees was relatively more in low-lying areas where cold air settled and remained for a longer time on ground. But temperate fruits such as apple, peach, plum and cherry gave higher yield due to extended chilling. Some major signs of cold stress are as follows, surface lesion, a water-soaked appearance of the tissue, discoloration, and desiccation, tissue breaks down, accelerated senescence, ethylene production, shortened shelf life and faster decay (Sharma et al. 2005). Most plants like rice, maize, tomato, soybean, banana, papaya, mango, grapes and orange are sensitive to low temperature and lose their quality and productivity (Larcher 1995). For instance a 30 to 40% yield reduction of rice was reported in temperate climate (Andaya and Mackill 2003). Similarly in low temperature when plant tissue temperatures fall below critical values, sensitive perennial crops such as grapevines and tender tree fruits can suffer irreversible cold injury, causing malfunction or death of plant cells.

Frost occurs when water vapour in the air freezes upon contact with an object that has a surface temperature below 0 °C. It is commonly seen as ice crystals on tree branches, grass stems, and car windscreens. Radiation frost occurs at night in cold temperatures when heat radiates into the atmosphere and leaves the ground and ambient air cooler. Drought years are conducive to radiant frost due to the lack of cloud cover and the increased difference between day and night time temperatures. When plant tissues freeze, ice crystals form and rupture cell walls. The resulting dead tissue leads to the familiar water-soaked and scorched appearance of frosted plants. Late, heavy frosts during spring can kill fruitlets at any stage of development. The severity of the damage varies with the phenological stage and type of fruit tree affected. The duration of exposure to frost and the rate of tissue defrosting also influence the amount of damage.

In addition, membrane damage occurs from ice formation inside the plant tissue, cellular dehydration and increased concentration of intracellular salts in frost stresses. Lipid peroxidation caused by accelerated reactive oxygen species resulting from freezing produce cold and frost injury in plants (Liang et al. 2008). Prolong decrease in temperature and change of cold stress into freezing stress affects plants growth to come into a critical stage. This causes visibility of damages on plants surface in some cases. Temperatures below 0 °C even for short time can be fatal for plant disease and death of cell viability (Raghavendera and Padmasree 1992). Severe, late frosts can also kill the buds responsible for the following year's crop and have an on-going effect on orchard productivity.

10.2.4.1 Management Strategies

Site Selection Where advection frosts occur, frost-sensitive crops need to be planted on slopes so that cold air can flow downhill and escape the orchard. Belts of trees, weedy fence lines and buildings can trap cold air. Low branches on wind breaks at the bottom of slopes need to be pruned to allow cold air to drain. Windbreaks uphill of an orchard can prevent cold airflows reaching the crop. Orchards located next to dams, lakes or rivers are usually less prone to frost damage. Air masses over water cool less rapidly at night than over land. Where radiant frost occurs, orchards need to be well irrigated. Moist, compact soil absorbs heat during the day and radiates this heat during the night. This heat may be enough to prevent damage from light frost. Dry, loose soils do not warm during the day and offer no protection during the night. Mulches need to be applied after flowering and kept moist. Inter-rows need to be slashed as low as possible to trap radiant heat during the day for release in the early mornings to reduce frost damage.

Netting Netting provides some protection from frost damage in light to moderate frost events. Nets hold heat radiating from the ground and raise the temperature in the orchard.

Wind Machines Wind machines can raise air temperatures around plants by about a third to half the temperature inversion difference. One protection method is to use wind machines in orchards of high value crops. Wind machines are tall, fixed-in-place, engine-driven fans that pull warm air down from at least 15 m above ground during strong temperature inversions, blowing it down and out, pushing away and replacing cold air near target crops. This raises air temperatures around cold-sensitive perennial crops such as grapes. Wind machines break up micro-scale air boundary layers over plant surfaces, improving sensible heat transfer from the air to the plants. Wind machines transfer heat by forced convection. The area protected covers 3–5 ha, depending on topography, field layout, strength of temperature inversion, time of year and drift due to slight winds.

Overhead Irrigation This method protects plant tissue. It does not raise the air temperature or warm plant parts. There is likely to be a build-up of ice on branches and the resulting weight can cause breakages. Sprinklers need to be turned on with the onset of freezing conditions and maintained at a rate of 1.5 to 4 mm per hour until temperatures have risen above freezing point. If irrigation is turned off prematurely, heat will transfer from the plant tissues to melt the ice, so freeze damage is likely.

Use of Chemicals Abscisic acid has been demonstrated to reduce chilling-induced injury in some crops (Wang 1993).

10.2.5 Hailstorm

Hot, humid, still days result in strong upward convection, creating massive thunderhead (cumulonimbus) clouds. The upper elevations of such clouds are below 0 °C, and droplets of water carried into the super cooled middle and uppercloud levels freeze. Enlargement of the ice ball occurs both from repeated trips up and down in the convective currents inthe cloud, and by continuous condensation of moisture in the supercooled regions of the cloud. Once the mass of the hail stone exceeds the upward force of the convective currents, the hailstone will fall to earth. Hailstones usually range in size from 0.5 to 1 cm in diameter, but sizes up to 2 to 4 cm are not uncommon (Schaefer and Day 1981).

The development of hailstones typically occurs 5 to 7 km/hr. above the earth's surface. Its size and shape depend on how fast the storm is moving and how strong the updrafts are inside the storm. A typical hail-streak is about 1.5 km wide and 8 km in length. However, these may vary from a few acres to large belts, about 16 km wide and 160 km long. The volume of hail reaching the ground falls at a speed of about 40 m/sec, and is usually less than 10 per cent of the volume of rain produced by a thunderstorm (Bal et al. 2014).

Hail damage to foliage, flowers, and tender stem tissues appears as bruising, shredding, defoliation, orphysical mangling. Tattered holes may be obvious in larger leaves. Fruit, twigs, and even larger stems may exhibit open, ragged-edged wounds in the skin or bark. Damage to fruits nearing maturity can resemble bird pecking damage. Such damage is easily identified immediately after the storm event, which seldom escapes the notice of people residing in the path of the hailstorm. Damaged fruits will mature with surface blemishes and malformations (Jones and Aldwinkle 1990;Whiteside et al. 1988).

In addition to the direct damage caused by hail, wounds caused by impact can serve as the infection court for fungal and bacterial diseases. A high velocity impact early in the season can cause deep depressions in fruit which persist to harvest. Later season damage can vary from bruised fruit to broken skin and holes in the fruit flesh. Fruit can also be stripped from the trees by the force of wind. Damage in mature fruit quickly become focal points for brown rot (stone fruit, cherries), blue rot (apples) and grey rot (all fruit). While the disease may initially be limited to damage tissue, it can quickly spread to intact fruit in warm, humid conditions.Twig canker fungi, bacterial and fungal soft rot of fruits, and bacterial fire blight of Rosaceae maytake advantage of wounds, resulting in even greater losses (Jones and Aldwinkle 1990). The permanent or temporary losses/damage occurred due to hail storm, high wind velocity and rainfall incidence in Maharashtra, Karnataka, M.P., in horticultural crops *viz.* pomegranate, grapes, citrus, mango, banana, papaya, onion and vegetable crops which experienced fruit injury, canopy damage besides uprooting and breakage of trees/plants.

10.2.6 Management Strategies

- In areas with higher probability of hail storm occurrence, shade nets can be a good option especially for high value crops.
- Planted shelter belts and wind breaks around orchard to avoid heavy damage to the main crop.
- Hail control mechanisms: Artificial hail control is an important measure in disaster prevention and mitigation. Cloud seeding for hail suppression is based on the cloud microphysical concept in which seeding is postulated to reduce hail severity. The natural and artificial ice crystals compete for the available super-cooled liquid cloud water within the storm. Hence, the hailstones that are formed within the seeded cloud volumes will be smaller and produce less damage if they should survive the fall to the surface. If sufficient nuclei are introduced into the new growth region of the storm, then the hailstones will be small enough to melt completely before reaching the ground. Another concept is to create shock waves which can prevent the formation and growth of hail by melting altogether. Shockwaves are produced using hail guns/cannons. The super-cooled water situated on the external layer of hailstone is transformed from liquid state to solid state. Therefore the hail nuclei are not able to melt anymore and remain at a small size which thus minimizes the damage when they hit the ground. It is be reiterated that hail control mechanisms cannot eliminate hail completely but the cloud seeding can be beneficial (Bal et al. 2014).
- After hail storm, clean up thedebris and trim or prune off broken stems, affected branchesand leaves.Remove damaged fruits, which will attract insects and helps in building secondary infection.
- Application of fungicide and/or insecticide to minimise chances of secondary infection by preventing fungus and/or insects from wounding parts of plants.The

spraying of 1% Bordeaux mixture was beneficial in healing of wounds. Application ofsupplemental nutrientsto the affected plants to help them to regrow and development of new foliage.

- Plants damaged in the spring season benefit from a 5–8 cm layer of mulch around the base of theplant to help it survive summer.
- Heavily affected plantsshould be removed and replaced.
- Bud breaking chemicals and growth regulators may be applied to induce the vegetative growth in orchard crop along with fertilizers.

10.2.7 Meteorological Drought

On an average India receives 118 cm annual rainfall which is considered to be highest anywhere in the world for the country of comparable size. But the uncertain, unreliable and erratic nature of rainfall by south-west monsoons creates drought conditions in different parts of the country. The meteorological drought is defined usually on the basis of the degree of dryness (in comparison to some "normal" or average amount) and the duration of the dry period. Definitions of meteorological drought must be considered as region specific since the atmospheric conditions that result in deficiencies of precipitation are highly variable from region to region. Drought results in wilting, reduction in photosynthesis, disturbances in physiological processes, cessation of growth, or even death of the plant or plant parts which ultimately reduce the potential yield. The occurrence of drought conditions during production of fruit and vegetable crops is becoming more frequent with climate change patterns (Whitmore 2000). Moisture stress during the production phase of crops may affect their physiology and morphology in such a manner as to influence susceptibility to weight loss in storage (Kumar et al. 2013). Another negative effect associated with water deficit is that, pre-harvest water stress (watering to 25–75% of soil water field capacity) can weaken the cells, resulting in higher membrane leakage (i.e. cell damage) and consequently greater weight loss in storage. Many aspects of plant growth are affected by moisturestressincluding leaf expansion which is reduced due to sensitivity of cell growth to water stress. Reduction in leaf areareduces crop growth and thus biomass production (Brown et al. 1985). In sugarcane, drought stress occurring when the leaf canopy is already well established has more serious impacts on final yield in terms of total biomass, stalk biomass, and stalk sucrose than when drought occurs earlier during the season (Robertson et al. 1999).

10.2.8 Management Strategies

Selection of Suitable Crops and Varieties Varieties having good root system and capacity to recoup after the alleviation of stress need to be selected and depending upon situation, it is recommended to use short duration varieties.The fruit crop selected for arid region must be such that its maximum growth period synchronizes with the period of maximum water availability and low vapour pressure deficit in the atmosphere. Fruits like ber, custard apple, phalsa, Cordiamyxa are suitable for such condition. The selected crop should be such that their reproductive cycle can be monitored to synchronize with maximum moisture availability periods e.g. guava, pomegranate and acid lime which bear fruits in distinct bahars and the bahar which coincides with rainy season can be encouraged. Since water is a limiting factor in arids, the crop selected should have drought tolerance mechanism like deep root system to draw water from deeper soil (ber, date palm), leaf shedding (ber, gonda), water binding mechanism (fig), presence of thorns (caronda), stomata at lower side (anona), wax coating (ber), leaf orientation (aonla), hairyness, sunken and covered stomata (fig, phalsa, ber and gonda).

Breeding Strategies Ittends to focus on reducing the risk of total yield loss. This can be achieved by decreasing cumulative transpiration, which often incurs a penalty in terms of yield potential and performance under mild or moderate water deficit conditions. Such strategies therefore often increase yield stability at the expense of yield potential. These strategies include (a) shortening the duration of the crop cycle (i.e. phenological adjustment) to escape from the drought period (b) reducing leaf area or stomatal conductance, traits which frequently increase the WUE.

Improved Method of Seedling Production Improved method of seedling production such as Portray grown seedling using coco peat, nylon net protection and bio-fertilizers/bio-pesticide inoculation at nursery stage has good potential for obtaining sturdy, uniform and healthy seedlings. These seedlings when transplanted in the main-field will establish better with less root damage and fare better in overcoming biotic and abiotic stresses particularly during water stress conditions.

Adoption of Soil and Moisture Conservation Techniques Contour cultivation, contour trip cropping, mixed Cropping, tillage, mulching, zero tillage, are some of the agronomical measures for the in-situ soil moisture conservation. Mechanical measures like contour bunding, graded bunding, bench terracing, vertical mulching etc. also need to be followed for effective soil and moisture conservation in dry lands. Trench planting (0.5–1.0 m) is recommended for ber, aonla and custard apple to conserve moisture. The trenches collect rain water along with silt and organic matter and thus promote tree growth. Planting in trenches is common for pineapple under dry conditions. Contouring, contour bunds, graded bunding, terraces, broad bed furrow system, curved land ploughing, loose boulders are some of the other agronomical measures used to conserve moisture.

Enrichment of Soil Organic Carboncontent Incorporation of crop residues and farm yard manure to soil improves the organic matter status, improves soil structure and soil moisture storage capacity. Organic matter content of the soil can also be improved by fallowing alley cropping, green manuring, crop rotation and agro forestry. Vegetables being short duration crop and having faster growth phases, the available organic matter needs to be properly composted. Vermi-composting can be followed for quicker usage of available organic matter in the soil and improving the soil moisture holding capacity.

Water Saving Irrigation Method In the midst of increasing urban and environmental demands on water, agriculture must improve water use efficiency generally. Adding climate change to this mix only intensifies the demands on water use in agriculture. With hotter temperatures and changing precipitation patterns, controlling water supplies and improving irrigation access and efficiency will become increasingly important.

In places with limited access to irrigation water, well timed 'deficit irrigation' or 'regulated deficit irrigation' or 'phase dependent limited irrigation' can make a substantial difference in productivity. Such deficit irrigation techniques like partial root zone drying (PRD), drip irrigation or sub-surface irrigation will become increasingly important. In non-irrigated areas, water conservation and water harvesting techniques may be farmers' only alternative to abandoning cultivation agriculture all together. Drip irrigation has proved its superiority over other conventional method of irrigation, in horticulture due to precise and direct application of water in root zone. A considerable saving in water, increased growth, development and yields of fruits and vegetables and control of weeds, saving in labour under drip irrigation are the added advantages. Drip irrigation can be adopted in fruit crops and also to all vegetable crops including closed spaced crops like onions and beans. The saving in water is to the tune of 30–50% depending on the crop and season. In capsicum, tomato, okra and cauliflower indicated that adopting alternate-furrow irrigation and widely-spaced furrow irrigation saved 35 to 40 per cent of irrigation water without adversely affecting yield.

Mulching Practices The technique of covering the soil with natural crop residues or plastic films for soil and water conservation is called mulching. Mulching can be practiced in fruits and vegetable crops using crop residues and other organic material available in the farm. Recently plastic mulches have come into use due to the inherent advantages of efficient moisture conservation, weed suppression and maintenance of soil structure. Wide variety of vegetables can be successfully grown using mulches. In addition to soil and water conservation, improved yield and quality, suppression of weed growth, mulches can improve the use efficiency of applied fertilizer nutrients and also use of reflective mulches are likely to minimize the incidence of virus diseases. For vegetable production generally polyethylene mulch film of 30micron thick and 1 to 1.2 m width is used. Generally raised bed with drip irrigation system is followed while laying the mulch film.

Application of Foliar Nutrition and Plant Growth Regulators The foliar application of nutrients during water stress conditions helps in the better growth by quick absorption of nutrients. The spraying of K and Ca induces drought tolerance in arable (cereals, pulses and vegetables) crops. Spraying of micronutrients and secondary nutrients improves crop yields and quality. Anti-transpiration coatings have been shown to be effective for maintaining quality through control of water loss. Several chemicals have been successfully used like Acropyl in grapes, polycot in banana and kaolinite (3–8%) in different fruit plants. The energy input can be reduced by increasing plant reflectivity by using effective chemicals like zinc oxide, kaolinite, chalk etc. alone or in combination with other anti-transpirants. Film forming compounds like wilt-pruf, mobileaf, vapour guard and folicoat can be used to reduce transpiration and water loss.Foliar sprays of 50 mMIAA, KNO_3, GA_3, or benzylaminopurine (BAP) partially counteracted the effect of water deficit on photosynthesis and transpiration. Growth regulator applications can also potentially enhance stress resistance in crops which are prone to show accelerated ripening or senescence in response to drought stress. As consequence anti-ethylene products such as aminovinylglycine (AVG) and 1-methylcyclopropene (1-MCP) could be used to mitigate drought stress. While foliar cytokinin (CK) application can prevent ABA-induced photosynthetic limitation, the effects can be transient and of little consequence in the long term. Paclobutrazole (10 mg/lit) is used to avoid moisture stress in mango. Accelerated ripening or senescence is most often mediated by ethylene production in response to the stress. As consequence anti-ethylene products such as aminovinylglycine (AVG) and 1-methylcyclopropene (1-MCP) could be used to mitigate drought stress.

10.2.9 Greenhouse Gases

Greenhouse gases are a group of compounds that are able to trap heat (long wave radiation) in the atmosphere, keeping the Earth's surface warmer than it would be if they were not present (Allison 2010). These gases are the fundamental cause of the greenhouse effect (Le Treut et al. 2007). An increase in the amount of greenhouse gases in the atmosphere enhances the greenhouse effect which is creating global warming and consequently climate change. Greenhouse gases allow sunlight (short-wave radiation) to pass through the atmosphere freely, where it is then partially absorbed by the surface of the Earth. But some of this energy bounces back out towards space as heat. The heat emitted back to space, some is intercepted and absorbed by greenhouse gases in the atmosphere. This is because these compounds are made of three or more atoms. This molecular structure allows them to absorb some of the escaping heat and then re-emit it towards the Earth which increases global temperatures.

According to The Ministry of Environment & Forests, Govt. of India estimates, agriculture sector in India contributes about 28% of total GHGs emissions and is likely to increase in the future due to increasing population and urbanistaion,. Emissions from soils account for about 12% while manure management and crop residues burning account for 6% of the emissions. Over 90% of N_2O emissions constituting 4% of India's GHGs emissions come from agriculture sector, largely due to fertilizer use.In order, the most abundant greenhouse gases in Earth's atmosphere are Water vapor (H_2O), Carbon dioxide (CO_2), Methane (CH_4), Nitrous oxide

(N_2O), Ozone (O_3), Chlorofluorocarbon (CFC_S). Recently the carbon dioxide (CO_2) and methane (CH_4) concentrations have reached from 280 ppm and 0.7 ppm in the pre-industrial period to 395.4 ppm and 1.76 ppm at present, respectively.Nitrous oxide (N_2O) concentration is increasing at the rate of 0.22% every year to attain a value of 310 ppb (IPCC 2001) and attained 324 ppb in 2014 (Blasing 2014).

Increased CO_2 Level and Plant Growth There is now general agreement that levels of atmospheric CO_2 will rise significantly over the next 50 to 100 years. It is important for us to consider these effects will have upon all aspects of the life cycle of plants, because the increase in CO_2 concentration will be of global nature.It is known that increased levels of CO_2 cause significant increases in fruit and seed yield in a wide range of C_3 plants but this may not always be true for flower production. Higher concentrations of atmospheric CO_2 due to increased use of fossil fuels, deforestation and biomass burning, can have a positive influence on photosynthesis under optimal growing conditions of light, temperature, nutrient and moisture supply, biomass production can increase, especially of plants with C_3 photo-synthetic metabolism above and even more below ground.TheC_3 species respond more too increased CO_2; C_4 species respond better than C_3 plants to higher temperature and their water-use efficiency increases more than for C_3 plants. Higher CO_2 values would also mitigate the plant growth damage caused by pollutants such as NO_x and SO_2 because of smaller stomatal openings.

10.2.10 Management Strategies

Reduced Tillage Soil tillage involves the physical disturbance of the upper soil layers for a variety of purposes, including seedbed preparation, weed control, and incorporation and mixing of crop residues, fertilizers or other amendments. The relationship between tillage, soil structure and soil organic matter (SOM) dynamics is integral to the C sequestration capacity of agricultural soils. Soil carbon sequestration can be used to mitigate the increase of atmospheric CO_2 concentration. Soil erosion can be prevented through conservation practices that may also sequester soil carbon and enhance methane (CH_4) consumption (Johnsona et al. 2007). Genetically modified crops such as Roundup Ready TM (herbicide resistant) soybean led to sequestration of 63,859 million tons of CO_2 (Kleter et al. 2008).

Reduced Use of Agricultural Chemicals The use of agricultural chemicals has led to the contamination of the environment with hazardous toxic chemicals which ultimately affect the biogeochemical cycles. Formation and release of greenhouse gases (particularly N_2O) from the soil to the atmosphere is mainly occurs due to the use of inorganic nitrogenous fertilizers namely ammonium sulphate, ammonium chloride, ammonium phosphates etc. (Brookes and Barfoot 2006). To avoid that effect biotechnology-based fertilizer is one of the solutions to reduce the adverse effects.The nitrogenfixing characteristics of Rhizobium inoculants were improved by using genetic engineering techniques (Zahran 2001).

10.3 Conclusion

Climate change is an ideal scenario of atmospheric and edaphic stresses in which there has been a change in the statistical distribution of weather (temperature, soil moisture, salinity, ecohydrology, soil fertility, emission of greenhouse gases, etc.) over periods of time that range from decades to centuries to millions of years. Agriculture crops do respond to these changes in the process of acclimation and acquiring tolerance– morphologically, structurally, physiologically, biochemical and molecular mechanisms.Thesestresses cause changes in soil–plant–atmosphere continuum which is responsible for reduced yield in several of the major agricultural crops in different parts of the world as well as in India.Therefore, the subject of crop adaptation and management interventions for climate resilient agriculture is gaining considerable significance to Indian agriculture. As climate change will amplify the thesestressors, especially temperature (low and high) and water availability (drought and flood) in many places, appropriate adaptive measures must be taken tomeet the future productivity demands by higher resourceuse efficiency. By incorporating various adaptation measures in the agriculture system one can increase the resilience and adaptive capacity of farmers. Applying climate resilient agricultural technologies that would increase farm production, productivity and qualityvia continuous management of natural and manmade resources constitute an integral part of sustaining agriculture in the era of climate change.

References

Aggarwal, P. K. (2008). Climate change and Indian agriculture: Impacts, adaptation and mitigation. *Indian Journal of Agricultural Sciences, 78*, 911–919.

Allison, I. (2010). *The science of climate change: Questions and answers*. Canberra: Australian Academy of Science.

Andaya, V. C., & Mackill, D. J. (2003). Mapping of QTLs associated with cold tolerance during the vegetative stage in rice. *Journal of Experimental Botany, 54*, 2579–2585.

Bal, S.K., Saha, S., Fand, B.B., Singh, N.P., Rane, J. & Minhas, P.S. (2014) Hailstorms: Causes, damage and post-hail management in agriculture. Technical Bulletin No. 5, ICAR-National Institute of Abiotic Stress Management, Malegaon, Baramati, Pune, Maharashtra. pp. 44.

Beck, E. H., Fetitig, S., & Knake, C. (2007). Specific and unspecific responses of plants to cold and drought stress. *Journal of Biological Sciences, 32*(3), 501–510.

Blasing, T. J. (2014). *Recent greenhouse gas concentrations*. https://doi.org/10.3334/CDIAC/atg.032. http://cdiac.ornl.gov/pns/current_ghg.html.

Brookes, G., & Barfoot, P. (2006). GM crops: The first ten years – global socio-economic and environmental impacts in the first ten years of commercial use. *ISAAA. Brief*, 36, 1–116

Brown, E., Brown, D., & Caviness, C. (1985). Response of selected of soybean cultivars to soil moisture deficit. *Journal of Agronomy, 77*(2), 274–278.

Craufurd, P. Q., & Wheeler, T. R. (2009). Climate change and the flowering time ofannual crops. *Journal of Experimental Botany, 60*, 2529–2539.

Dhawan, A. K., Singh, K., Saini, S., Mohindru, B., Kaur, A., Singh, G., & Singh, S. (2007). Incidence and damage potential of mealybug, *Phenacoccussolenopsis* Tinsley, on cotton in Punjab. *Indian Journal of Ecology, 34*, 110–116.

Farooq, M., Basra, S., Wahid, A., Cheema, Z., Cheema, M., & Khaliq, A. (2008). Physiological role of exogenously applied glycine betaine to improve drought tolerance in fine grain aromatic rice (*Oryza sativa* L.). *Journal of Agronomy and Crop Science, 194*, 325–333.

IPCC. (2001). *Climate change 2001: The scientific basis- contribution of working group I to the third assessment report of the intergovernmental panel on climate change (IPCC)*. Cambridge: Cambridge University Press.

IPCC. (2013). Summary for policymakers. In T. F. Stocker, D. Qin, G.-K. Plattner, M. Tignor, S. K. Allen, J. Boschung, A. Nauels, Y. Xia, V. Bex, & P. M. Midgley (Eds.), *Climate change 2013: The physical science basis. Contribution of working group I to the fifth assessment report of the intergovernmental panel on climate change*. Cambridge, UK/NY: Cambridge University Press.

Jalota, S. K., Kaur, H., Ray, S. S., Tripathi, R., Vashisht, B. B., & Bal, S. K. (2013). Past and general circulation model-driven future trends of climate change in Central Indian Punjab: Ensuing yield of rice-wheat cropping system. *Current Science, 104*(1), 105–110.

Johnsona, J. M. F., Franzluebbersb, A. J., Weyersa, S. L., & Reicoskya, D. C. (2007). Agricultural opportunities to mitigate greenhouse gas emissions. *Environmental Pollution, 150*(1), 107–124.

Jones, A. L., & Aldwinkle, H. S. (Eds.). (1990). *Compendium of apple and pear diseases* (pp. 85–86). St. Paul: American Phytopathological Society Press.

Kleter, G. A., Harris, C., Stephenson, G., & Unsworth, J. (2008). Comparison of herbicide regimes and the associated potential environmental effects of glyphosate-resistant crops versus what they replace in Europe. *Pest Management Science, 64*, 479–488.

Kotak, S., Larkindale, J., Lee, U., von Koskull-Döring, P., Vierling, E., & Scharf, K. D. (2007). Complexity of the heat stress response in plants. *Current Opinion in Plant Biology, 10*, 310–316.

Kumar, P. S., Minhas, P. S., Govindasamy, V., & Choudhary, R. L. (2013). Influence of moisture stress on growth, development, physiological process and quality of fruits and vegetables and its management strategies. In R. K. Gaur & P. Sharma (Eds.), *Approaches to plant stress and their management* (pp. 125–148). New Delhi: Springer.

Lang, P., Zhang, C. K., Ebel, R. C., Dane, F., & Dozier, W. A. (2005). Identification of cold acclimated genes in leaves of Citrus unshiu by mRNA differential display. *Gene, 359*, 111–118.

Larcher, W. (1995). *Physiological plant ecology-ecophysiology and stress physiology of functional groups*. Berlin/Heidelberg/New York: Springer.

Le Treut, H., Somerville, R., Cubasch, U., Ding, Y., Mauritzen, C., Mokssit, A., Peterson, T., & Prather, M. (2007). Historical overview of climate change. In S. Solomon, D. Qin, M. Manning, Z. Chen, M. Marquis, K. B. Averyt, M. Tignor, & H. L. Miller (Eds.), *Climate change 2007: The physical science basis. Contribution of working group I to the fourth assessment report of the intergovernmental panel on climate change*. Cambridge, UK/NY: Cambridge University Press.

Liang, Y., Zhu, J., Li, Z., Chu, G., Ding, Y., Zhang, J., & Sun, W. (2008). Role of silicon in enhancing resistance to freezing stress in two contrasting winter wheat cultivars. *Environmental and Experimental Botany, 64*, 286–294.

Mengel, K., & Kirkby, E. A. (2001). *Principles of plant nutrition* (5th ed.). Dordrecht: Kluwer Academic Publishers.

Mittler, R. (2006). Abiotic stress, the field environmentand stress combination. *Trends in Plant Science, 11*(1), 15–19.

Monteith, J. L., Ong, C. K., & Corlett, J. E. (1991). Microclimatic interactions in agroforestry systems. *Forest Ecology and Management, 45*(1–4), 31–44.

Oerke, E. C. (2006). Crop losses to pests. *The Journal of Agricultural Science, 144*, 31–43.

Palta, I. P., & Li, P. H. (1980). Alteration in membrane transport properties by freezing injury in herbaceous plants. *Plant Physiology, 50*, 169–175.

Porter, J. R., & Semenov, M. A. (2005). Crop responses to climatic variation. *Philosophical Transactions of the Royal Society B: Biological Sciences, 360*, 2021–2035.

Prasad, P. V. V., Craufurd, P. Q., Kakani, V. G., Wheeler, T. R., & Boote, K. J. (2001). Influence of high temperature during pre- and post-anthesis stages of floral development on fruit-set and pollen germination in peanut. *Australian Journal of Plant Physiology, 28*, 233–240.

Raghavendera, A. S. S., & Padmasree, K. (1992). Short term interaction between photosynthesis and respiration in leaves and protoplasts. *Plant Physiology, 3*, 12–18.

Robertson, M. J., Inmam-Bamber, N. G., Muchow, R. C., & Wood, A. W. (1999). Physiology and productivity of sugarcane with early and mid-season water deficit. *Field Crops Research, 64*, 211–227.

Sage, R. F., Way, D. A., & Kubien, D. S. (2008). Rubisco, rubisco activase, and globalclimate change. *Journal of Experimental Botany, 59*, 1581–1595.

Sakamoto, A., & Murata, N. (2000). Genetic engineering of glycinebetaine synthesis in plants: Current status and implications for enhancement of stress tolerance. *Journal of Experimental Botany, 51*, 81–88.

Samra, J. S. (2003). Impact of climate and weather on indian agriculture. *Journal of the Indian Society of Soil Science, 51*, 418–430.

Schaefer, V. J. & Day, J. A. (1981). A field guide to the atmosphere, Peterson field guide series, no. 26. Houghton Mifflin Company, Boston pp. 197, 244–245.

Sharma, P., Sharma, N., & Deswal, R. (2005). The molecular biology of the low-temperature response in plants. *BioEssays, 27*, 1048–1059.

Vara Prasad, P. V., Craufurd, P. Q., Summerfield, R. J., & Wheeler, T. R. (2000). Effectsof short-episodes of heat stress on flower production and fruit-set of groundnut (*Arachishypogaea* L.). *Journal of Experimental Botany, 51*, 777–784.

Wang.C.Y. (1993). Approaches to reducing chilling injury of fruits and vegetables. *Horticultural review, 15*, 63–95.

Waraich, E. A., Ahmad, R., Ashraf, M. Y. S., & Ahmad, M. (2011). Improving agricultural water use efficiency by nutrient management in crop plants. *Acta Agriculturae Scandinavica Section B Soil and Plant Science, 61*(4), 291–304.

Whiteside, J. O., Garnsey, S. M. and Timmer, L. W. (Eds.) (1988). *Compendium of citrus diseases* (pp. 5, 19, 59). St. Paul: American Phytopathological Society Press.

Whitmore, J. S. (2000). *Drought management on farmland*. Dordrecht: Kluwer Academic Publishers.

Zahran, H. H. (2001). Rhizobia from wild legumes: Diversity, taxonomy, ecology, nitrogen fixation and biotechnology. *Journal of Biotechnology, 91*, 143–153.

Chapter 11
Impact of Climate Change on Tropical Fruit Production Systems and its Mitigation Strategies

Vishal Nath, Gopal Kumar, S. D. Pandey, and Swapnil Pandey

Abstracts Scientists are almost of unanimous opinion that tropics will be the first and most to suffer due to climate change even though the magnitude of projected change as well as past climate trend is moderate as compared to other part of the world. Reasons are though complex, include poor economic conditions of majority of population, higher dependence on natural resources and ecosystem service and relatively narrow temperature ranges. Projection of climate change for tropics indicates rise in temperature of 0.4–1 °C, 0.8–3.2 °C and 1.2–6 °C by the 2020, 2050 and 2100 respectively as compared to base line period (1961–1990). Atmospheric CO_2 concentration is likely to reach 550–800 ppm from the present level of 400 ppm by the end of this century. The change associated with climate has never been smooth and likely to be expressed in terms of increased uncertainties and extremities of weather. Number of heat waves witnessed during last 25 years (1990–2015) in tropics has exceeded as compared to preceding 70 years. Similarly the number of extreme rainfall events as well as drought has also increased in recent past. These changes are likely to impact natural resources, societies and their interdependence.

Fruit production being one of the important activities in tropical regions has been largely ignored from the systematic impact analysis of climate change and adaptation studies. Tropical fruits are attuned to the prevailing weather conditions marked by high temperature. There are delineated and definite ranges of temperature at different phenophases for optimum production of different fruits. Any deviation from the optimum is likely to affect production and quality drastically. The complex interaction of altered temperatures, corresponding phenophases of different fruits, relative suitability of species and cultivars, elevated CO_2, reduced water availability, pollinators, pest and disease and management practices have demonstrated impact and thus will largely determine the tropical fruit production. Tropical fruits where production is presently limited by high temperature are likely to suffer most and probably have to shift to new areas. Some tropical fruits growing in elevated/fringe

V. Nath · G. Kumar · S. D. Pandey
ICAR-NRC on Litchi, Musahari, Muzaffarpur, Bihar, India

S. Pandey (✉)
Research Scholar, PAU, Ludhiana, Punjab, India

S. Sheraz Mahdi (ed.), *Climate Change and Agriculture in India: Impact and Adaptation*, https://doi.org/10.1007/978-3-319-90086-5_11

areas and production is limited by law temperature are likely to be benefited from elevated temperature. For rest of the large section of tropical areas, a strong adaptation strategy needs to be developed. In all above three cases, preparedness which is essential part of adaptation is must to deal with extreme weather events and increased uncertainties. In general, higher resilience of tropical perennial fruits against climate change as compared to annuals needs to be properly harnessed through adaptation mechanisms.

Suitable and improved cultivars, alteration in cultural practices including plant architectural management, water management, micro climate modifications, soil organic carbon built-up etc. are possible adaptation strategy. Shifting tropical fruit cultivation to new areas as an option is still debatable but seems necessary owing to extending tropics Adaptation strategy must have imbedded some mitigation potential in long run.

Keywords Climate change · Tropical fruits · Impact and management

11.1 Introduction

The year 2016 has been declared as warmest year in the recorded history of climate data. There is almost unanimous agreement among scientist that tropics will be first and most to bore brunt of climate change even though the magnitude of projected change as well as past climate trend is moderate as compared to other part of the words. Reason are complex comprising economic, geographic and meteorological factors including relatively narrow temperature ranges means even the small deviation is likely to have significant effect (Martin 2015). Estimated changes in terrestrial metabolic rates in the tropics are large, because of warm base line temperature (Dillon et al. 2010) even though tropical temperature change has been relatively small. Carbon cycle in the tropics has becomes more sensitive to temperature change in last 50 years and around two billion tons of more carbon per year is likely to be released from tropical forests and savannahs, in response to 1 °C rise in temperature compared with the 1960s and 1970s (Ross 2013). In addition, warming has extended the tropical region over last several decades' (Lovett 2007) resulting newer areas are getting importance for tropical fruits.

Global warming is inevitably happening and affects many aspect of life on earth. More fluctuation in temperature, rainfall, occurrence of frequent drought, floods, storms are likely expressions of climate change. There are many projections for future climate and temperature projections which show an increase of up to 6 °C by the year 2100 while CO_2 concentration might increase to 850 ppm. Temperature is known to affect both photosynthesis and respiration and their ratio must be high in order to achieve high yield (Moretti et al. 2010). Photosynthetic activity increases with temperature until certain level but further increase in temperature results inactivation of enzymes thus reduces the ability to cope with heat stress. Temperatures above 35 °C are considered to stop ripening process in climacteric fruit as it suspends ethylene production. Zhang et al. (2011) reported very high response of heat stress

(43%) in terms of protein spots in a hot water treatment in peach. Higher temperatures during growing season also resulted in earlier harvest.

Developing countries account for about 98% of tropical fruit production. Tropical areas is though endowed with great diversity of tropical fruit trees, out of about 2700 species very few are commercially grown and even lesser number is traded (Paull and Duarte 2011). Presently the impact of climate change on tropical fruits has been somewhat neglected probably with assumption that these fruits are already adapted to hot and humid conditions and also due to difficulty attached in experimentation with perennial, however, there are reports of impacts, most notably on tree phenology, especially flowering and fruiting.

Fruit crops have longer period of flowering thus stay exposed to climate variability. Temperature brings about changes in hormones needed for growth and development of trees. In these fruit species temperature play important role in flower bud differentiation as well as flowering and fruit set. Symptoms like early blooming and advanced crop harvest has already been seen (Hribar and Vidrih 2014) in some crops including mango and litchi. Tropical fruit production is not only directly affected by weather modification as expression of climate change but also by the changed over all production scenario and competition for resources. Developing effective adaptation strategy along with mitigation efforts are must for sustainable and quality production in these fruit crops.

11.2 Global Climate Change Scenario

Globally the increase in greenhouse gas concentration in the atmosphere has been considered as the prime causes for climate change. Major greenhouse gases such as carbon dioxide, methane, nitrous oxide, sulpherdioxide, etc., are increasing in the atmosphere. Burning fossil fuels, deforestation, industrial processes, some agricultural practices have contributed to increased greenhouse gases into the atmosphere. After fossil fuel use, land use change and forestry, especially deforestation and degradation, are the next largest emitters of CO_2 (Baumert et al. 2009). CO_2 alone is responsible for 77% of global warming (Climate Analysis Indicators Tool 2011). The CO_2 concentration in atmosphere has sharply increased from 280 ppm (preindustrial period) to 400 ppm in 2014 (Vati and Ghatak 2015). Global air temperature increased by 0.74 °C between 1906 and 2005 (IPCC 2007a) which is further projected to increase by 0.5 to 1.2 °C by 2020, 0.88 to 3.16 °C by 2050 and 1.56 to 5.44 °C by 2080 (IPCC 2007b). The rate of sea level rise was 1.8 mm per year over 1961 to 2010 (Chadha 2015).

Global warming which is predicted by general circulation models and supported by mounting evidence will lead to an excessive change in climate conditions and thereby crop production. Global temperature is projected to increase up to 6 °C by the year 2100 while CO_2 concentration is projected to increase in the range from 550 to 850 ppm (Hribar and Vidrih 2014). Increase in temperature by the end of the twenty-first century and under the GHGs emission scenario SRES-A2 (IPCC 2000) is likely to be 1.8–6 °C for Asia (IPCC 2007) even after accounting most of the variability to the different global climate models (GCMs) used in the IPCC Fourth Assessment Report (AR4), and with warm periods showing greater increases as compared with cold periods. Though least increase is projected for South Asia (except for the Himalayas), but nevertheless it also ranges between 1.8–5 °C (IPCC 2007). Rainfall in cold season is predicted to change between −5 to 20% and for summer season between −40 to 15% (IPCC 2007). These changes will certainly affect agricultural production mainly due to over reliance of production systems on favorable climatic conditions (Jarvis et al. 2010).

Regional climate trends and projections are important for adaption strategy. For Indian condition, high-resolution climate change scenarios using regional climate models, predicted a change of 2.5 to 5.0 °C in annual mean surface temperatures over a period of 100 years across different scenarios (Rupa Kumar et al. 2006). MoEF (2004) has mentioned projected increase of 2–4 °C in maximum and minimum temperature by 2050. There is large scale uncertainties associated with monsoon rainfall projections (Mishra 2015). An increase in monsoon rainfall by 20%, in most of the Indian states and reduction in rainfall in some others over a period of 100 years are also projected (Rupa Kumar et al. 2006). Mishra (2015) reported declining trend of monsoon rainfall in India. Projections show increase in number of rainy days in north-eastern parts (by 5–10 days), and decrease in western and central parts of the country. Rainfall intensity and weather extremes are also likely to increase. Kumar et al. (2010, 2010a) comparing PRECIS downscaled base line (1961–1990) with A2a scenario for the period of (2071–2100) projected 3.96 °C increase in average annual maximum temperature in Gujarat, Maharashtra and Madhya Pradesh with highest mercury level in month of November. For North West part of the country, Kumar et al. (2010a, 2010) indicated more increase (to the tune of 5.6 °C) in average annual temperature which may affect productivity and seasonality of many crops and perennials. Using the CMIP5 climate projections, Chaturvedi et al. (2012) reported 3–4 °C increase in projected temperature under the representative concentration pathways (RCP) 8.5 by the end of twenty-first century and also found large uncertainty in precipitation projections. By using high resolution multimodal climate change projection Kumar, et al. (2013) reported that India is likely to be warmer by 1.5 °C and 3.9 °C by the end of 2050 and 2100 respectively as compared to 1970–1999.

In past Thailand has experienced country-wide significant warming over the last four decades and the extreme events associated with both the cold and warmth have changed remarkably (Limjirakan and Limsakul 2012). During 1980–2012, annual rainfall, number of rainy days, relative humidity, maximum and minimum temperatures significantly increased by 29.5 mm/year, 0.83 day/year, 0.116% /year, 0.033 and 0.035 °C/ year, respectively. In tropical part of Australia, mean annual temperature in all production regions has been increased and under projected climate change condition, winter temperature rise is likely to be more as compared to summer temperature rise.

Sea level rise as a result of the melting of ice glaciers is likely to increase floods and cyclones, sea water intrusion into fresh ground water, acceleration of coastal erosion, submergence of coastal and island horticulture crops. Due to melting of glaciers, many rivers in tropical regions are at the risk of losing perennial nature thereby effecting water availability in long run. By the end of twenty-first century, sea level is projected to be 40 cm higher than present level and associated flood is likely to affect additional 80 million coastal residents in Asian countries particularly Bangladesh and India.

Although the perennial fruit trees have number of survival mechanism to cope with stress but it comes at cost of productivity and quality. Climate change impact exceeding the buffer mechanism of perennial may be un imaginable/disastrous for grower. While determining the potential area of any crops specially horticulture/ perennials, the projected climate change must be considered.

11.3 Climate Optimums for Major Tropical Fruits

Mean temperature range for optimum growth of most of tropical fruits are about 24 to 30 °C (Tables 11.1 and 11.2), however some crops like mango and litchi can tolerate more than 45 °C for short period. Mango grows at temperatures as low as 0 °C and as higher as 45 °C whereas, the ideal temperature is 24–30 °C with high humidity during the growing season. Temperatures below 10 °C and above 42 °C, affect the growth. Well distributed precipitation of 900 to 1000 mm in a year is ideal.

Litchi and Longan is suitable in warm subtropical to tropical climate but frost free condition with high summer heat, rainfall and humidity are essential. Warm humid summer and cool dry winter is considered best but temperature less than 0 °C is harmful. Rambutan grows well under optimum temperature of 22–30 °C but temperature below 10° drastically reduces the growth. Citrus though a subtropical fruits is well adapted to tropical condition, the temperature of 16–20 °C is optimum but it can grow well up to 30 °C. A very high temperature is detrimental for growth of citrus. In arid and semiarid condition, best quality citrus fruits can be grown with supplemental irrigation. At least 700 mm well-distributed rainfall is necessary for these crops. Pomelo prefers warm climate (27° to 30 °C). Good quality of Guava with higher yield is produced in areas having distinct winter season. Temperature range of 23 °C to 28 °C is optimum for guava. It grows best with an annual rainfall around 1000 mm restricted between June and September. Young plants are susceptible to drought and cold. As such guava is considered tolerant to environmental stress. For custard apple, the temperatures above 40 °C and less than 0 °C are harmful. Poor fruit set in custard apple in northern India is attributed to high temperature and dry condition. For the growth and fruiting of custard apple, an annual rainfall of 800 mm is considered adequate. High rainfall, high temperature and humidity are favorable for Mango steen. Its growth is slow if temperature is below 20 °C and above 35 °C which creates stress on Mango steen.

Aonla grow well under tropical conditions with dry summer and frost free winter. It is adaptable to a wide range of climatic conditions from sea coast to an altitude of 1800 m (Bose et al. 1999). Tamarind grows well in tropical climate with hot

Table 11.1 Optimum climatic condition and latitudinal distribution of some important tropical fruits

Fruits	Optimum temp	Optimum annual rainfall	Lat. Range (distribution)	References
Mango	24–30	900–1100 well distributed, range(900–2500)	30°N -27°S	Mukherjee (1953), Davanport (2009)
Litchi	25–35	1200–2500	28°N -27°S	Tindall (1994), Menzel et al. (1989)
Citrus	16–27	1400–1800 (well distributed: About 700 mm)	44°N -27°S	Verjeij and Coronel (1992)
Guava	23–28	About 1000 mm (from June to Sept)	26°N -30°S	Verjeij and Coronel 1992
Mangosteen	25–35	(1200–1800)	12°N -10°S	Osman and Milan (2006), Cox (1976)
Tamarind	18–25	100–1400	15°N -23°S	
Rambutan	24–32	1500–2500	18°N -17°S	Tindall (1994), UDPMS (2002), Verjeij and Coronel (1992)
Ber	22–36	700–1600	22°N -18°S	
Aonla	20–34	700–2000 (dry summer, frost free winter)	35°N -18°S	Bose et al. (1999)
Pomelo	27–30	1800–2200	35°N -35°S	Verjeij and Coronel 1992, Gaffar et al. (2008)
Jackfruit	16–28	1000–1500	28°N -30°S	Ghosh (2000), Haq (2006)

Table 11.2 Optimum temperature, precipitation and latitude condition for other fruits gaining popularity in tropical condition

Fruits.	Optimum temp (°C)	Annual Rainfall (mm)	Latitude	Climate description
Sapota	14–38	900–1500	25°N -25°S	Tropical semiarid to arid with moderate humidity.
Custard apple	18–27	800–1800	22°N-22°S	.
Papaya	18–27	800–1500	25 N- 29S	Tropical, semiarid with mild winter
Banana	20–30	500–3000	28 N- 18S	Semiarid to humid subtropical
Pomeranate	18–36	600–1200	22 N - 29S	Semiarid tropics with hot dry summer and cold winter.
Longan	20–25	1400–1600	14°N-30°S	Warm subtropical to tropical with frost free winter.

summers and dry but mild winters. It is drought resistant but susceptible to frost. Ripening of fruits do not commence under cold climate (Singh 1992).

Though the optimum climatic conditions are defined for different tropical fruits but it can sustain deviation in day to day weather variation to certain extent. However this variation of weather is likely to increase under climate change condition and therefore its impact need to be studied and adaptation must be explored.

11.4 Impact of Climate Change on Tropical Fruits

Climate change may have negative as well as some positive impacts on tropical fruits however even for full actualization of positive impact, there is need to develop suitable strategy. Warming has been core of many study regarding climate change impact, elevated CO_2 and changed precipitation are also considered important.

Change in time to harvest, reduced irrigation water, increased irrigation cost, changed suitability and availability of cultivars for current and future production, increased physiological disorders, inferior qualities, increased pest (extra generation of insect pest and more period of activities) and disease, as well as outbreak of new pests and diseases, extreme event damage, negative effect on soil due to extreme rainfall and temperature are most important factors which can be associated with impact of climate change on tropical fruits.

Phenological shifts in response to climate change have been reported from different parts of the globe, Europe- Sparks and Menzel (2000), United States - Parmesan and Yohe (2003), Australia - Chambers (2009) and East Asia - Sthapit et al. (2012).

Changes in rainfall distribution affects year-to-year variations in flowering, productivity and quality of tropical fruits (Sdoodee et al. 2010). Flowering period of tropical fruits in Thailand tends to shift to second-half of the year (Apiratikorn et al. 2014). Climate variability in southern Thailand may cause off-season flowering of longkong and change in rainfall distribution may shift flowering of longkong to the second-half of year Uraipan (2009). Chen Q (2012) reported total crop failure in banana production due to extreme climate. Warmer night deteriorated the fruit quality particular fruit flavor in China. Reduced crop duration and yield due to dry spell during flower emergence & fruit sets has also been observed in banana. Increase of 1–2 °C temperature beyond 25–30 °C promotes vegetative in place of flowering flushes in citrus whereas untimely winter rain affects flower initiation and favors Psylla incidence.

Rise in a temperature of above 1 °C may shift a major area of potential tropical fruits. Many suitable areas of fruit crops may become marginally suitable where as new suitable area may come up. Increased temperature is likely to have more prominent effect on reproductive biology of these crops. Rapid development of fruit and maturity is predicted in citrus, grapes and litchis due to increased temperature. The faster maturity, quick ripening may reduce fruit availability period (Kumar 2015). Floral abortion and poor pollination due to warming is also anticipated. Soil and canopy temperature increase earlier in spring may adversely affect grafting time and callus formation. Increased irrigation requirement due to higher evaporation coupled with faster development of trees due to faster accumulation of heat unit reduces fruit sizes and anthocyanin pigmentation in litchi.

Both early and delayed flowering may be characteristic features in mango under changed climate. Low temperature (4 to 11.5 °C), high humidity (>80%) and cloudy weather in January has been observed to delay panicle emergence whereas low temperatures during inflorescence development reduces number of perfect flowers (Chadha 2015) At temperature regime of around 27/13 °C, perfect flowers were significantly higher than in a temperature regime of 21/14 °C in mango. In the regions experiencing severe winter, mango malformation is a common problem (Sankaran et al. 2008) and thus may get benefited under warmer climate. Unseasonal rain coupled with variation in temperature and high humidity may result to altered flowering trend, delayed panicle emergence and fruit set (Chadha 2015). Increased temperature during panicle development in mango may speed up the growth and reduces the number of days available for effective pollination. Desiccation of pollen and poor pollinator activities are also likely to worsen the situation. Flower bud may change into vegetative one under warm night condition. Area optimally suitable for Dussaheri, a popular mango variety drastically reduced with 1 °C increase in temperature and Alphanso (a mango cultivar) is likely to be confined to Ratnagiri area due to its suitability under changed climate (Dinesh and Reddy 2012) Increased frequency of early rain for some mango area may result in to blackening. In Southern India, even marginal increase of temperature in Nov and Dec may reduce flowering in mango. Elevated temperature is attributed for fruitlet and flower dropping in mango. Rains at flowering are harmful, washout of pollen grains during flowering and poor pollination due to reduced pollinator activities has been reported (Rajan 2012). Under continued high humidity due to unseasonal rain, severe attack of mango hoppers and certain fungi and heavy shedding of flowers and fruits take place. In case of frequent hail storms, mango cultivation is not feasible (Bhriguvanshi 2010). In mango cv. Chausa the rate of development of fruit fly increased with the increase in temperature from 20–35 °C (Kumar and Shukla 2010). Dussaheri , a popular mango variety of Uttar Pradesh produces 40–50 days early crop when grown in Andhra Pradesh. At high soil temperature, more "spongy tissue' a physiological disorder has been observed in Alphonso, a popular mango variety from Konkan region, India.

Citrus plants grow well in tropical and subtropical climates, with very high temperatures being detrimental. Warm days and cool nights are good for color development. Soil moisture stress coinciding with maturation improves TSS and reduces acidity. Cool weather in subtropics and moisture stress in tropics are known condition that favours conversion of major part of the shoots to flower at one time during the year. Increased temperature may increase water requirement in tropical condition and reduced flowering in subtropical condition where plant puts to flower under cold stress. Warming may be favorable for scale and mite pest in citrus because of drier and dustier condition. Increase in temperature beyond 30 °C promotes vegetative instead of flowering flushes.

Good quality of guava is produced in area having distinct winter season. Guava is though more tolerant to environmental stresses, young plants are susceptible to drought and cold. Red color development on the peel of guava requires cool nights during fruit maturation. Warner night temperature may impair the colour development. Varieties like Apple Color guava, which have attractive apple skin color under subtropical conditions of North India, have red spots on the skin under warmer conditions. Observation in areas suitable for production of red color guava) revealed negative correlation of total soluble solids, fruit firmness and percentage dry mass with temperature during fruit growth (Rajan 2008). However, the relationship varied with the cultivar (Hoppula and Karhu 2006). Even few daily events of extremely high temperatures (above 40 °C) which are often encountered in northern India is harmful. With projected change of 1 to 5 °C, with seasonal contrast, the area suitable is likely to reduce. Although plants can tolerate extreme climatic conditions, yet for good fruiting, high humidity, occasional rains and warm temperatures are required (Bose et al. 1999).

Aonla is adaptable to a wide range of climatic conditions (Bose et al. 1999). However in the event of frost, it is severely affected and in young budded plants scion portion and young grafts are dried. It is one of the fruit considered having greater resilience under climate change condition as it can sustain draught. Varieties that matures before occurrence of frost, produce maximum (More and Rakeshbhargava 2010).

Though high temperature in general improves fruit quality but excessive temperature has found to delay maturity in grapes. Color development has also been found to reduced due to excessive temperature. Damage due to downy mildew in grapes increased with temperature coupled with unseasonal rains. Advancement is cherry blossom by 4 days was observed in response to 1 °C rise in March temperature in Japan and Korea (Yoshino and Sook 1996). Higher temperature also results into flower drop of female and hermaphrodite plants in papaya. In banana, dry spell during flower emergence and fruit set reduces crop duration.

The predicted changes of rainfall and temperature are also likely to affect tropical fruits and regions through changes in enterprise structure and location. Increased irrigation demand and altered reliability of irrigation schemes and water availability is likely to be core issue for tropical fruit under changed climate. Redistribution of existing pests, diseases and weeds and increased threat of incursions into new crops, prominence of otherwise minor pest and disease, more physiological disorders like tip burn, sun burn, fruit cracking (Kumar and Kumar 2007), blossom end rot, hail damage and soil erosion etc. are likely impacts for which evidences are mounting. Drought and floods may promote soil borne disease. Phytophthora infection may increase in flood condition and likely to spread more. Poor color development, quick desiccation, less residence time on tree due to fast maturity are likely impact on tropical fruits. Severe pressure on food security in South Asia and Africa is predicted under climate change condition (Lobell et al. 2008).

Though impacts of climate change are likely to be adverse, but a careful adaptation strategy and preparedness for harnessing good opportunities provided by elevated CO_2 and temperature may compensate to great extent. Kimball et al. (2007)

reported that sour orange (*Citrus aurantium*) plants grown for 17 years in elevated CO_2 (300 ppm above ambient), produced 70% more fruit instead of acclimation. It also increased biomass accumulation (70% extra) due to higher wood growth. Altered carbon availability to mango fruit affected both dry mass and water mass of peel, pulp and stone (Léchaudel and Joas 2007).

Using Eco Crop data set, base line data of GCM (Had CM3) and projected data using DIVA GIS model, area suitability of some important tropical fruits was determined (Beebe et al. 2011). It shows an average increase of 1.6% in suitable area of mango globally, with 26% of the global suitable areas for mango production being negatively impacted. For coconut, some 40% of the global suitable areas seem to be decreasing their climatic suitability whereas increase of about 1% suitable area is predicted. Similarly in case of orange some (<1%) area is added to suitable area but suitability is reduced in about 46% of suitable area (Mathur et al. 2012).

11.5 Adaptation Strategy to Mitigate Ill Effects of Climate Change in Tropical Fruits

Adaptation is important but mitigation must be part of adaptation because even if we stop all emissions, the temperature will continue to rise for many decades before it gets stabilized. If left unattended, the rate of increase of will cross many dangerous tipping points which further accelerate climate change to the extent that adaptation may become inadequate (Sthapit and Scherr 2012). Mixing adaptation and mitigation is the best strategy to achieve greater levels of both adaptation and mitigation (IPCC 2007). There is sufficient scope in agriculture, land use and forestry sector, to attend adaptation and mitigation jointly as one provides opportunities for the other.

The simplest adaptation strategy to maintain tropical fruit is development and use of adaptable cultivars and cultural practices (Singh 2012). Adapting to changed climate in current location is main option at farmer's level where as shifting current production system to new favorable location is also an option and global scale. Due to perennial nature, tropical fruit trees are considered to be more resilient therefore provide important adaptive values to climate change (Chadha 2015) and can be an important substitute for more vulnerable annuals. However, if impacts exceed adaptation capacity of current location, it can be catastrophic and then northward shift of agro climatic zones in tropical region of northern hemisphere and south ward shift in southern hemisphere may be inevitable. Climate change is already considered to be shifting the habitat ranges of plants and animals (Pereira et al. 2010).

The most important adaptation and mitigation strategy in tropical production system is enrichment of soil organic carbon, as it not only has the biggest agriculture based mitigation potential but also provides resilience against climate change. Minimizing tillage, erosion and chemical use, incorporation of organic inputs and use of perennials are key strategy. Perennial tropical fruit is good substitute for annual.

11.5.1 Change of Crop and Varieties

Change in crop and variety in response to changing climate in one of the greatest adaptation strategy. Crop varieties which are able to withstand biotic and abiotic stresses offer greater climate resilience. Identifying more adaptable cultivars and a range of cultural practices may help growers maintain current production. Drought tolerant varieties and varieties that can skip stress loaded period though modified fruiting period need to be developed/ identified. Pomegrante hybrid Ruby (drought tolerant), Annona hybrid Arka Sahan (drought tolerant) and Grape root stock Dogridge (*Vitis champine*) has been found promising. Different races and genetic base need to be explored and experimented. Place of origin of fruits need to be explored for climate tolerance. For example, mono embryonic type mango may be tested for land logged climate in northern India with well-defined winter and poly embryonic type may be tested for coastal climate for adaptation under changed climate.

11.5.2 Good Management Practices for Land and Plants

Good management practices comprising different aspect of production and post-production need to be adapted to provide better resilience to tropical fruit production. Adapting improved irrigation technology targeting water for plat rather than for soil, erosion control measures, drainage facilities, retaining residue infield, reducing fallow period, canopy management for optimum light utilization in different orchards are adaptation options. Developing and practicing a monthly/biweekly calendar of practices in any tropical fruits and adjusting time of operation according to prevailing weather will be advantageous.

11.5.3 Canopy Management

Canopy management for improved leaf and fruit exposure help improve yields and fruit quality (Kumar and Nath 2013). If done properly, it minimizes resource competition and improves physiological processes of crops. It is essential in many cases including Litchi. The main aim of training in younger plants is to give desired shape to canopy and developing strong framework which also offers greater resilience against strong wind. In mature plants training and pruning aims upkeep of desired shape and maximize bearable surface area. In old plants/orchards, pruning helps maintain vigour of plant which may otherwise develop senility. It has been established that pruning stimulates shoot initiation. Rising levels of atmospheric CO_2 is likely to boost canopy development which may offers conducive microclimate, more susceptible tissue, more interception of inoculums, more opportunity for infection, more polycyclic infection and radiation shield for inoculums. If all leaves which comprises 2–5% of total biomass removed in training or/and pruning is incorporated in soil, and 50% of hard biomass is used as timber with more than 30 years of life, the pruning and training is estimated to have net mitigation effect. Schaffer and Gaye (1989) found increased in light utilization in mango by removing 25% of canopy.

11.5.4 Microclimate Modifications

There is scope of minimizing impact of extreme weather event by modifying micro climate. Water has been mostly used for microclimate modification both against heat and cold stress. Overhead irrigation, mid canopy sprinkler, shade net, reflecting materials, water channels, canopy management, etc. offers adaptation options against extreme weather. The microclimate modification may also be manipulated to reduce pest and disease.

11.5.5 Micro Site Modifications

The ideal site conditions are though defined for all tropical fruits but rarely found in actual field situation. There is wide scope of micro site modification for new plantation as well as in existing orchards. Favorable soil and filling materials including manures and nutrients are used for the same. In coarse texture soil, clay enrichment is used to improve water holding capacity. Micro site improvement has been successfully used on difficult site like gravely soil, saline or alkaline soils. Micro site modification not only improves resilience against climate change but also has mitigation effects. Drying of litchi orchards planted on shallow soil was observed near Muzaffarpur, Bihar, India due to heat and water stress during 2016, but were less on adjacent site where micro site improvement though incorporation of pond soil was done.

11.5.6 Sound Weather Forecasting System

In day to day management of production affair, farmers need to tackle extreme weather probably with higher uncertainties. Though long term larger scale and very short term at smaller scale weather forecasting system has matured and performing satisfactory but there is much to be done for taluka or district level medium term weather forecasting at which is mainly used in production management. A close to accurate forecasting is very important for adjusting field operation to minimise damage to the crops.

11.5.7 Efficient Water Management

Among the crops, maximum adoption of drip system has been in coconut, followed by banana, grape, papaya, pomegranate, mango, and sapota (Chadhga 2015). Water availability is likely to be reduced even though rain fall increased marginally as more intense rain produce more runoff. Efficient irrigation system like drip irrigation will save significant water that can be used for expanding irrigated area. More favorable soil moisture coupled with elevated CO_2 may encourage more biomass production and soil carbon enrichment.

11.5.8 Water Harvesting and Artificial Ground Water Recharge

Considering more intense rainfall, in-situ as well as ex-situ water harvesting will be inevitable to meet multi sector demand including tropical fruits which usually comes lower in priority. Surface water harvesting with suitable structure and protection against evaporation loss is an important adaptation against projected drought. Artificial ground water recharge is necessary because of over dependence on ground water. Artificial recharge usually reduces the time of evaporation thus make more water available for sector like fruit production.

11.5.9 Soil and Water Conservation Measures

Warmer climate, poor soil organic matter, poor vegetation cover under tropical condition and more intense rain under changed climate make soil highly vulnerable to erosion and degradation. Soil and water conservation measures like, leveling, bunding, terracing, trenching, and including in situ moisture conservation measures like mulching are important adaptation against, frequent drought and foods. Tropical fruits grown with suitable soil and water conservation measures offer greater resilience against climate change and also have higher mitigation potential through soil protection and soil carbon builtup.

11.5.10 Degraded and Wasteland Opportunity

Significant area in the tropics is under degraded and wastelands. About 120.72 million hectare of degraded and wastelands are in India alone (Maji et al. 2010). Out of which 16.53 million hectare is under open forest (<40% canopy cover). With proper treatment approximately 50% this area can be converted to perennial fruit belt.

Greening degraded land has added advantage as it has larger carbon sink potential due to usually very poor soil carbon base line.

11.5.11 High Density Planting (HDP)

Tropical fruits have great adaptation as well as climate change mitigation potential as it not only produce more but also accumulates biomass and soil carbon per unit area. In a well-managed high density planting of perennials, terrestrial carbon accumulation can be as high as 1.5 times as compared to normal planting in one life cycle. Selective adoption of HDP is already practice in mango, guava, banana, citrus, pineapple, pomegranate, papaya, cashew and coconut. These success experience need to be utilized for other tropical fruits like litchi, longan etc. Mango, Guava and citrus can be planted at 2.5 m × 2.5 m, 3 m × 3 m and 1.8 m × 1.8 m respectively in place of 10 m × 10 m, 6 m × 6m and 6 m × 6 m respectively. Canopy management is the key to the success but future high-density plantings have to be developed by using dwarfing rootstock, inter stocks, and scion varieties. A combination of canopy management and growth regulators is important for success of HDP.

11.5.12 Protected Cultivation

Protected cultivation using net house or poly house has advantage in terms of modified micro climate and crop is less exposed to weather extremities. Increase in CO_2 build up in the growing condition can also be beneficial in enhancing photosynthesis. Reduction in pest and disease and wild animal damage is added advantage of protected cultivation. About 40,000 hectare area is used of protected cultivation in India but mostly of annuals or for nursery in case of perennial fruits.

11.5.13 Eco Fortification in Tropical Fruit Production System Through Perennial Fruit Based Cropping Model

Crops are less affected by frost if grown under the canopy of other cops. Growing different crops under partial shade in younger orchard have multiple advantages including advantage of modified microenvironment. Intercropping of coffee and bananas, growing tuber crops in litchi and mango orchards, growing shade loving medicinal plants in full grown orchards not only improves income frpm cultivation but also encourage ecological condition by reducing soil erosion, soil heating and by adding nitrogen in case of leguminous crop. This multitier cropping system including fruits has greater resilience against climate change in tropics. More plants and residue recycling under these multi cropping system has high potential for soil carbon enrichment. Adjusting management calendar as per changed cropping system and prevailing climate is another common adaptation to climate variability at the farm level.

11.6 Conclusion

With mounting evidences of climate change there is need of preparedness to minimise adverse impact as well as to harness from the opportunities. Though the predicted warming is less for tropics as compared to higher latitude, nevertheless the impact is predicted to be worse. Tropical fruits are not only exposed to changed climate but also will be impacted by changes in other sector. Early maturity, accelerated growth, altered flowering, more insect pest, new pests and diseases, increased irrigation requirement, higher physiological disorder like fruit cracking, tip burn, poor colour development, reduced fruit size are likely impact on tropical fruits. For day to day management, grower need to tackle extreme weather and increased uncertainties. Mitigation must be part of every adaptation strategy. Improved irrigation technique, microsite and climate modification, change in crop and varieties, protected cultivations, canopy management, use of growth chemicals are adaptation options. High density planting, multitier cropping, soil and water conservation, enriching soil organic carbon are mitigation options with adaption potential in tropical fruits.

References

Apiratikorn, S., Sdoodee, S., & Limsakul, A. (2014). Climate-related changes in tropical-fruit flowering phases in Songkhla Province, Southern Thailand. *Research Journal of Applied Sciences, Engineering and Technology, 7*(15), 3150–3158.

Baumert, K. A., Herzog, T., & Pershing, J. (2009). *Navigating the numbers, greenhouse gas data and international climate policy*. Washington, DC: World Resources Institute.

Beebe, S., Ramirez, J., Jarvis, A., Rao, I. M., Mosquera, G., Blair, M., & Bueno, J. M. (2011). Genetic improvement of common beans and the challenges of climate change. In S. S. Yadav, B. Redden, J. L. Hatfield, & H. Lotze-Campen (Eds.), *Crop adaptation to climate change* (pp. 356–369). Hoboken: Wiley-Blackwell Publishing.

Bhriguvanshi, S. R. (2010). Impact of climate change on mango and tropical fruits. In H. P. Singh, J. P. Singh, & S. S. Lal (Eds.), *Challenges of climate change-Indian horticulture*. New Delhi: Westville Publishing House.

Bose, T. K., Mitra, S. K., Farooq, A. A., & Sadhu, M. K. (1999). Tropical Fruits. 1, Nayaprakash, 206, Bidhan Sarani.

Chadha, K. L. (2015). Global climate change and Indian horticulture, Climate Dynamics in Horticultural Science. In M. L. Chaudhary, V. B. Patel, M. W. Siddiqui, & R. B. Verma (Eds.), *Impact, Adaptation and Mitigation* (Vol. 2, p. 351). Hoboken: Apple Academic Press.

Chambers, L.E., (2009). *Evidence of climate related shifts in Australian phenology*. Proceeding of the 18th World IMACS/MODSIM Congress. Cairns, Australia, Jul 13–17, pp 2597–2603.

Chaturvedi, R. K., Joshi, J., Jayaraman, M., Bala, G., & Ravindranath, N. H. (2012). Multi-model climate change projections for India under representative concentration pathways. *Current Science, 103*(7), 791–802.

Chen, Q. (2012). Adaptation and mitigation of impact of climate change on tropical fruit industry in China. *Acta Hort (ishs), 928*, 101–104. http://www.actahort.org/books/928/928_10.htm.

Climate Analysis Indicators Tool. (2011). CAIT version 8.0 [online]. http://cait.wri.org. Accessed 10 Dec 2016.

Cox, J. E. K. (1976). Garcinia mangostana-Mangosteen. In R. J. Garner & S. Ahmed Chaudhari (Eds.), *The propagation of tropical fruit trees, Horticultural Review* (Vol. 4, pp. 361–375). East Malling: Commonwealth Bureau of Horticulture and Plantation Crops.

Davanport, T. L. (2009). Reproductive physiology. In R. E. Litz (Ed.), *The mango. botany, production and uses* (2nd ed.). Wallingford: CABI.

Dillon, M. E., Wang, G., & Huey, R. B. (2010, October 7). Global metabolic impacts of recent climate warming. *Nature, 467*, 704–706. https://doi.org/10.1038/nature09407.

Dinesh, M. R., & Reddy, B. M. C. (2012). Physiological basis of growth and fruit yield characteristics of tropical and sub-tropical fruits to temperature. In B. R. Sthapit, V. Ramanatha Rao, & S. R. Sthapit (Eds.), *Tropical fruit tree species and climate change*. New Delhi: Bioversity International.

Gaffar, M. B. A., Osman, M. S., & Omar, I. (2008). Pomelo (Citrus maximaL.). In C. Y. Kwok, T. S. Lian, & S. H. Jamaluddin (Eds.), *Breeding horticultural crops* (pp. 67–82). Malaysia: MARDI.

Ghosh, S. P. (2000). In B. Govindasamy, B. Duffy, & J. Coquard (Eds.), *Status report on genetic resources of jackfruit in India and SE Asia*. New Delhi: IPGRI.

Haq, N. (2006). Fruits for the Future 10. Jackfruit (Artocarpus heterophyllus) Southampton Centre for Underutilised Crops. p. 192.

Hoppula, K. B., & Karhu, S. T. (2006). Strawberry fruit quality responses to the production environment. *Journal of Food, Agriculture and Environment, 4*(1), 166–170.

Hribar, J. Vidrih, R. (2014) Impacts of climate change on fruit physiology and quality, proceedings of 50th Croatian and 10th international symposium on agriculture. Opatija. Croatia (42–45).

Intergovernmental Panel on Climate Change. (2000). *IPCC special report on emissions scenarios*. Geneva: IPCC.

IPCC. (2007a). *Climate change 2007, impacts, adaptation and vulnerability*. Cambridge/New York: Cambridge University Press.

IPCC, (2007b). A report of working group one of the intergovernmental panel on climate change summary for policy makers. Intergovernmental Panel on Climate Change.

Jarvis, A., Ramirez, J., Anderson, B., Leibing, C., & Aggarwal, P. (2010). Scenarios of climate change within the context of agriculture. In M. P. Reynolds (Ed.), *Climate change and crop production*. CAB International. isbn:13: 978 1 84593 633 4.

Kimball, B. A., Idso, S. B., Johnson, S., & Rillig, M. T. (2007). Seventeen years of carbon dioxide enrichment of sour orange trees, final results. *Global Change Biology, 13*, 2171–2183.

Kumar, R. (2015). Climate issues affecting sustainable litchi (Litchi chinensis Sonn) production in eastern India, Climate Dynamics in Horticultural Science, Vol 2, Impact, adaptation and mitigation, Eds by Chaudhary, M. L., Patel, V. B., Siddiqui, M. W., and Verma, R. B., Apple Academic Press, pp 351.

Kumar, R., & Kumar, K. K. (2007). Managing physiological disorder in Litchi. *Indian Horticulture, 52*(1), 22–24.

Kumar, R., & Nath, V. (2013). In H. C. P. Singh, N. K. S. Rao, & K. S. Shivakumar (Eds.), *Canopy management in a book entitled "Climate Resilient Horticulture-Adaptation and Mitigation Strategy"* Springer Science & Business Media/Technology & Engineering 303.

Kumar, R., & Omkar Shukla, R. P. (2010). Effect of temperature on growth, development and reproduction of fruit fly Bractocera dorsalis Hendel (Diptera Tephritidae) in mango. *Journal of Ecofriendly Agriculture, 5*(2), 150–153.

Kumar, R. K., Sahai, A. K., Kumar, K., Patwardhan, S. K., Mishra, P. K., Ravadekar, K. K., & Pant, G. B. (2006). High-resolution climate change scenarios for India for the 21st century. *Current Science, 90*(3), 334–345.

Kumar, G., Chakravarty, N. V. K., Kurothe, R. S., Sena, D. R., Tripathi, K. P., Adak, T., Haldar, D., & Anuranjan. (2010). Effect of projected climate change on mustard. *Journal of Agrometeorology, 12*(2), 168–173.

Kumar, G., Kurothe, R. S., Sena, D. R., Vishwakarma, A. K., Madhu, M., Rao, B. K., Tripathi, K. P., & Anuranjan. (2010a). Sensitivity of wheat crop to the projected climate change in non-traditional areas. *Journal of Agrometeorology, 12*(2), 161–167.

Kumar, P., Wiltshire, A., Mathison, C., Asharaf, S., Ahrens, B., Philippe Lucas-Picher, P., Christensen, J. H., Gobiet, A., Saeed, F., Hagemann, S., & Jacob, D. (2013). Downscaled climate change projections with uncertainty assessment over India using a high resolution multi-model approach. *Science of the Total Environment, 468–469*, S18–S30.

Léchaudel, M., & Joas, J. (2007). An overview of preharvest factors influencing mango fruit growth, quality and postharvest behavior. *Brazilian Journal of Plant Physiology, 19*(4), 287–298.

Limjirakan, S., & Limsakul, A. (2012). Observed trends in surface air temperatures and their extremes in Thailand from 1970 to 2009. *Journal of the Meteorological Society of Japan, 90*, 647–662.

Lobell, D. B., Burke, M. B., Tebaldi, C., Mastrandrea, M. D., Falcon, W. P., & Naylor, R. L. (2008). Prioritizing climate change adaptation needs for food security in 2030. *Science, 319*, 607–610.

Lovett, R. A. (2007). *Climate change pushing tropics farther*, Faster, National Geographic News http://news.nationalgeographic.com/news/2007/12/071203-expanding-tropics.html

Maji, A. K., Reddy, G. P. O., & Sarkar, D. (2010). *Degraded and Wasteland of India – Status and spatial distribution* (pp. 1–167). New Delhi: ICAR.

Martin, R. (2015). Climate change: Why the tropical poor Will Suffer most https://www.technologyreview.com/s/538586/climate-change-why-the-tropical-poor-will-suffer-most/

Mathur, P. N., Ramirez-Villegas, J., & Jarvis, A. (2012). The impacts of climate change on tropical and sub-tropical horticultural production. In B. R. Sthapit, V. Ramanatha Rao, & S. R. Sthapit (Eds.), *Tropical fruit tree species and climate change* (p. 137). New Delhi: Bioversity International.

Menzel, C. M., Rasmussen, T. S., & Simpson, D. R. (1989). Effects of temperature and leaf water stress on growth and flowering of lychee (Litchi chinensis Sonn.). *Journal of Horticultural Science, 64*, 739–752.

Mishra, V. (2015). Climatic uncertainty in Himalayan water towers. *Journal of Geophysical Research-Atmospheres, 120*(7), 2689–2705.

MOEF, Govt. of India (2004). Initial National Communication Report of India to UNFCCC 2004, http://unfccc.int/resource/docs/natc/indnc1.pdf accessed on 8-7-2016.

More, T. A., & Bhargava, R. (2010). Impact of climate change on productivity of fruit crops in arid regions. In H. P. Singh, J. P. Singh, & S. S. Lal (Eds.), *Challenges of climate change-Indian horticulture*. New Delhi: Westville Publishing House.

Moretti, C. L., Mattos, L. M., Calbo, A. G., & Sargent, S. A. (2010). Climate changes and Po-tential impacts on postharvest quality of fruit and vegetable crops, a review. *Food Research International, 43*, 1824–1832.

Mukherjee, S. K. (1953). The mango-its botany, cultivation, uses and future improvement, especially observed in India. *Economic Botany, 7*, 130–162.

Osman, M. B., Milan, A. R. (2006). Mangosteen (Garcinia mangostana L.). DFID, FRP, CUC, World Agroforestry Centre and IPGRI.

Parmesan, C., & Yohe, G. (2003). A globally coherent fingerprint of climate change impacts across natural systems. *Nature, 421*(6918), 37–42.

Paull, R. E., & Duarte, O. (2011). Introduction. In R. E. Paull & O. Duarte (Eds.), *Tropical Fruits* (2nd ed., pp. 1–10). London: CAB International.

Pereira, H. M., Leadley, P. W., Proença, V., Alkemade, R., Scharlemann, J. P.W., and Fernandez-Manjarrés, J. F. (2010). Scenarios for Global Biodiversity in the twenty-first century. Science, 330 1496–1501.

Rajan, S. (2008). Implications of climate change in mango. In *Impact Assessment of Climate Change for Research Priority Planning in Horticultural Crops* (pp. 36–42). Shimla: Central Potato Research Institute.

Rajan, S. (2012). In B. R. Sthapit, V. Ramanatha Rao, & S. R. Sthapit (Eds.), *Phenological responses to temperature and rainfall: A case study of Mango, tropical fruit tree species and climate change*. New Delhi: Bioversity International.

Ross, P. (2013) Climate Change Will Affect Tropics First, Long before Arctic Sees Shift International business times 1-9-2013 (http://www.ibtimes.com/climate-change-will-affect-tropics-first-long-arctic-sees-shift-1419618)

Sankaran, M., Jaiprakash Singh N. P., and Datta, M. (2008). Climate change and horticultural crops. In Climate change and food security (Datta, M., Singh, N. P., & Daschoudari, D., eds.) New India Publishing House, New Delhi, 243–265.

Schaffer, B., & Gaye, G. O. (1989). Season effects of pruning on light penetration, specific leaf density, and chlorophyll content of mango. *Scientia Horticulturae, 41*, 55–61.

Sdoodee, S., Lerslerwong, L., & Rugkong, A. (2010). *Effects of climatic condition on off-season mangosteen production in Phatthalung Province*. Songkhla: Department of Plant Science, Prince of Songkla University.

Singh. (1992). *Fruit crops for wasteland* (pp. 215–221). Jodhpur: Scientific publisher.

Singh, H. P. (2012, January). Adaptation and mitigation strategies for climate resilient horticulture. In *Key Note Address (28–29th)* (pp. 1–20). Hessarghatta/Bangalore: Indian Institute of Horticultural Research.

Sparks, T. H., & Menzel, A. (2000). Observed changes in seasons: An overview. *International Journal of Climatology, 22*, 1715–1725.

Sthapit S R. and Scherr, S J. (2012) Tropical Fruit Trees and Climate Change, Tropical Fruit Tree Species and Climate Change. Bioversity International, New Delhi, India, by Sthapit BR, Ramanatha Rao V, Sthapit SR. 2012.

Sthapit B. R., Rao, R. V., Sthapit S. R. (2012). Tropical fruit tree Species and climate change. Bioversity International, New Delhi. 137.

Tindall, H. D. (1994). Rambutan cultivation. Food and Agriculture Organization of the United Nations. ISBN 9789251033258.

UDPSM. (2002). Plant propagation and planting in uplands. Training Manual for Municipal Extension Staff. Upland Development Programin Southern Mindanaao, Oct 2002, Davao, Philippines.

Uraipan, P. (2009). *The impact of climate changes on phenology of Longkong (Lansium domesticum Corr.)*. M.Sc. Thesis, Prince of Songkla University, Thailand.

Vati, L. and Ghatak, A. (2015). Phytopathosystm modification in response to climate change, Climate Dynamics in Horticultural Science, Vol 2, Impact, Adaptation and Mitigation, Eds by Chaudhary, M. L., Patel, V. B., Siddiqui, M. W., and Verma, R. B., Apple Academic Press, pp 351.

Verheij, E. W. M., & Coronel, R. E. (1992). *Edible fruits and nuts. Plant resources of South-East Asia* (Vol. 2, pp. 128–131). Bogor: Prosea Foundation.

Yoshino, M., & Sook, P. O. H. (1996). Variations in the plant phenology affected by global warming. In K. Omasa (Ed.), *Climate change in plants in East Asia* (pp. 93–107). Tokyo: Springer Verlag.

Zhang, L., Yu, Z., Jiang, L., Jiang, J., Luo, H., & Fu, L. (2011). Effect of post-harvest heat treatment on proteome change of peach fruit during ripening. *Journal of Proteomics, 74*, 1135–1149.

Chapter 12
Tackling Climate Change: A Breeder's Perspective

P. K. Singh and R. S. Singh

Abstract The threat of climate change is well evident by the fact of increasing temperature and more frequent severe drought and floods in recent times, and higher incidence of insects-pest and diseases impacting agriculture and food production. This situation has aggravated the scarcity of food and hunger around the world. To mitigate the ill effects of climate change, developing climate resilient varieties for heat, cold, drought and flood stresses is one of the options, where breeders can play major role. Several Institutions in the world are engaged in developing viable strategies. This will require a much better understanding of our genetic resources, the underlying mechanism of gene interactions and pyramiding multi-stress related genes for developing new variety or improving the already cultivated variety. The most suited approaches should involve conventional breeding as well as new emerging technologies like doubled haploidy, marker-assisted selection, high throughput phenotyping and bioinformatics to hasten the crop improvement. For breeders, ample opportunity lies in developing climate resilient high-yielding varieties, resistant/tolerant to biotic and abiotic stresses that help increasing food production and productivity, thus ease the cultivation under climate change regime. In this direction, several international institutes have initiated work on developing climate resilient crops, for example, the International Rice Research Institute (IRRI) has released 44 varieties of rice that are resilient to the effects of climate change and work is underway on a tripartite rice variation to cope with stresses like droughts, floods and saltiness. Even, the International Crops Research Institute for the Semi-Arid Tropics (ICRISAT) identified 40 germplasm lines of chickpea with resistance to extreme weather conditions such as drought, high temperature and salinity. In India, various ICAR institutes and state agricultural universities, under National Innovations on Climate Resilient Agriculture (NICRA) programme, made the concerted efforts to develop different high yielding cultivars with enhanced tolerance to heat, drought, flooding, chilling and salinity stresses for different agro-climatic zones. Thus, effect of climate change can be withstand to a greater extent with a

P. K. Singh (✉) · R. S. Singh
Department of Plant Breeding and Genetics, Bihar Agricultural University, Bhagalpur, Bihar, India

S. Sheraz Mahdi (ed.), *Climate Change and Agriculture in India: Impact and Adaptation*, https://doi.org/10.1007/978-3-319-90086-5_12

suitable genetic blue print in our cultivars and that need more focussed research and development from breeder's side.

Keywords Climate change · Plant breeding · Genetic resources · Climate resilient varieties

12.1 Introduction

Climate change is very obvious now, we see increasing temperature, increasing CO_2 concentration, melting of glacier and rising global average sea level. The threat of climate change is well evident also by the fact of more frequent and severe drought and floods in recent times, and higher incidence of insects-pest and diseases impacting agriculture and food production. For breeders, ample opportunity lies in developing eco-friendly and high-yielding varieties, resistant to pests and diseases that could increase food productivity and production, thus ease the cultivation under climate change regime. Modern breeding methods enable faster and more efficient breeding of crops adapted to specific regions. Breeding a particular variety for a specific situation gets further boost with the help of genomics-based approaches.

The International Rice Research Institute (IRRI) has released 44 varieties of rice that are resilient to the effects of climate change and work is underway on a tripartite rice variation to cope with stresses like droughts, floods and saltiness. Even, the International Crops Research Institute for the Semi-Arid Tropics (ICRISAT) identified 40 germplasm lines of chickpea with resistance to extreme weather conditions such as drought, high temperature and salinity. Similarly, International Centre for Agricultural Research in the Dry Areas (ICARDA) has started evolutionary participatory programmes for barley and durum wheat. In our country, various ICAR institutes and state agricultural universities, under National Innovations on Climate Resilient Agriculture (NICRA) programme, made the concerted efforts to develop different high yielding cultivars with enhanced tolerance to heat, drought, flooding, chilling and salinity stresses for different agro-climatic zones. Thus, effect of climate change can be withstand to greater extent with a suitable genetic blue print in our cultivars and that need more focussed research and development from breeder's side.

12.2 Climate Change Impact on Agriculture

Increased intensity and frequency of storms, drought and flooding, altered hydrological cycles and precipitation variance have implications for future food availability. The potential impacts on rainfed agriculture *vis-à-vis* irrigated systems are still not well understood. The developing world already contends with chronic food problems. Climate change presents yet another significant challenge to be met.

While overall food production may not be threatened, those least able to cope will likely bear additional adverse impacts (WRI 2005). In developing countries, 11 percent of arable land could be affected by climate change, including a reduction of cereal production in up to 65 countries and about 16 per cent of agricultural GDP (FAO, Committee on Food Security Report 2005).

Enhanced frequency and duration of extreme weather events such as flood, drought, cyclone, cold and heat wave as a result of climate change may adversely affect agricultural productivity. This is projected to reduce the crop yield in India, like in irrigated rice by ~4% in 2025 and rainfed rice yield by 6%; wheat by 6 to 23%; mustard by 2%; potato by ~2.5%, rainfed sorghum by 2.5%; maize yield in Kharif season by 18% (www.nicra.iari.res.in).

The growing concentration of CO_2 in the atmosphere, global warming and a higher frequency of extreme weather events mean increasing fluctuations in yields from agricultural crops and growing yield losses. The initial effects of the shift in climatic zones, vegetation zones and habitats are already visible: Pest and disease pressure on crops is growing. Global population is increasing while arable land area reducing day-by-day due to industrialization and human encroachment, and at the same time water shortages like situation, decreases the efficiency of agricultural production. In the short to medium term, agriculture can adapt by means of agronomical measures, such as tilling, crop rotation and optimized fertilization methods, and plant protection measures to safeguard the plants.

Each and every sphere of agriculture get affected by climate change be it crops, pasture, forests and livestock, land, soil and water resources, weed and pest, besides socio-economic impacts. The increased temperature (2–4 °C) by 2100, rise in CO_2 concentration, droughts and floods might be frequent events of future therefore, emphasis needs to be on climate smart agriculture with the aim of reduction of greenhouse gas emissions enhanced resilience and reduced wastes with the increase in the productivity of small and large scale farmers (Ahmed et al. 2013).

12.3 How Can Plant Breeding Help?

Increasing frequency of biotic stress is also highly probable. Plant breeding has addressed both abiotic and biotic stresses for a long time. Strategies in the face of climate changes may be based more on plant architectural changes to plant nutrient use efficiency, photoperiod and temperature responsiveness, with different maturity period to escape or avoid the unfavourable situation crop life cycles. Development of new more climate-resilient crop varieties will always be critical. All the cereals, oilseed, pulses, vegetables, flowers and fruits that is part of our day today life come from varieties developed by plant breeders and grown by farmers across the world. Now, with the intervention of modern genomics and biotechnology tool, the development of new adapted varieties has become a more precise and rapid process that contributes in feeding the world and tackling Climate Change at the same time. In the present scenario, the most suited approaches should involve conventional breeding as well as new emerging technologies like doubled haploidy, marker-assisted selection, high throughput phenotyping and bioinformatics to hasten the crop improvement.

For long time, plant breeders have been applying mutagenesis to induce genetic variation for increasing crop yield and improving the adaptability of crop plants. In the beginning, the X-ray radiation and gamma-ray radiation were used to induce point mutations and chromosomal mutations (Muller 1927, Devreux and Scarascia Mugnozza 1964. Designing crop breeding system suited for climate change with the help of superior combinations of genes into new varieties for new cropping systems and new environments. Environmentally friendly varieties such as improved varieties resistant to pests require fewer pesticides. Increase food production per unit area and alleviate pressure to add more arable land to production systems. The new improved varieties should be environmentally friendly, ensuring food security, while conserving the environment. Designing the crop breeding systems (1) to study the useful traits in response to various climatic factors such high CO_2 and high temperature, (2) a non-invasive phenotyping methods for specific traits, and (3) simulation of phenotypes will be helpful in plant breeding programs.

12.4 Exploring Plant Genetic Resources

Utilizing our plant genetic resources as source of traits/genes for resilience to changing environmental conditions and stresses could be a good option. Biodiversity itself is capable of providing genetic resources resilience to changing environmental conditions and stresses. The concerted effort on their bio-prospection is needed. The variability is the essence of plant breeding and the exploitation of plant ecosystems has resulted in loss of many valuable genetic resources. Plant breeders in recent time can take the advantage of genetics and genomics in characterizing the genetic resources so that enhanced breeding materials could be developed. The institution like National Bureau of Plant Genetic Resources (ICAR-NBPGR) does the task of several jobs related to plant genetic resources from introduction, exploration and collection, documentation, quarantine, maintenance and sharing. The gene bank maintained at NBPGR is good resource for exploring the genetic resources for traits suitable for high temperature, drought, and flooding, high salt content in soil, pest and disease resistance have been harnessed for new crop varieties by national and International programmes.

Developing genetic responses to climate change is incumbent on effective evaluation and exploitation of the existing genetic variability. Use of molecular markers to predict adaptive variability is rather ineffective because little correlation exists between molecular genetic diversity and quantitative genetic variation (Gilligan et al. 2005). Genomics possesses the potential to increase the diversity of alleles available to breeders through mining the gene pools of crop wild relatives. Genomics tools also enable rapid identification and selection of novel beneficial genes and their controlled incorporation into novel germplasm. In the genomics era, this technology will be used to safeguard the future through improved food security. Taken together, the application of genomics for crop germplasm enhancement offers the greatest potential to increase food production in the coming decades. With continued rapid advances in genome technologies, the application of genomics to identify and transfer valuable agronomic genes from allied gene pools and crop relatives to elite crops will increase in pace and assist in meeting the challenge of global food security.

12.5 Conventional Breeding Approaches

One of the strategies for introgressing traits to mitigate the effect of climate change on plants performance for sustainable food production is identifying and incorporating suitable alleles by means of conventional breeding approaches. For this purpose, crop relatives have been used for decades for breeding, in particular to transfer genes of resistance or tolerance to pests, diseases or abiotic stress to the cultivated species (Hajjar and Hodgkin 2007; Honnay et al. 2012). In case of non-availability or no known source of suitable alleles, mutation breeding is one of the choice upon which breeder's can rely. By exposing seeds to mutagenic chemicals or radiation the mutants are generated with desirable traits to be bred with other cultivars. In radiation-based approaches, FAO/IAEA reported in 2014 that over 1000 mutant varietals of major staple crops were being grown worldwide.

In India, Bhabha Atomic Research Centre (BARC) has contributed a lot by induced mutation (mostly using gamma rays)-based varietal development programme in different crops for example, groundnut, mungbean, blackgram, soybean, redgram, cowpea, mustard, rice and jute. Mutation breeding using radiation technology for crop improvement is an active area of research at the Bhabha Atomic Research Centre (BARC), Trombay, Mumbai. (http://dae.nic.in). Using radiation induced mutation and cross-breeding, 39 new crop varieties (Trombay varieties) developed at BARC have been released. These include 20 in oil seeds (14-groundnut, 3-mustard, 2 soybean, 1 sunflower), 17 in pulses (8-greengram, 4-blackgram, 4-pigeonpea, 1-cowpea) and one each in rice and jute for high yield, some with additional desirable characters like disease resistance, early maturity, suitability to rice fallows, improved quality parameters etc. (Table 12.1).

ICAR-National Rice Research Institute, Cuttack, has development blast tolerant mutants of rice (CRM 49, CRM 51 and CRM 53) from IR 50 (a blast susceptible variety) through mutation breeding (chemical-based mutagenesis). CRM 49 and CRM 51 were isolated from sodium azide-treated populations, while CRM 53 from ethyl methane sulfonate (EMS)-treated populations. These mutants possessed semi-dwarf stature, long slender grains and with a yield potential of 5 t/ha.

In mutational programme, the work on calmodulin-binding protein gene family involved in abiotic stress responses in rice is going on at BAU, Sabour. More than 1, 50, 000 M_3 mutant rice (Gamma irradiated) plants of Rajendra Mahsuri-1 and Rajendra Kasturi were grown and screened for drought stress in field condition.

Table 12.1 Varieties developed by mutation breeding for various stresses

Crop	Recommended for states	Features (including stress tolerance)
Groundnut		
TG 26	Gujarat North Maharashtra Madhya Pradesh	Earliness, high harvest index, 20 days seed dormancy, smooth pods, salinity tolerance, high harvest index
TG 37A	Haryana, Rajasthan, Punjab, Uttar Pradesh, Gujarat, Orissa, West Bengal, Assam, North Eastern states	High yield, smooth pods, wider adaptability, collar rot and drought tolerance
Sunflower		
TAS-82	Maharashtra	Black seed coat, tolerant to drought
Pigeonpea		
TJT- 501	Madhya Pradesh, Gujarat, Maharashtra, Chhattisgarh	High yielding, early maturing, tolerant to Phytophthora blight
Mungbean (Green gram)		
TJM-3	Madhya Pradesh	Resistant to powdery mildew, yellow mosaic virus and Rhizoctonia root –rot diseases.
TM 2000–2	Chhattisgarh	Suitable for rice fallows and resistant to powdery mildew
TMB-37	Eastern Uttar Pradesh, Bihar, Jharkhand, West Bengal, Assam	Tolerant to yellow mosaic virus
Urdbean (Black gram)		
TU 94–2	Andhra Pradesh, Karnataka, Kerala, Tamil Nadu	Resistant to yellow mosaic virus

Source: http://dae.nic.in

Rice mutants (R. Mahsuri-1) were showing different phenotypes like early flowering, increased L/B ration and any marked visual changes were selected.

12.6 Breeding Strategies for Major Climate Change Effect

The breeding strategies to tackle the adverse climate change effect on crops should encompass the development of stress tolerant genotypes together with sustainable crop and natural resources management along with sound implementation of policies. Meanwhile, development of resistance genotypes to biotic/abiotic stresses, choice of crops, change in the cropping patterns, rotation, time of planting and avoidance, nutrient use efficiency and the approaches like wide hybridization, mutagenesis, genomics-assisted breeding and transgenic will be very helpful to achieve the goal of yield sustainability.

Under "National Inovation in Climate Resilient Agriculture" (NICRA), IARI has developed new crop varieties with higher yield potential and resistant to multiple stresses (heat, drought, flood, salinity) that will be the key to maintain yield stability. Two germplasm, Nerica L44 and N22 were identified as novel sources of heat tolerance in rice by the Institute. Further, in an effort to map the QTLs governing heat tolerance recombinant inbred line mapping populations are being generated involving heat tolerant genotypes namely L44 and N22, which are in F4 generation. Marker assisted backcross breeding was carried out using molecular marker linked to the QTL governing drought tolerance, qDTY1.1 into Pusa Basmati 1 and qDTY3.1 into Pusa 44 and 41 (in Pusa Basmati 1 background), 36 (in Pusa 44 background). Similarly, in wheat the standardized physiological trait-based phenotyping protocol for screening for heat and drought tolerance in wheat and also, maize genotypes tolerant to low and high temperature tolerance were identified.

Maize inbred lines were also screened for multiple disease resistance. In tomato, two wild species i.e., *L. peruvianum* and *L. pimpinellifoium* crossable to cultivated tomato have been identified for temperature stress tolerance. Among cultivated genotypes (Pusa Sadabahar and TH-348) and hybrids (DTH-9 and DTH-10) were identified for heat stress tolerance.

Under Cereal Systems Initiative for South Asia (CSISA) programme, at BAU Sabour, with rigorous effort of scientists, Rajendra Mansuri has been identified as suitable for direct seeded rice (DSR) cultivar that would save labour and water. This variety can be recommended for water deficit region and would help in climate change regime.

Plant breeding programmes at ICAR-IARI has led to release of several varieties in different crops for stress tolerance (Table 12.2).

12.6.1 Flood and Salinity Tolerant

Climate induced weather extremes, such as flooding, submergence and salinity impacts crops adversely. About 20 million hectares of rice land is prone to flooding in Asia, which is major rice growing continent. Of this, India and Bangladesh share more than 5 million hectares of rice field, flooded during most of the planting seasons (http://irri.org/). Progressive salt accumulation due to excessive irrigation with poor water quality coupled with poor or improper drainage results in high salt levels (Ismail et al. 2007). Excess water in the soil reduces oxygen availability to the plant (Kozlowski 1984). The extended deep submersion can cause plant death because of a lack of oxygen required to sustain plant growth and an accumulation of toxic substances, such as organic acids, NO^{2-}, Mn^{2+}, Fe^{2+}, and H_2S (Kozlowski 1984; Janiesch 1991).

Generally, the traits like the vigorous roots and less evapo-transpiration could be developed in plants that can help maintaining water balance up to some extent. The past decade has witnessed an increase in studies related to detection of quantitative trait loci (QTL) for drought-related traits, and the first encouraging results in QTL cloning have been reported (Salvi and Tuberosa 2005). Thus, development of more flood-tolerant cultivars is critical for enhancing sustainable production of crops. For

Table 12.2 List of varieties released by ICAR-IARI for various stresses

Crops	Variety	Stress
Maize	Pusa Composite3, Pusa Composite4	Moisture stress
Karan rai	Pusa Swarnim (IGC01), Pusa Aditya (NPC9)	
Pearl millet	Pusa Composite 383, Pusa Composite 443,	
	Pusa Composite 612	
Pigeonpea	Pusa 991	Salinity
Indian mustard	Pusa Vijay (NPJ93)	
Wheat	Kaushambi (HW 2045), Pusa Gold (WR544),	High temperature tolerance
	Pusa Basant (HD2985), Pusa Wheat111, (HD2932), Harshita (HI1531)	
Chickpea	Pusa 547	
Mungbean	Pusa 9531	
Indian mustard	Pusa Agrani (SEJ2), Pusa Mahak (JD6), Pusa Vijay (NPJ93), Pusa Tarak (EJ13), Pusa Mustard25, (NPJ112), Pusa Mustard26, (NPJ113), Pusa Mustard27 (EJ17)	
Mungbean	Pusa 0672	Cold tolerance
Maize	AH421 (PEHM5)	Water logging
Cotton	PSS2 (Arvinda)	Hot and humid conditions
Wheat	VSM (HD 2733), Kaushambi (HW 2045)	Diseases-pest
Chickpea	Pusa 1088 (Kabuli)	

Source: http://www.iari.res.in

example, rice dies within 5–6 days of complete submergence, resulting in total crop loss. These losses affect rice farmers in rainfed and flood-affected areas where alternative livelihoods are limited. After the gene called Sub1 gene was found, it was infused into popularly grown rice varieties in rice-growing countries in Asia ((http://irri.org/)). IRRI has developed a rice variety using Sub1 gene through marker-assisted backcrossing, Swarna-sub1 in 2005 that can withstand being submerged under water for two weeks.

Soil salinity is a major cause of concerns in rice growing areas. In India nearly 6.72 million ha of total land are salt affected out of which 2.95 million ha are saline (including coastal) and 3.77 million ha are alkaline (IAB 2000).In Bangladesh's coastal areas, salinity affects about one million hectares of land that can otherwise be used for rice farming. Rice productivity in salt-affected areas is very low less than 1.5 tons per hectare. But this can potentially increase by at least two tons per hectare with improved varieties that can withstand soil salinity.

IRRI scientists have identified a major region of the rice genome called Saltol that gives the rice plant tolerance to salinity. Saltol is being used to develop varieties that can cope with exposure to salt during the seedling and reproductive stages of the plant. Recent work at IRRI has shown that the SUB1 gene and Saltol can be combined in the same variety of rice, increasing the rice plant's tolerance to salinity and submergence. Plant breeders have incorporated Saltol in popular rice varieties such as the BRRI Dhan 11, 28, 29 varieties released in Bangladesh. To date, IRRI, with the help of its national partners, has developed more than 100 salinity-tolerant elite lines. These elite lines possess superior traits such as high yield, good eating quality, resistance to pests and diseases, and tolerance of stresses, and are ready for testing in farmer's fields.

Central Soil Salinity Research Institute, Karnal, India (CSSRI) has identified several promising salt-tolerant crops such as salt-tolerant varieties of rice, for example, CSR 10, CSR 11, CSR 13, CSR 27 for inland situations and CST 7–1, CSR 4 and CSR 6 for coastal areas. Likewise, salt-tolerant varieties for wheat, KRL 1–4 and mustard, CS-52, CS-330 have been developed and released (Dagar 2005).

12.6.2 Cold, Heat and Drought tolerance

Tolerance to freezing temperatures is the most important component for winter survival, but also of considerable importance is the capability to withstand combinations of stresses due to desiccation, wind, ice-encasement, low light, snow cover, winter pathogens, and fluctuating temperatures. Resistance to desiccation through the maintenance of cell membrane integrity and retention of cellular water is essential, and it is unsurprising that the same genetic response to the onset of freezing temperatures is often observed with drought or salinity stress (Yue et al. 2006). Indeed, cold acclimation can frequently improve adaptation to a mild drought stress and vice versa (Seki et al. 2002). Higher temperatures are speculated to reduce rice grain yields through two main pathways: (i) high maximum temperatures that in combination with high humidity cause spikelet sterility, and (ii) increased night time temperatures, which may reduce assimilate accumulation (Wassmann and Jagadish 2009).

Cold tolerance is a complex trait controlled by many genes. IRRI scientists have identified three regions of the rice genome that have a direct link to cold tolerance at the plant's reproductive stage. Cold stress at critical times of reproduction hinders the formation of fertile pollen that is crucial for fertilization and consequently the rice plant may fail to produce grains. In pigeonpea at BAU, Sabour, it was found that the pigeonpea lines starts dropping flowers and newly developed pod when temperature goes below 10 °C. In this context, 120 pigeonpea genotypes planted in augmented design with four checks. Genotypes categorized into cold escape, cold tolerant and susceptible. It has been found that the early group of pigeonpea genotypes which completed their reproductive cycle before temperature down below 10 °C was not affected by temperature considered cold escape such as, ICP-13359, ICP-11627, ICP-11059, ICP-11477. Other long to medium duration genotypes ICP-15382, ICP-7076, ICP-7076 and ICP-14229, i.e., also developed pods normally since temperatures were not critically low during winter season. Hence conclusion cannot be made this year due to normal winter season.

IRRI scientists are looking for rice that can tolerant high temperatures by screening improved and traditional rice varieties. These donors are used in a crossing program to incorporate tolerance of high temperature into elite rice lines that are then tested for heat tolerance in 'hot and dry' and 'hot and humid' countries. Some of the development that can help fight against climate change, for example, discovery of a cold-tolerant breeding line called IR66160–121–4-4-2 that inherited cold tolerance genes from Indonesia's tropical japonica variety Jimbrug and northern China's temperate japonica variety, Shen-Nung89–366 by IRRI's and South Korea's Rural Development Administration.

Drought is the most widespread and damaging of all environmental stresses, affecting 23 million hectares of rainfed rice in South and Southeast Asia. In some states in India, severe drought can cause as much as 40% yield loss, amounting to $800 million. In all cases, the emphasis will be on identifying and using sources of genetic variation for tolerance/ resistance to a higher level of abiotic stresses. The two most obvious sources of novel genetic variation are the gene banks (ICARDA has one of the largest gene banks with more than 120,000 accessions of several species including important food and feed crops such as barley, wheat, lentil, chickpea, vetch, etc.) and/or the farmers' fields. Currently, there are several international projects aiming at the identification of genes associated with superior adaptation to higher temperatures and drought. At ICARDA, as elsewhere, it has been found that landraces and, when available, wild relatives harbour a large amount of genetic variation some of which is of immediate use in breeding for drought and high temperature resistance (Ceccarelli et al. 1991; Grando et al. 2001).

Sabour Ardhajal, a drought tolerant variety of rice developed at BAU, Sabour under Stress-Tolerant Rice for Africa and South Asia (STRASA) programme, which contains QTL for reproductive stage drought tolerance would be important climate change scenario. Further, under the same project at BAU Sabour, several multi-environmental trials (MET) underway on aspect of breeding for drought, heat and cold tolerance, in H-MET total 14 entries were evaluated including local check R. Mahsuri-1. Yield differences were found to be significant and varied from 2827 kg/ha (CGZR-1) to 4971 kg/ha (IR 91953–141–2-1-2(R-119). Two entries IR 91953–141–2-1-2(R-119) and IR 92937–178–2-2(R-155) recorded yield significantly superior to R. Mahsuri-1 (3752 kg/ha). While in MET-2 (Loc-1), total seven entries were evaluated including local check Prabhat. Yield differences were found to be significant and varied from 2183 kg/ha (Prabhat) to 3638 kg/ha (MTU1010). Entry R-RHZ-7 recorded yield significantly superior to Prabhat (2183 kg/ha). In MET-2 (Loc-3), total seven entries were evaluated including local check Prabhat. Yield differences were found to be significant and varied from 2123 kg/ha (Prabhat) to 3437 kg/ha (MTU1010). Entry R-RHZ-7 recorded yield significantly superior to Prabhat (2123 kg/ha). In Swarna Sub-1+ drought (control), total 16 entries were evaluated including two checks namely S. Sub-1 and Swarna. Yield differences were found to be varied from 6696 kg/ha (IR 96321–315-294-B-1-1-1) to 9494 kg/ha (IR 96321–558-563-B-2-1-3). However, none of the entry could statistically surpass the best check Swarna (9063 kg/ha). In Swarna Sub-1+ drought (Drought), total 16 entries were evaluated including two checks namely

S. Sub-1 and Swarna. To identify suitable drought donors, a total 24 entries were evaluated including checks. Yield differences were found to be varied from 949 kg/ha (Koi Murali) to 4348 kg/ha (Binuhangin). Top three yielder donors were Binuhangin, Dular and Uri. Donors were crossed with the locally adapted varieties.

Under Heat stress Tolerant Maize (HTMA) for Asia project at BAU, Sabour, 20 entries (in AMDWTC-17″ trial) were tested and significant differences were observed among the entries for grain yield. Grain yield ranged from 2400–6930 kg/ha but none of the entry showed significantly higher yield than the best check "PIO3396" (6930 kg/ha).

In another programme on improvement of pigeonpea for plant type, early maturity, pod borer resistance and moisture stress tolerance at BAU, Sabour, 250 pigeonpea germplasm received from NBPGR were planted with four checks (Asha, Patam, Pusa 9 and Bahar) to screen genotype against pod borer at flowering and podding. For this experiment was left without chemical treatment (insecticide/pesticide). Plants (5/row) were tagged to monitor damage by pod borer. Before harvest pods were plucked from individual plant and counted. The damage pods were counted separately and percent infestation were recorded on plant basis. It has been found that out of 250 entries 23 genotypes showed less infestation (< 30%) with minimum 7%. However, maximum infestation went up to 90%.

In chickpea programme at BAU, Sabour, the evaluation of advanced generation chickpea genotypes for yield and abiotic / biotic stress tolerance (Desi) were done involving nine chickpea advanced breeding lines including PG 186 as check were evaluated in RCBD (Timely sown condition). Varietal difference with respect to grain yield was found significantly superior which ranged from 1555 kg/h to 2043.2 kg/ha. Highest grain yield (2043.2 kg/ha) was recorded by cross no 14 (ICCV 10 x ICCV 97105) which was 15% superior to check PG 186, while two entries were at par with check. The trials were planted under normal moisture condition; however disease and insects infestation was less than 10%. Also characterization and evaluation of eighty chickpea genotypes (Desi) including four checks for heat tolerance (late sown and normal sown condition) is underway. The varietal difference with respect to grain yield was found significantly superior which ranged from 300 kg/h to 1905.2 kg/ha. Highest grain yield was recorded by ICCV 4958 followed by IPC 10–59, JG 14, PG 186, JG 18 under late sown condition.

The work on understanding heat and drought tolerance mechanism in lentil and its improvement by over-expression of antioxidant genes is also being carried out keeping in view climate change at BAU, Sabour. Under *in vitro* condition shoot cultures of four genotypes viz. Noori, HUL 57, Arun and SBO Local were established. Shoot regeneration for Noori and HUL 57 have been standardized from embryonal axis. Pot experiments were conducted using 14 contrast genotypes under heat and drought stress condition.

The evaluation of linseed germplasm under utera (sowing in standing paddy crop) condition is being carried out that may help climate resilience by early harvest of linseed. One hundred eight entries were evaluated along with two checks (T-397 & R-552), twenty five entries were found promising as they recorded more yield than the best check, T-397 (825 kg/ha). The top ranker five entries were BRLS-112-2

(1497 kg/ha), BRLS-113 (1454 kg/ha), BRLS-110-2 (1453 kg/ha), BRLS-109-2 (1428 kg/ha), BRLS-111–2 (1312 kg/ha). One of the variety released recently Linseed variety "Sabour Tisi-1" has been released for utera condition.

In rapeseed-mustard breeding programme, to combine traits of interest viz. early and late sowing, bold seed, semi compact canopy, lodging and shattering resistant, TW high, MR to AB & aphid, high yield, profuse branch, semi-apprressed, non-glossy stem and upright branches in popular varieties is being carried out. Also, inter-specific crosses (*Brassica juncea* X *Brassica carinata*) for drought tolerance and better growth under late sown condition were done.

In safflower, Brassinosteroid-mediated increase in seed yield and enhanced abiotic stress tolerance Safflower (A1 variety) were transformed with CaMV:DWF4:NOS gene construct by *Agrobacterium* transformation. The presence of transgene in transgenic safflower lines (T1 and T2) were confirmed using PCR analysis. RT-PCR was carried out to confirm the expression of transgene. Sub-cellular localization experiment showed that the expression of DWF4 in cytoplasm.

12.6.3 Diseases and Pest

To date, research on the impact of climate change on plant diseases has been limited, with many studies focusing on the effects of a single atmospheric constituent or meteorological variable on the host, pathogen, or the interaction of the two, under controlled conditions. Rising temperature and atmospheric CO2 are also indirectly affecting crops through their effects on pests and diseases. These interactions are complex, and their full impact on crop yield is yet to be fully appreciated. Impacts of warming or drought on resistance of crops to specific diseases may be through the increased pathogenicity of organisms or by mutations induced by environmental stresses (Gregory et al. 2009). The influence of climate change on plant pathogens and their consequent diseases has been reviewed extensively (Coakley et al. 1999; Chakraborty 2005). Different individual parameters associated with climate change, such as warming, increased levels of CO_2, decreased rainfall, and erratic pattern of rainfall, have been studied for their influence on different aspects of pathogens and diseases across various crops (Chakraborty 2005).

At BAU, Sabour, the engineered resistance in rice against fungal pathogens is underway. Rice calli of Rajendra Kasturi were transformed using *Dwf1* gene construct using *Agrobacterium* mediated transformation. The presence of transgene in transgenic rice lines (T_1 and T_2) were confirmed using PCR analysis. RT-PCR was carried out to confirm the expression of transgene. While in lentil, for biotic stress the work on identification of donor parents resistant to *Fusarium* wilt is going on. In this, a trial consisting 16 released varieties including check from different states were tested for the performance for its adaptability in different ecological regions of Bihar. HUL 57 (1250 Kg/ha) was the best check, three entries were found significantly superior than the check. Out of 16 entries BRL-3 has yielded out highest (1860 Kg/ha) followed by BRL-1 (1629 Kg/ha) and BRL-2 (1539 Kg/ha), other varieties were at par with the best check HUL 57 (1250 Kg/ha).

In wheat, for biotic tolerance the work on development of spot blotch resistant genotypes of spring wheat for eastern Gangetic plain of India using double haploid (DH) technology was undertaken keeping in view climate change. For this purpose, spot blotch resistant genotypes viz. Chirya 3, Chirya 7, MACS 2496, GW373, HD 3118, RAJ 3765, WH730, BH1146 were procured. Crosses were made involving Spot blotch resistance genotypes and high yielding locally adapted genotypes. Procedures and physical conditions were optimized for pseudo seeds production. Nutrient media and physical conditions were standardized for haploid production.

12.7 Genomics Assisted-breeding and Biotechnological Intervention for Climate Change

Potential of Genomics-assisted Breeding in Producing Climate Resilient Crops Genomics offers tools to address the challenge of increasing food yield, quality and stability of production through advanced breeding techniques. Advances in plant genomics provide further means to improve the understandings of crop diversity at species and gene levels, and offer DNA markers to accelerate the pace of genetic improvement (Muthamilarasan et al. 2014). A genomics-led breeding strategy for the development of new cultivars that are "climate change ready" (Varshney et al. 2005) commences by defining the stress that will likely affect crop production and productivity under certain climate change scenarios.

The ideotype breeding coupled with DNA fingerprinting, and gene/ quantitative trait loci (QTL) will assist in selecting screening promising accessions against specific stress. Similarly, precise phenotypic assessments and appropriate biometric analysis will assist in identifying unique responses of a set of genotypes in a given physiological stage influenced by variation of weather patterns. This information will be further used in genomics aided breeding approaches such as genome-wide selection of promising germplasm for further use in crop breeding aiming at both population improvement and cultivar release. Genetic mapping and QTL analysis, via bi-parental or association mapping (AM) populations, have accelerated the dissection of genetic control of agricultural traits, potentially allowing MAS, QTL, and AM studies or direct calculation and genomic selection (GS) of high value genotypes to be made in the context of breeding programs (Kulwal et al. 2011).

With the advent of next-generation sequencing (NGS) methods has facilitated the development of large numbers of genetic markers, such as single nucleotide polymorphisms (SNP), insertion-deletions (InDels), etc. even in relatively research-neglected crop species. Discovery of novel genes/alleles for any given trait could be then performed through genotyping-by-sequencing (GBS) approaches. Similarly, genome-wide association studies (GWAS) could be used to identify the genomic regions governing traits of interest by performing statistical associations between

DNA polymorphisms and trait variations in diverse collection of germplasm that are genotyped and phenotyped for traits of interest.

Compared to radiation methods, chemical mutagens tend to induce SNPs rather than chromosomal mutations. Currently, chemical mutagens, such as Ethyl methanesulfonate (EMS) are being used to induce random mutations into the genome and have become a useful complement to the isolation of nuclear DNA from mutated lines by TILLING (Targeting Induced Local Lesions in Genomes) technology and screening of the M_2 population at the DNA level using advanced molecular techniques.

At BAU, Sabour, the work on architectural modification of Katarni rice through marker assisted selection is underway. The work will bring down the height of Katarni rice resulting in lodging resistance and help the crop stand against storm or high wind and rain. Survey of parental polymorphism between Katarni, IR-64, Rajendra Sweta and BPT 5204 was undertaken through the set of 25 additional SSR markers in rice were conducted. Validation of F_1 plants of Katarni/BPT5204 through parental polymorphic SSR was done in PCR. Backcrossing of 8 validated Katarni/ BPT5204 F_1s with recurrent parent Katarni was done to obtain the BC_1F_1 seeds. 388 BC_1F_1 plants of Katarni/R.Sweta//Katarni were raised out of which 107 plants were selected based on the foreground selection using *badh2* gene specific primer (Bradbury 2005). Out of 107 fragrant allele positive plants, 87 were found to be positive for *sd1* gene through PCR using *sd1* gene specific primer as suggested by Spielmeyer et al. (2002). On the basis of grain and leaf aroma KOH sensory test, 15 plants were selected out of the 87 positive plants for *sd1* gene. These 15 plants will be evaluated for the presence of aroma, *sd1* gene with early flowering trait in next segregating generation. Out of about 10,000 F_2 plants of Katarni/R.Sweta, 410 plants were selected based on KOH sensory test, plant height and early date of flowering. Out of about 6000 F_2 plants of Katarni/IR64, 350 plants were selected on the basis of plant height and date of flowering.

For the first time, the germin-like protein multi-gene family in tomato has been identified in our lab at Plant Breeding and Genetics lab, BAU Sabour. Through detailed bioinformatics analysis, a potent candidate gene (annotated in this study as *SlGLPH*) has been identified. The *SlGLPH* transcript has been detected in leaf, stem, flower and fruit tissues of tomato. Relative abundance of the *SlGLPH* transcript has been found to be significantly increased under abiotic stress conditions. The coding DNA sequence of the *SlGLPH* gene has been amplified from tomato (cultivar: Pusa Ruby) genomic DNA and verified through custom sequencing. A genetic construct for over-expression of the *SlGLPH* gene has been prepared, mobilized to *Agrobacterium* and plant transformation (tomato and brinjal) has been initiated.

12.8 Conclusion

Plant breeders need to focus on traits with the greatest potential to increase yield. Hence, new technologies must be developed to accelerate breeding through improving genotyping and phenotyping methods and by increasing the available genetic diversity in breeding germplasm. Crop improvement through breeding brings immense value relative to investment and offers an effective approach to improving food security. Many new improved varieties are environmentally friendly, ensuring food security, while conserving the environment. The holistic approach in tackling climate change should encompass the climate resilient genotypes coupled with a suitable crop and natural resources management and sound implementation policies that could led to climate-smart agriculture.

References

Ahmed, M., Asif, M., Sajad, M., Khattak, J. Z. K., Ijaz, W., Fayyaz-ul-Hassan, W.,. A., & Chun, J. A. (2013). Could agricultural system be adapted to climate change? Review. *Australian Journal of Crop Sciences., 7*(11), 1642–1653.

Bradbury, L. M. T., Fitzgerald, T. L., Henry, R. J., Jin, Q. S., & Waters, D. L. E. (2005). The gene for fragrance in rice. *Plant Biotechnology, 3*(3), 363–370.

Ceccarelli, S., Valkoun, J., Erskine, W., Weigand, S., Miller, R., & Van Leur, J. A. G. (1991). Plant genetic resources and plant improvement as tools to develop sustainable agriculture. *Experimental Agriculture, 28*, 89–98.

Chakraborty, S. (2005). Potential impact of climate change on plant-pathogen interactions. *Australasian Plant Pathology, 34*, 443–448.

Coakley, S. M., Scherm, H., & Chakraborty, S. (1999). Climate change and plant disease management. *Annual Review of Phytopathology, 37*, 399–426.

Dagar, J. C. (2005). Salinity Research in India: An Overview [In: Gupta et al. (Editors), Ecology and Environmental Management: Issues and Research Needs] Bulletin of the National Institute of Ecology 15: 69–80.

Devreux, M., & Scarascia Mugnozza, G. T. (1964). Effects of gamma radiation of the gametes, zygote and proembryo in *Nicotiana tabacum* L. *Radition Botany, 4*, 373–386.

FAO. (2005). Impact of climate change, pests and diseases on food security and poverty reduction. Special event background document for the 31st Session of the Committee on World Food Security. Rome. May 2005 p. 23–26

Gilligan, D. M., Briscoe, D. A., & Frankham, R. (2005). Comparative losses of quantitative and molecular genetic variation in finite populations of Drosophila. *Genetical Research, 85*, 47–55.

Grando, S., Von Bothmer, R., & Ceccarelli, S. (2001). Genetic diversity of barley: use of locally adapted germplasm to enhance yield and yield stability of barley in dry areas. In H. D. Cooper, C. Spillane, & T. Hodgink (Eds.), *Broadening the Genetic Base of Crop Production* (pp. 351–372).

Gregory, P. J., Johnson, S. N., Newton, A. C., & Ingram, J. S. I. (2009). Integrating pests and pathogens into the climate change/food security debate. *Journal of Experimental Botany, 60*, 2827–2838.

Hajjar, R., & Hodgkin, T. (2007). The use of wild relatives in crop improvement: a survey of developments over the last 20 years. *Euphytica, 156*, 1–13.

Honnay, O., Jacquemyn, H., & Aerts, R. (2012). Crop wild relatives: more common ground for breeders and ecologists. *Frontiers in Ecology and the Environment, 10*, 121–121.

IAB. (2000). *Indian Agriculture in Brief* (27th ed.). New Delhi: Agriculture Statistics Division, Ministry of Agriculture, Govt. of India.

Ismail, A. M., Heuer, S., Thomson, M. J., & Wissuwa, M. (2007). Genetic and genomic approaches to develop rice germplasm for problem soils. *Plant Molecular Biology, 65*, 547–570.

Janiesch, P. (1991). Eco-physiological adaptation of higher plants in natural communities to water logging. In J. Rozema & J. A. C. Verkleij (Eds.), *Ecological Responses to Environmental Stresses* (pp. 50–60). The Netherlands: Kluwer Academic Publishers.

Kozlowski, T. T. (1984). Plant responses to flooding of soil. *Bioscience, 34*, 162–167.

Kulwal, P. L., Thudi, M., & Varshney, R. K. (2011). Genomics interventions in crop breeding for sustainable agriculture. In R. A. Meyers (Ed.), *Encyclopedia of Sustainability Science and Technology* (pp. 2527–2540). NewYork: Springer.

Muller, H. J. (1927). Artificial transmutation of the gene. *Science, 66*, 84–87. https://doi.org/10.1126/science.66.1699.84. New York/Rome: CABI//FAO/IPRI.

Muthamilarasan, M., Venkata Suresh, B., Pandey, G., Kumari, K., Parida, S. K., & Prasad, M. (2014). Development of 5123 intron-length polymorphic markers for large-scale genotyping applications in foxtail millet. *DNA Research, 21*, 41–52.

Salvi, S., & Tuberosa, R. (2005). To clone or not to clone plant QTLs: present and future challenges. *TRENDS in Plant Science, 10*(6), 297–304.

Seki, M., Narusaka, M., Ishida, J., Nanjo, T., Fujita, M., Oono, Y., et al. (2002). Monitoring the expression profiles of 7000 Arabidopsis genes under drought, cold and high-salinity stresses using a full-length cDNA microarray. *The Plant Journal, 31*, 279–292.

Spielmeyer, W., Ellis, M. H., & Chandler, P. M. (2002). Semidwarf (sd-1), "green revolution" rice, contains a defective gibberellin 20-oxidase gene. *Proceedings of National Academy of Sciences USA, 99*, 9043–9048.

Varshney, R. K., Graner, A., & Sorrells, M. E. (2005). Genomics-assisted breeding for crop improvement. *Trends in Plant Science, 10*, 621–630.

Wassmann, R., & Jagadish, S. (2009). Regional vulnerability of climate change impacts on Asian rice production and scope for adaptation. *Advances in Agronomy, 102*, 91–133.

World Resources Institute (WRI) in collaboration with United Nations Development Programme, United Nations Environment Programme, and World Bank, (2005): World Resources 2005: The Wealth of the Poor—Managing Ecosystems to Fight Poverty. Washington, DC

Yue, B., Xue, W., Xiong, L., Yu, X., Luo, L., Cui, K., et al. (2006). Genetic basis of drought resistance at reproductive stage in rice: separation of drought tolerance from drought avoidance. *Genetics, 172*, 1213–1228.

Chapter 13
Decreasing the Vulnerability to Climate Change in Less Favoured Areas of Bihar: Smart Options in Agriculture

Anshuman Kohli, Sudhanshu Singh, Sheetal Sharma, S. K. Gupta, Mainak Ghosh, Y. K. Singh, B. K. Vimal, Vinay Kumar, and Sanjay Kumar Mandal

Abstract Anthropogenic climate change results from developmental activities across various sectors including agriculture. Threats resulting from these have variously been proposed to be manageable with mitigation and adaptation mechanisms by the stakeholders. The population inhabiting the less favoured environments is much more vulnerable to climate change. Despite the potential to produce under designated management, these environments have not received the due attention for development initiatives. This could partially be because of the lack of infrastructure facilities and partly also due to the tendency to concentrate in the comfort zone. Hence management adaptations that can directly influence the responsive indicators of climate change such as the concentration of green house gases in the atmosphere could be promising. System intensification is an unambiguous choice keeping in view the increasing population. Subtle climate smart technologies are also available as simple production techniques that can potentially reduce the vulnerability to climate change such as competent cultivars and cropping systems apart from time tested modifications in production practices. System diversification is a key component of disaster risk reduction and seen as a tool for reducing vulnerability to climate change. Precision nutrient management for smallholders aided by IT enabled tools helps in filling the deficit between the crop needs and indigenous nutrient

A. Kohli (✉) · Y. K. Singh · B. K. Vimal · V. Kumar
Department of Soil Science and Agricultural Chemistry, Bihar Agricultural University, Sabour, Bhagalpur 813 210, Bihar, India

S. Singh · S. Sharma
International Rice Research Institute (IRRI), New Delhi, India

S. K. Gupta · M. Ghosh
Department of Agronomy, Bihar Agricultural University, Sabour, Bhagalpur 813 210, Bihar, India

S. K. Mandal
Krishi Vigyan Kendra, Jagatpur, Banka, Bihar Agricultural University, Bhagalpur, Bihar, India

S. Sheraz Mahdi (ed.), *Climate Change and Agriculture in India: Impact and Adaptation*, https://doi.org/10.1007/978-3-319-90086-5_13

supply of the soil for rational yield targets and results in saving fertilizers, increase fertilizer use efficiency and reduce greenhouse gas emissions vis-a-vis conventional management. Sustainable biochar technology can trap atmospheric carbon dioxide in the soil for a time scale of the order of thousands of year and at the same time improve crop productivity and soil physical conditions.

Keywords Climate change · Ahar-pyne system · Biochar

13.1 Introduction

Climate change is a global environmental and humanitarian concern and with increasing awareness and scientific efforts to monitor and combat climate change, the threat of changing climate appears more real (Solomon et al. 2007). Agriculture is the precursor for economic growth and more vulnerable sector than others because of its inherent larger dependence on weather and climate. The challenge and threat from climate change in agriculture has been thought to be manageable with a host of adaptive and mitigating options (Makuvaro et al. 2018), most of which are directed by farmers themselves and not pointed out by government institutions. The mitigation options can be exercised by all consumers and producers; however, adaptive role is largely limited to the stakeholders such as the primary cultivators and others engaged in primary occupations. Masud et al. (2017) revealed that education, experience as well as access to resources and societal support systems have significant impacts on adaptation practices. However, it is pertinent to assume that the adaptation has to be in proportion with their vulnerability or their stake.

13.2 Possible Indicators of Global Climate Change

There is a general consensus among the scientific community that the threat of climate change is real on a global scale. There are indicators such as a generalised global warming characterised by a gradual increase in the mean temperature, increased atmospheric concentration of green house gases, sea level rise, erratic weather patterns, and the increased frequency of extreme weather events. Though the climate of the globe has never been constant on a geological scale, the scientific data available has confirmed beyond doubt that the climate of the globe is changing on scales far shorter than the geological scale, largely because of anthropogenic activities.

13.3 Vulnerability of the Less Favoured Areas

Communities living in the less favoured areas are more vulnerable because they have lesser access to resources. They need to adapt to a greater degree in order to decrease their vulnerability to the effects of climate change. Climate change directly affects the vulnerable by damaging physical resources through the increased frequency of extreme weather events (Garnaut 2013; Nelson and Shively 2014). Ismail et al. (2013) has commented that rural poverty and food insecurity prevail in rainfed and flood-prone ecologies and these areas are expected to increase substantially as a consequence of climate change (Coumou and Rahmstorf 2012). With complete submergence, most of the rice cultivars die within days, resulting in total crop loss (Mackill et al. 2012). Hence the vulnerability of rice farmers in such areas is much more to climate change. The flood prone areas on the banks of Ganges, Kosi, Gandak, Budhi Gandak, Mahananda and their tributaries; the rainfed ecosystems of south Bihar; the *tal* and *diara* lands; *chaur* lands; and the areas fed by the *ahar pyne* systems predominant in Bihar have characteristics of the less favoured areas. Despite their potential to produce under designated management, these areas have not received attention in terms of development initiatives, largely because these are not the ideal choice of the development professionals. Development professionals usually prefer more remunerative postings and even the private sector enterprises do not give an equitable share to these areas. This could partially be because of the lack of infrastructure facilities and partly also due to the tendency to concentrate in the comfort zone.

Another conviction in asserting that the communities in the less favoured areas are more vulnerable to climate change is that the less favoured areas are so largely neglected because of their agro-climatic locations or agro-ecological setting. These areas are less favoured simply because of the harsh agro-climatic conditions that make the agricultural enterprises in these conditions more risky than in the competing areas. Although the communities have adapted to the harsh agro-climate of these areas and there are technologies (both traditional as well as modern) for sustained productivity from these locations, this call for an extra effort and a greater risk; which weans the stakeholders from these ecologies at the very first opportunity in the comfort zone. To illustrate the vulnerability of communities in the less favoured areas to climate change, let us consider the behaviour of irrigation production function in the scenario of climate change (Fig. 13.1). If suppose the climate change scenario involves the shift from a normal to a drier regime, though the irrigation production function would show a southward shift meaning a decrease in the yield of all crops at the similar level irrigation, the maximum decrement in relative terms would be observed in case of rainfed crops. Contrastingly, if the change involves the shift from an existing to a wetter regime, the irrigation production function would show a northward shift meaning an increase in the yield of all crops at the same level of irrigation with a maximum increment in case of rainfed crops. In the first scenario, obviously the most vulnerable communities would be those inhabiting the rainfed areas. In the second scenario, the vulnerability of the rainfed communities

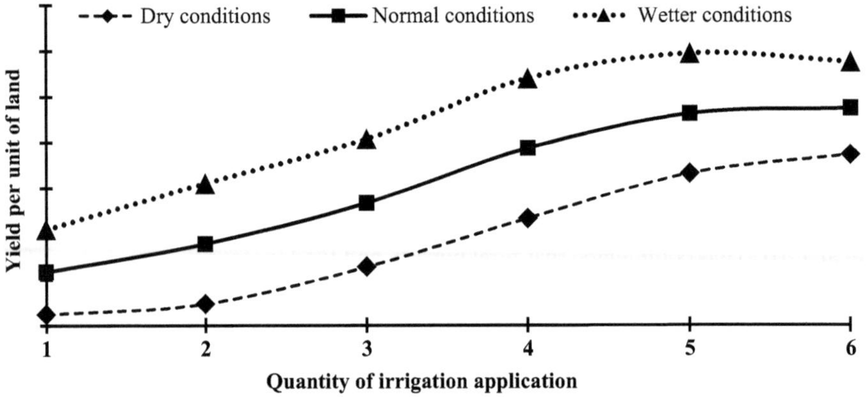

Fig. 13.1 Production function for irrigation under climate change

would be no less because they are likely to encounter an extreme of the precipitation events which would call for a greater adaptation given their original adaptation for the water scarce environments. Another community at the receiving end in a scenario of shift to a wetter regime would be those inhabiting the flood prone areas which have the increased risk of flooding. These areas, having fewer and poorer infrastructure facilities, have the danger of facing greater losses with the advent of similar inclement weather conditions.

13.4 Management Adaptations for Mitigating Climate Change

Adaptations in agronomic practices and adoption of agro-advisories at the farm level can be key components in improving the adaptation of agriculture to climate change (Byjesh et al. 2010; Naresh Kumar et al. 2014; Naresh Kumar and Aggarwal 2013). These options can significantly improve crop yields, increase input-use efficiencies and net farm incomes, and reduce greenhouse gas emissions (Smith et al. 2007). Many of these interventions have been successful in increasing production, income and building resilience among farming communities in many areas such as the Indo-Gangetic plains (Parihar et al. 2016).

Amongst the various possible indicators of global climate change discussed above, the indicator that is likely to respond directly to management adaptations in agriculture is the concentration of green house gases in the atmosphere. The management adaptations that can contribute to decreasing the emission of greenhouse gases from the agricultural soils are primarily reducing or eliminating tillage along with a host of management decisions that can increase crop intensification and plant production efficiency. To illustrate, let us consider a hypothetical case where there is a certain level of greenhouse gas emission under a traditional management regime.

These emissions will usually be far greater than those under a fallow system but fall short of those under an intensified production system. This is the scenario when we consider the emissions per unit of land area per se. With a consideration that there is a need to grow more food for the increasing demand due to increasing population as well as to meet the requirements of improved lifestyles, there is no alternative to intensification. Fallowing cannot be allowed when taking decision based on economic sense. Hence, in our analysis, if we just try to focus on emission per unit land area per unit of output or production, we can see that the emissions scenario will be the best for an intensified system followed by that of a traditional cultivation system or a fallow, which could vary with the season. So this means that the whole set of practices that characterise a system as intensive can contribute to a decreased overall load of greenhouse gases in the environment.

13.4.1 Subtle Climate Smart Technologies

Climate smart technologies are simple production techniques or apart from the techniques that can potentially reduce the vulnerability to climate change. The introduction of SUB1 gene into varieties that are already popular among farmers considerably shortened the time for further evaluation and adoption of flood tolerant varieties (Ismail et al. 2013). Another option is preferring cultivation of resilient crops. One example of such resilient crops is millets in rainfed environments during the *kharif* (rainy) season. Another is *lathyrus* during the *rabi* (dry) season. Such resilient crops are least dependent on external inputs and have a lower incidence of pests and diseases. Traditionally these crops have been grown by broadcasting the traditional seeds and can compete with the native weeds in the rainfed environments. They are generally capable of completing their life cycle from seed to seed solely on the soil moisture residual from the previous crop. The root system is robust to capture the soil moisture from the lower soil layers even when the upper soil layers become too dry to allow any vegetation. When roots transgress the lower layers of the profile for moisture, they inadvertently access the nutrients held up in these soil layers and also enrich those layers with the root exudates and associated organic carbon. During broadcasting of seeds, the traditional practice is to use slightly higher seed rates which takes care of the lower germination under rainfed environments. In case of a rainfall event during the initial stages there is an increased plant population which also gets an extra bout of nutrients from the upper profile till the time it has sufficient moisture. Research into sustainable intensification of these practices under such resilient systems has provided resilient cultivars of various crops that can compete with the water limited environments. This has been made possible by breeding of varieties that have a more robust and extensive root system, those which can be cultivated in a bigger planting window that makes them resilient across cropping systems as well as vagaries of weather. This is just another expression of a cultivar resilient to climate change. An example that is coming out of research initiatives in eastern India is that planting of medium to short duration rice varieties such as

Fig. 13.2 Drought resistant rice cultivar Sahbhagi dhan cultivated in rainfed tracts of Banka amidst traditional rice cultivars depict a distinct advantage in terms of the availability of an early window for planting rabi season crops that can utilize soil moisture residual in the rice fields for crop establishment and a substantial portion of the early growth

Sahbhagi dhan make it possible to use residual soil moisture for establishment of the subsequent dry season crop (Fig. 13.2).

Scientific research has demonstrated such practices as zero or minimum tillage to be more resilient in the unfavourable environments than the conventional practices. The traditional crop establishment practices such as *bhokha* (direct drilling) or *paira* (relay cropping) have an advantage over the conventional tillage in terms of an early crop establishment that enables the crop to complete a portion of its life cycle on the residual soil moisture. Lack of tillage further reduces the soil moisture loss during the subsequent stages of the crop. This is akin to resilience against climate change. The improved resilient crop establishment practices employ the same principles involving an early establishment of the crop and minimizing tillage so as to retain as much residual soil moisture as possible. Another subtle technology is showing respect to and following the time tested traditions. For instance, the *ahar-pyne* systems prevalent in the plains of south Bihar, present a lively example of a tradition so suited for a fragile ecosystem. An *ahaar* is a traditional water harvesting structure which is a common sight in the rainfed lowland tracts of several districts in south Bihar. Structurally, an *ahaar* is a catchment basin embanked on three sides and the fourth side consisting of a natural gradient of the topography. Substantial areas along the highways can be seen in Gaya that have evolved such systems with a thorough understanding of the local topography and agro-climatic conditions. The lifeline of *ahaars* are the *pynes* which simply are the channels leading water to or from an *ahaar*. The crops being cultivated on the pyne bed include chickpea/ lentil, vegetable peas and wheat. The conventional practice is to broadcast the seeds and then till the soil with a bullock driven plough to cover the seeds. The seedbed is usually very cloddy in case of vegetable peas and chick pea, which are being grown in the area without any supplemental irrigation; however the seed bed for wheat is comparatively smoother. Probably the indigenous communities recognize the presence of water in the profile and so the existing practice is of tilling the soil leaving a cloddy surface with much surface roughness and breaking the capillaries to restrict the movement of profile soil moisture to the surface soil. These indigenous practices are an essential part of the rainfed cultivation of rabi season crops in the ecology. The *ahaar* bed is usually flooded during the *kharif* season and so there is limited

Fig. 13.3 Rabi cultivation in an *Ahaar* of Khizarsarai block of district Gaya, Bihar shows a distinct gradient in the time of planting of *rabi* season crops based on drying of the *Ahaar* bed starting from the highest point in the watershed to the lower points

Fig. 13.4 Based on visual interpretation key, the *Ahaar-Pyne* systems can be seen densely existing in the rainfed tracts of district Gaya, Bihar

scope of cultivation. However, the advent of the *rabi* season enables cropping in phases as visible in distinct bands as the water level in the *ahaar* recedes (Fig. 13.3). The *Ahaars* can be seen very densely scattered in the area as identified in the archived post monsoon LANDSAT-TM imagery (Fig. 13.4).

13.4.2 Diversification and Intensification of Rainfed Systems Reduces Vulnerability to Climate Change

Drought is a widespread abiotic stress and drastically reduces rice yield. Technologies that reduce the production risk caused by drought will favor input use and have a major impact on system productivity in good and bad years. In addition, improved technologies that reduce labor and land requirements for crops are needed to allow these resources to be released for other income-generating activities. Several varieties of rice have been evaluated for tolerances to abiotic stresses such as drought and submergence. Drought affects rice at morphological, physiological, biochemical and molecular levels and thereby affects its yield. Good opportunities for diversification through post-rice crops are provided by the development of new drought-tolerant short-duration rice varieties/lines such as *Sahbhagi dhan* and *Susk Samrat.* Most farmers practice limited or no cropping during the *rabi* season in drought-prone areas of Bihar, thus failing to realize the full production potential of their land. New area specific systems need to be developed to improve the post-rice production. Timely and appropriate planting techniques allow earlier-maturing crops, maximizing the use of residual water for post-rice *rabi* crops. New, improved varieties of *rabi* season crops, notably pulses, allow for a greater range of options for *rabi* season cropping and land productivity. Broadcasting of seeds of pulses like lentil or lathyrus into the standing rice crop before harvest is practiced in these areas if sufficient moisture is available. There is a strong need of managing midterm and terminal droughts of 2–3 weeks to have a capacity for intensifying cropping in hitherto single crop situations; more so because of the prevailing low cropping intensity in these areas (e.g. the cropping intensity of Bihar is just 146 per cent). Besides weather, soil moisture availability in a situation also depends on land use practices. There is a need to identify agricultural practices that maximize precipitation utilization and minimize evaporation. Reduced tillage, raising the height of farm bunds, rational residue retention and reducing the time of fallow between crops can allow better utilization of precipitation. Soil incorporation of biochar like substances can improve the water holding capacity, infiltration rate, microbial ecology and nutrient relations. System diversity can be increased by introducing interventions such as growing short to medium duration legumes like chickpea, green gram, grass pea, vegetable peas and lentils or short duration oilseeds like linseed. A host of practices that aim to increase the infiltration opportunity time, minimize evaporation, decrease soil bulk density and erosion susceptibility, increase the access of the rainfed crops to the stored water and nutrient resources in the lower layers, staggered planting in conjunction with diverse crop rotations, weed management, controlling grazing and human interference and avoidance of terminal heat stress can contribute towards obtaining greater yields from rainfed agriculture on a sustainable basis. However, the effects of each of management option needs to be quantified on a uniform scale for the effects to be additive and comparable.

13.4.3 Targeting Precision Nutrient Management Using IT Enabled Tools for Smallholders in Less Favoured Areas for Mitigating Climate Change

Precision nutrient management aims at satisfying the total mineral nutrient requirements of the crop by filling the deficit between the total needs to the crop and the soils' indigenous supply by first ensuring the effective use of the indigenous nutrients and then assessing the crops additional needs based on an attainable yield target. The benefits towards mitigating climate change by adopting precision nutrient management are expected in terms of the savings in fertilizers and increased fertilizer use efficiencies that can reduce the emissions of green house gases. The conventional approach for implementing a precision nutrient management programme is through soil testing based management. There is no doubt about the validity of this approach, but still it is a fact that despite the best efforts of government agencies, agricultural universities and others, the soil testing programme has had little visibility. This is obvious considering the high cost of analysis and difficulty in availability of timely results of analysis. Of late, several organizations have made concerted efforts to develop IT enabled tools to recommend a balanced fertilizer dose for various crops considering the system as a whole. Use of SSNM-based fertilizer recommendations for rice were shown to increase yields, increase net income of farmers, and provide positive impacts on the environment when compared to existing fertilizer practices. Field experiments were conducted at Bihar Agricultural University, Sabour with tools such as Nutrient Expert™ for Hybrid Maize, Nutrient Expert™ for Rice, Rice Wheat Crop Manager, Rice Crop Manager for stressed conditions (drought) and with Crop manager for rice based systems, all of which have shown a potential for considerable fertilizer saving, improving yields and profitability.

13.4.4 Sustainable Biochar Technology for Mitigating Climate Change in Less Favoured Areas

Scientific literature is multiplying exponentially with new benefits and applications of biochar as a soil conditioner and as a carbon negative technology. No other technology has a potential to trap atmospheric carbon dioxide in the soil for a time scale greater than that offered by biochars, whose mean residence time in the soil is of the order of thousands of years. During biochar production, up to a large fraction of the carbon in the original organic residue is retained in the crystalline biochar structure. Simultaneously, there is production of energy in the process. Thus our organic residues / wastes can be used for energy production using vessels which allow pyrolysis. The byproduct of these energy production systems can be a useful material known as biochar. Biochar production can be achieved during energy production systems at household to industrial scale. For instance, electricity co-generation in several rice mills is achieved by pyrolysis of rice husk, and this simultaneously produces rice husk biochar. The communities in the less favoured areas are dependent on fuel wood or other agricultural waste biomass for cooking activities with traditional stoves. At a household scale, there are top lit up draft (TLUD) stoves that can be used for cooking with any organic residue as feedstock and result in generation of clean energy for cooking and simultaneously producing biochar as a byproduct. TLUD technology needs to be promoted as a low external input technology which can be used easily in the far flung areas and hinterlands as against the current impetus of the governments on taking LPG connections to each household. Use of biochar in soils not only increases crop productivity, but also improves soil tilth, fertility, water holding capacity, reduces risk of soil erosion and the need for fertilizer inputs.

References

Byjesh, K., Naresh Kumar, S., & Aggarwal, P. K. (2010). Simulating impacts, potential adaptation and vulnerability of maize to climate change in India. *Mitigation and Adaptation Strategies for Global Change, 15*, 413–431. https://doi.org/10.1007/s11027-010-9224-3.

Coumou, D., & Rahmstorf, S. (2012). A decade of weather extremes. *Nature Climate Change, 2*, 491–496.

Garnaut, R. (2013). Removing climate change as a barrier to economic progress: twentysecond Colin Clark Memorial Lecture November 2012. *Journal of Economic Analysis and Policy, 43*(1), 31–47. https://doi.org/10.1016/S0313-5926(13)(50002-6).

Ismail, A. M., Singh, U. S., Singh, S., Dar, M. H., & Mackill, D. J. (2013). The contribution of submergence-tolerant (Sub1) rice varieties to food security in flood-prone rainfed lowland areas in Asia. *Field Crops Research, 152*, 83–93. https://doi.org/10.1016/j.fcr.2013.01.007.

Mackill, D. J., Ismail, A. M., Singh, U. S., Labios, R. V., & Paris, T. R. (2012). Development and rapid adoption of submergence-tolerant (Sub1) rice varieties. *Advances in Agronomy, 115*, 303–356.

Makuvaro, V., Walker, S., Masere, T. P., & Dimes, J. (2018). Smallholder farmer perceived effects of climate change on agricultural productivity and adaptation strategies. *Journal of Arid Environments, 152*, 75. https://doi.org/10.1016/j.jaridenv.2018.01.016.

Masud, M. M., Azam, M. N., Mohiuddin, M., Banna, H., Akhtar, R., Alam, A. S. A. F., & Begum, H. (2017). Adaptation barriers and strategies towards climate change: Challenges in the agricultural sector. *Journal of Cleaner Production., 156*, 698. https://doi.org/10.1016/j.jclepro.2017.04.060.

Naresh Kumar, S., & Aggarwal, P. K. (2013). Climate change and coconut plantations in India. Impacts and potential adaptation gains. *Agricultural Systems, 117*, 45–54.

Naresh Kumar, S., Aggarwal, P. K., Swaroopa Rani, D. N., Saxena, R., & Chauhan, N. A. J. (2014). Vulnerability of wheat production to climate change in India. *Climate Research, 59*, 173–187. https://doi.org/10.3354/cr01212.

Nelson, G. C., & Shively, G. E. (2014). Modeling climate change and agriculture: An introduction to the special issue. *Agricultural Economics, 45*(1), 1–2. https://doi.org/10.1111/agec.12093.

Parihar, C.M., Jat, S.L., Singh, A.K., Kumar, B., Pradhan, S., Pooniya, V., Dhauja, A., Chaudhary, V., Jat,M.L., JAt, R.K. and Yadav, O.P. (2016) Conservation agriculture in irrigated intensive maize-based systemsofnorth-western India: effects on crop yields, water productivity and economic profitability. Field Crop Res., 193,104-116. https://doi.org/10.1016/j.fcr.2016.03.013

Solomon, S., Qin, D., Manning, M., Chen, Z., Marquis, M., Averyt, K. B., Tignor, M., & Miller, H. L. (2007). *Climate change: the physical science basis. In: Contribution of Working Group I to the Fourth Assessment Report (FAR) of the Intergovernmental Panel on Climate Change* (p. 996). United Kingdom/New York: Cambridge University Press, Cambridge.

Smith, P., Martino, D., Cai, Z., Gwary, D., JAnzen, H., Kumar, P., McCarl, B., Ogle, S.O'MAra, F., Rice, C. andScholes, B. (2007) Policy and technological constrints to implementation of greenhouse gas mitigation options inagriculture. Agric. Ecosyst. Environ., 118(1), 6-28. https://doi.org/10.1016/j.agee.2006.06.006

Chapter 14
Nanotechnology in the Arena of Changing Climate

Nintu Mandal, Rajiv Rakshit, Samar Chandra Datta, and Ajoy Kumar Singh

Abstract Technological interventions in the changing climatic conditions seem to be imperative in formulating adaptation and mitigation strategies. Nanoscience, the study of matter at atomic or nano (1 nm = 10^{-9} m) scale can be one of the viable options/ interventions in increasing input use efficiency, effective pest control and drought management in agriculture to address the ill issues of climate change in agriculture. The chapter discussed about the Nanotechnology to play important role in development of intelligent delivery of agrochemicals, development of novel superabsorbent polymers, real time monitoring of agrochemicals via nanobiosensors, increased resilience and resilience of microbes under heat stress which are very relevant in changing climate scenario.

Keywords Nanotechnology · Climate change · Impact · Mitigation

14.1 Introduction

Climatic aberrations are increasing day by day causing adverse impact on essential ecosystem functions. Spatial and temporal shifting of amount and frequency of rainfall is affecting agricultural operations in a massive way. Climatic extremities (Heat wave, cold wave *etc*) are affecting agricultural production system in a bigger way.

Surface temperature is projected to rise over the twenty-first century under all assessed emission Scenarios (IPCC 2014). It is very likely that heat waves will occur more often and last longer, and that extreme precipitation events will become

N. Mandal (✉) · R. Rakshit
Department of Soil Science and Agricultural Chemistry, Bihar Agricultural University, Sabour, Bhagalpur, Bihar, India

S. C. Datta
Division of Soil Science and Agricultural Chemistry, Indian Agricultural Research Institute, New Delhi, India

A. K. Singh
Bihar Agricultural University, Sabour, Bhagalpur, Bihar, India

S. Sheraz Mahdi (ed.), *Climate Change and Agriculture in India: Impact and Adaptation*, https://doi.org/10.1007/978-3-319-90086-5_14

more intense and frequent in many regions. The ocean will continue to warm and acidify, and global mean sea level to rise (IPCC 2014). There are multiple mitigation pathways that are likely to limit warming to below 2 °C relative to pre-industrial levels (IPCC 2014). These pathways would require substantial emissions reductions over the next few decades and near zero emissions of CO_2 and other long-lived greenhouse gases by the end of the century. Implementing such reductions poses substantial technological, economic, social and institutional challenges, which increase with delays in additional mitigation and if key technologies are not available. Limiting warming to lower or higher levels involves similar challenges but on different timescales (IPCC 2014).

Technolological interventions in the changing climatic conditions seem to be imperative in formulating adaptation and mitigation strategies. Nanoscience is the study of matter at atomic or nano (1 nm = 10^{-9} m) scale. Nanotechnology deals with fabrication of materials at nanoscale. Nanomaterials are materials having at least one dimension in 1–100 nm scale as per USEPA (United State Environment Protection Agency). Nanotechnological interventions in increasing input use efficiency, effective pest control and draught management in agriculture are of utmost importance in present day agriculture and in future also.

14.2 Intelligent Delivery of Agrochemicals

14.2.1 Rhizosphere Controlled Release Nutrient Formulations

14.2.1.1 Major Nutrient Formulation

Montmorlonitic nanoclay separated from Vertisol was employed for controlled release N, P formulation by Sarkar et al. 2014. They prepared series of nanoclay polymer composites (NCPCs) by using various clay minerals *viz.* Kaolinite (Alfisol), Smectite (Vertisol) and Mica (Inseptisol) and concluded that smectitic type of clays were most suitable for NCPC preparation owing to its higher specific surface area and high aspect ratio. The types of composites were exfoliated types were confirmed by disappearance of typical bentonitic peak within polymer matrixes (Fig. 14.1).

In case of the NCPC incorporated with clay I (the kaolinite dominated clay) and clay II (the mica-dominated clay), the reaction occurred on the surface of the clays. However, the polymer layer penetrated into silicate layers, and the clay was exfoliated when clay III (the smectite-dominated clay) was introduced. The equilibrium water absorbency and nutrient release rate decreased (Fig. 14.2) with the incorporation of clay into the polymer matrix because of the increase in crosslinking points and the decrease in the mesh size of the NCPCs as compared to those in the pure polymer.

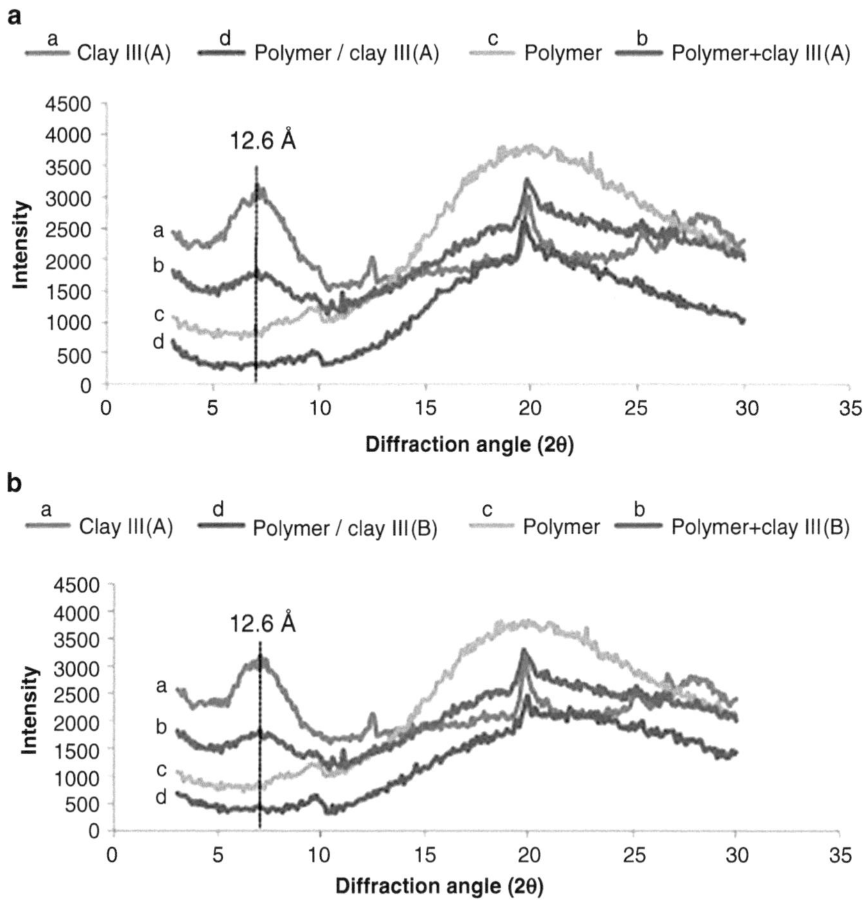

Fig. 14.1 Randomly oriented powder XRD patterns of the clay, polymer/ clay composite, polymer, and polymer1clay physical mixture for (A) clay III(A) and (B) clay III(B)

14.2.1.2 Micronutrient Formulations

Zincated nanoclay polymer composites (ZNCPCs):
NCPC based micronutrient formulation was reported by Mandal and his coworkers (Mandal et al. 2015; Mandal et al. 2018). Bentonitic nanoclay were separated through ultracentrifugation and used as a diffusion barrier in the acrylic acid (AA) and acrylamide (Am) copolymer masteries using ammonium persulphate(APS) as initiator and N, N, Methylene Bis acryalmaide (NNMBA) as crosslinker. Zn was loaded as Zn-citrate. Laboratory release study revealed that nanoclay were more effective as compared to clay in slow release behaviour (Fig. 14.3). Olsen-P content in soil also increased in ZNCPC treatments owing to citrate Solubilization of native soil P (Fig. 14.4).

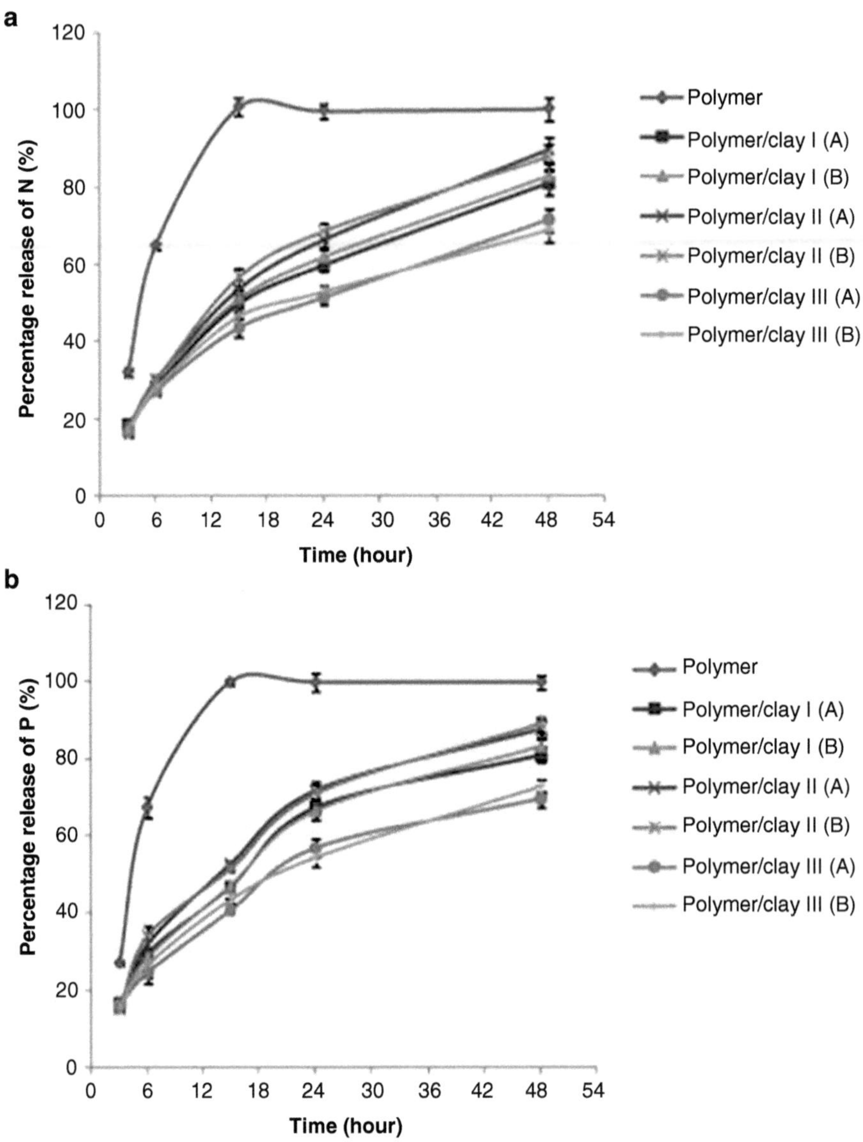

Fig. 14.2 Release of N and P from the DAP-loaded NCPCs incorporated with different kinds of clay (10 wt %) in distilled water. I, II and III represents clay separated from Alfisol, Inseptisol and Vertisol respectively. A and B indicates with and without aluminosilicates respectively

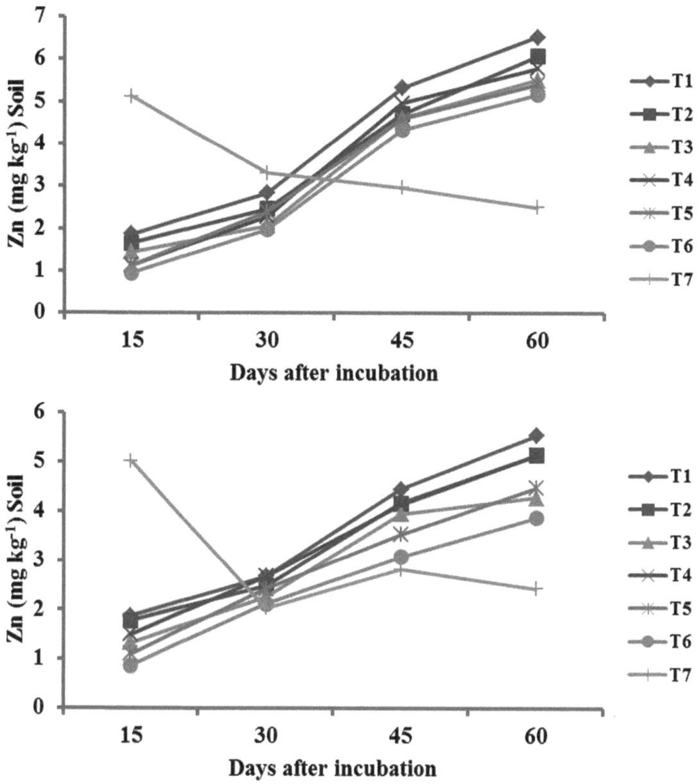

Fig. 14.3 Release of DTPA extractable Zn in soils during incubation experiment
T1: **8% clay;** *T2*: **10% Clay;** *T3*: **12% Clay;** *T4*: **8% Nanoclay;** *T5*: **10% Nanoclay;** *T6*: **12% Nanoclay and** *T7*: **$ZnSO_4$. 7 H_2O**

14.3 Superabsorbent Nanocomposites for Soil Moisture Conditioner Novel Superabsorbent Nanocomposites: Improvement in Moisture Retention and Available Moisture Content in Soil

Novel superabsorbent nanocomposites are recently being reported as promising materials in improving moisture retention characteristics in soil. Singh et al. 2011 prepared a novel nanosuperabsorbent composite (NSAPC) by in situ grafting polymerization and cross-linking on to a novel biopolymer of plant origin (complex heteropoly saccharide in nature) in the presence of a clay mineral using a green chemistry technique. The inorganic clay mineral acted as additional network point, resulting in increase in crosslinking with increase in clay content, manifested in decreased water absorbency.

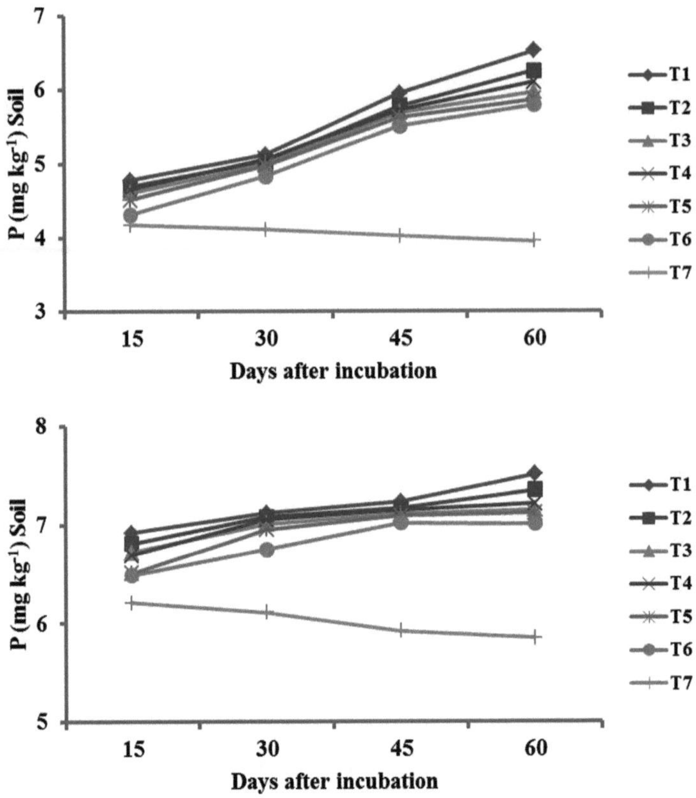

Fig. 14.4 Release of Olsen- P in soils during incubation experiment
T1: **8% clay;** *T2*: **10% Clay;** *T3*: **12% Clay;** *T4*: **8% Nanoclay;** *T5*: **10% Nanoclay;** *T6*: **12% Nanoclay and** *T7*: **$ZnSO_4.7H_2O$**

Addition of test hydrogels to soil and soil-less media significantly increased the availability of water to plant compared with control (Fig. 14.5). In case of soil-less medium, the lower rate of application (0.5%) was as effective as higher rate (0.75%) for both the gels. This observation was substantiated by the values of onset wilting point in amended and unamended plant growth media (Fig. 14.6). As is clear, because of more availability of water in gel amended treatments, the permanent wilting point approached in the amended soil in 4.7–6.3 days compared with 2.4 days in control. Delay by 1.4–3.6 days was observed in soil-less media.

Effect of NSAPC on water absorption and retention characteristics (Fig. 14.6) of sandy loam soil and soil-less medium was also studied as a function of temperature and tensions. Addition of NSAPC significantly improved the moisture characteristics of plant growth media (both soil and soil-less), showing that it has tremendous potential for diverse applications in moisture stress agriculture

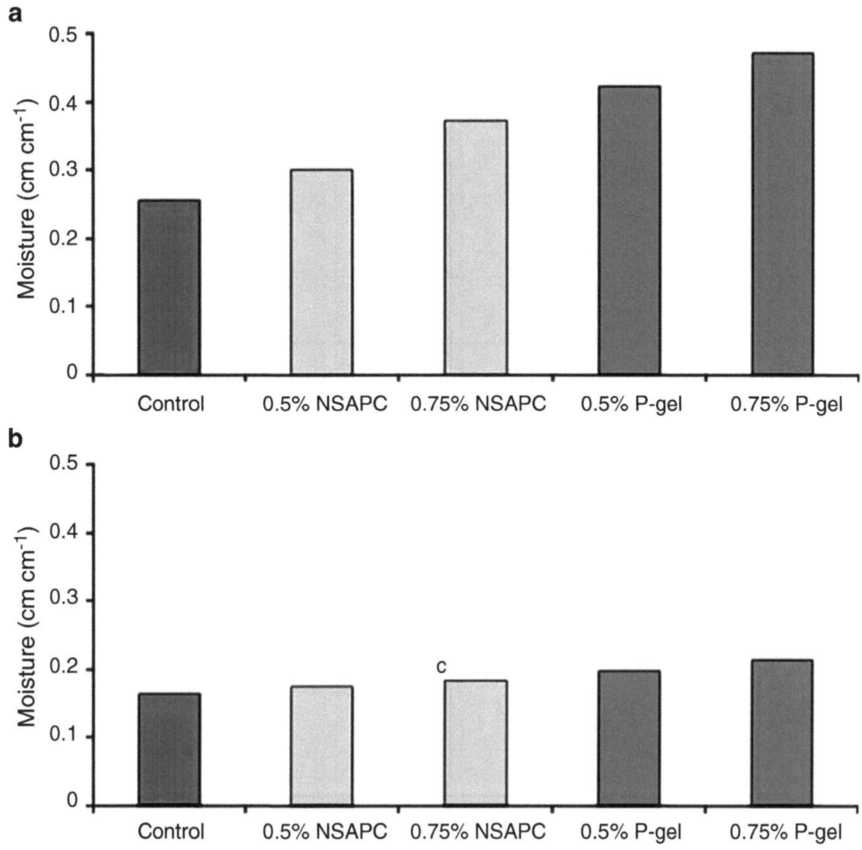

Fig. 14.5 Effect of gel addition on available water from soil (**a**) and soil-less media (**b**)

14.4 Interaction of Microbes with Nanomaterials Under Changing Temperatures

Resistance and resilience are considered as the ecological concepts of high policy relevance. Almost any study could be considered in terms of resilience, in that the common experimental format – the effect of X on Y – will have information about the effect of a disturbance on the system if X is a disturbance. To get an understanding of resilience, there needs to be a measurement soon after the disturbance to gauge resistance and then several subsequent measurements to assess the pattern of resilience. The time period can be a matter of days in a laboratory incubation, or even minutes for some physical measurements (Zhang et al. 2005), through to years for field-based observations and is generally related to the nature of the disturbance. Studies have investigated the resilience of microbial communities to disturbance due to human activities such as land use and agricultural practices. But, impact assessment of

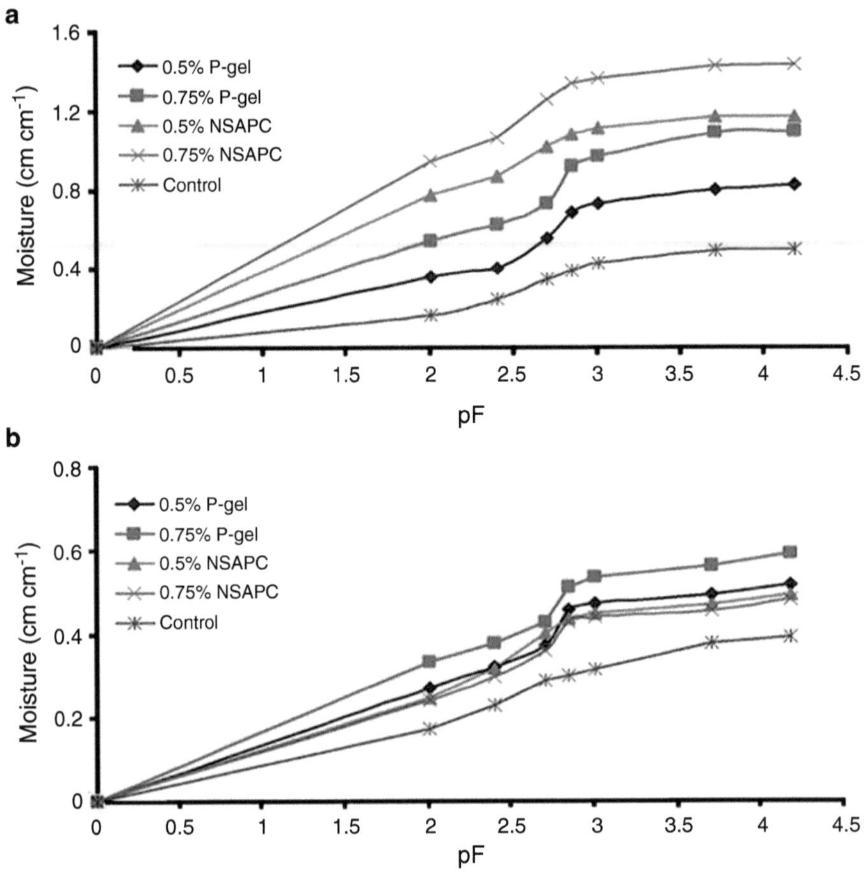

Fig. 14.6 Moisture release curves of soil (**a**) and soil-less media (**b**) amended with P-gel and NSAPC

nanomaterial addition (Zn and Fe nanomaterial) as extraneous substances added to soil system on soils resistance and resilience were still unidentified under heat stress and thus the hypothesis of the current experiment was that the addition of nanomaterials could have considerable implications on the resistance and resilience of soil organisms against abiotic stress, such as high temperature.

Kumar et al. (2016) showed that Fluorescien Diacetate Hydrolysing (FDA) activity was significantly reduced when the concentration of nano Fe was increased from 10 ppm (22.37 μg fluorescein released g^{-1} dry soil h^{-1}) to 40 ppm (18.08 μg fluorescein released g^{-1} dry soil h^{-1}) at $P < 0.05$. Data showed no significant differences between nano Zn @ 10 ppm (25.00 μg fluorescein released g^{-1} dry soil h^{-1}) and nano Zn @ 40 ppm (24.74 μg fluorescein released g^{-1} dry soil h^{-1} reflecting no changes in FDA activity at different doses of nano Zn. With respect to resistance and resilience indices of FDA activity against heat stress, it was observed that the nano Zn @ 10 ppm and nano Fe @ 10 ppm treatment had the greatest stress resistance,

Fig. 14.7 Resistance index of FDA activity in soil after heat stress (48 °C for 24 h) under various doses of nanomaterial. T_1 represents control; T_2 represents nano Zn@ 10 ppm; T_3 represents nano Zn@ 40 ppm; T_4 represents nano Fe@ 10 ppm; T_5 represents nano Fe@ 40 ppm. Bars indicate the CD values

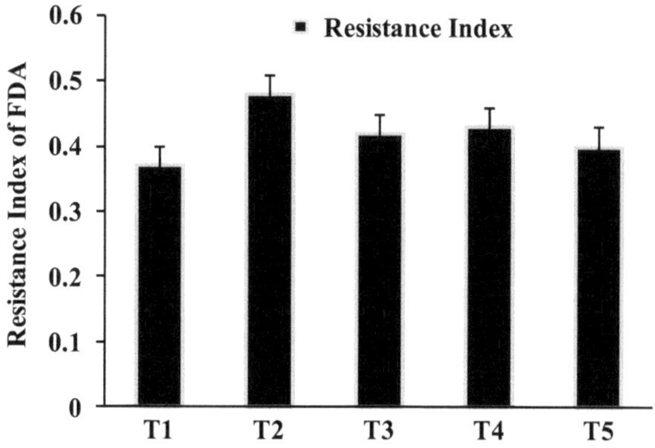

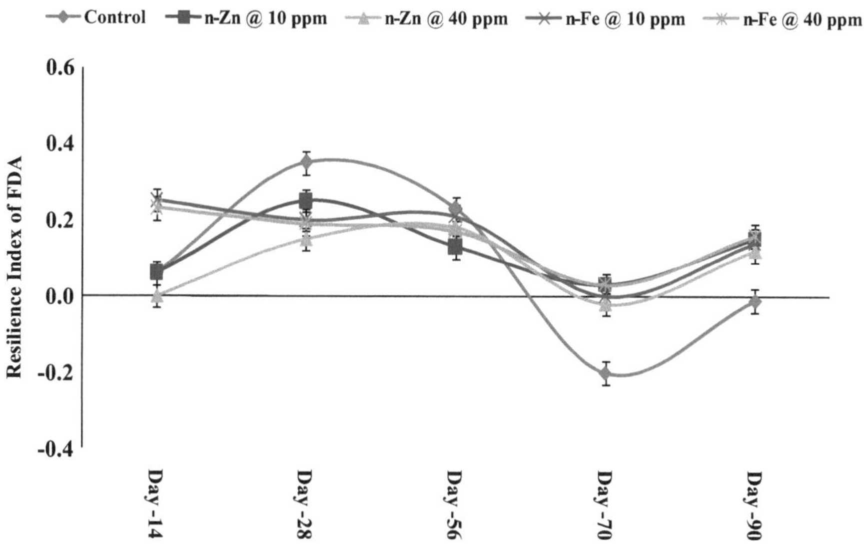

Fig. 14.8 Resilience index of FDA activity in soil after heat stress (48 °C for 24 h) under various doses of nanomaterial

with an index rating of 0.48 and 0.43 respectively (Fig. 14.7). Resistance index values of control was 0.37, there was no significant difference between treatments supplied with higher doses of nanomaterial (Zn or Fe @ 40 ppm) ($P < 0.05$).

Resilience indices of all the treatments varies from −0.01 to 0.16 after 90 days of incubation. It was observed that all the treatments supplied with nanomaterials showed a statistically similar resiliency against the heat stress after 90 days (Fig. 14.8). Although the control treatment showed a higher resilience index upto 56 days, but there was sharp decline in the resilience indices after 70 (−0.20) and 90 days (−0.01).

With respect to resistance indices of microbial biomass carbon against heat stress, it was observed that the nano Fe @ 40 ppm treatment had the greatest stress

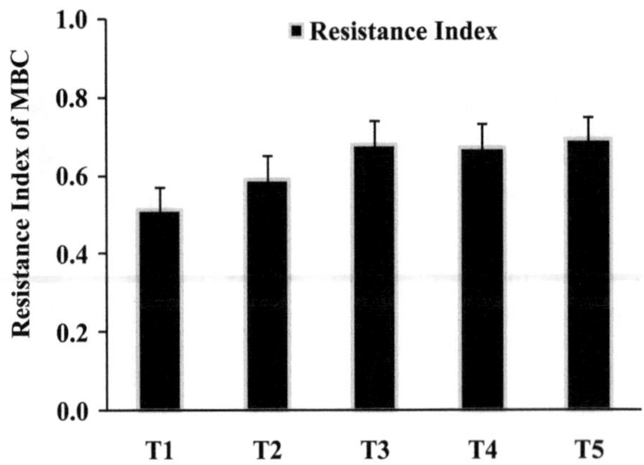

Fig. 14.9 Resistance index of MBC in soil after heat stress (48 °C for 24 h) under various doses of nanomaterial. T_1 represents control; T_2 represents nano Zn@ 10 ppm; T_3 represents nano Zn@ 40 ppm; T_4 represents nano Fe@ 10 ppm; T_5 represents nano Fe@ 40 ppm. Bars indicate the CD values

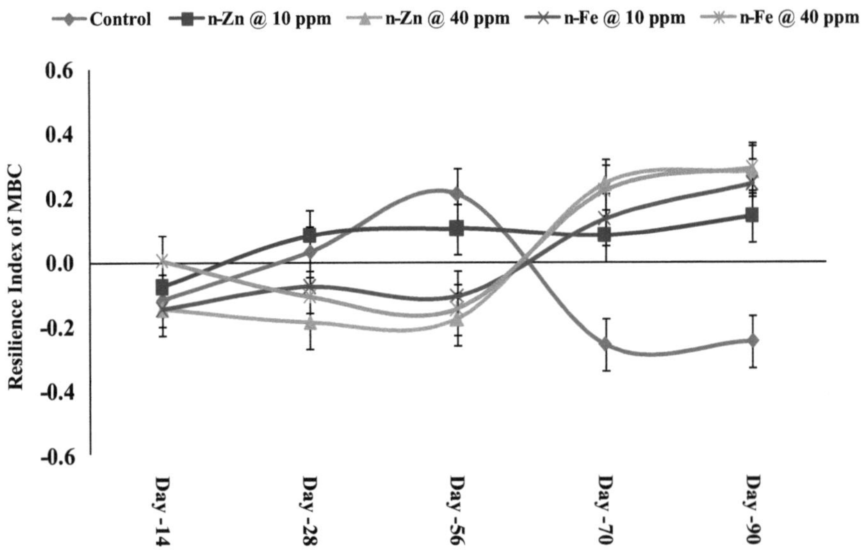

Fig. 14.10 Resilience index of MBC activity in soil after heat stress (48 °C for 24 h) under various doses of nanomaterial

resistance, with an index rating of 0.69, which was statistically comparable to nano Zn@ 40 ppm with an index rating of 0.68 and nano Fe@ 10 ppm with an index rating of 0.67 (Fig. 14.9). Recovery rate of microbial biomass carbon showed a similar pattern as in the enzymatic activity. Although the resilience indices were lower during the first 56 days, but the recovery indices were maximum after 90 days of incubation. Nano Zn @ 40 ppm and nano Fe @ 40 ppm had the greatest stress resilience with an index rating of 0.28 and 0.29 respectively after 90 days of incubation. The lower dose (10 ppm of nano Zn and nano Fe) of nanomaterials used in this experiment also showed a positive resilience index (index rating of 0.14 and 0.24 for nano Zn and nano Fe respectively) (Fig. 14.10).

14.5 Thermal and Hydrolytic Stability of Soil Enzymes: Immobilization of Soil Enzymes on Double Layer

Acid phosphatase was immobilized on layered double hydroxides of uncalcined- and calcined-Mg/Al-CO_3 (Unc-LDH-CO_3, C-LDH-CO_3) by the means of direct adsorption by Zhu et al. 2010. Optimal pH and temperature for the activity of free and immobilized enzyme were exhibited at pH 5.5 and 37 °C. The Michaelis constant (Km) for free enzyme was 1.09 mmol mL^{-1} while that for immobilized enzyme on Unc-LDH-CO_3 and C-LDHCO$_3$was increased to 1.22 and 1.19 m molmL^{-1}, respectively, indicating the decreased affinity of substrate for immobilized enzymes. The residual activity of immobilized enzyme on Unc-LDH-CO_3 and C-LDH-CO_3at optimal pH and temperature was 80% and 88%, respectively, suggesting that only little activity was lost during immobilization.

The deactivation energy (Ed) for free and immobilized enzyme on Unc-LDH-$CO_3$and C-LDH-CO_3 was 65.44, 35.24 and 40.66 kJ mol^{-1}, respectively, indicating the improving of thermal stability of acid phosphatase after the immobilization on LDH-CO_3 especially the uncalcined form. Both chemical assays and isothermal titration calorimetry (ITC) observations implied that hydrolytic stability of acid phosphatase was promoted significantly after the immobilization on LDH-CO_3especially the calcined form (Fig. 14.11).

Reusability investigation showed that more than 60% (Fig. 14.12a, b) of the initial activity was remained after six reuses of immobilized enzyme on Unc-LDH-CO_3 and C-LDH-CO_3. A half-life (t1/2) of 10 days was calculated for free enzyme, 55 and 79 days for the immobilized enzyme on Unc-LDH-CO_3 and C-LDH-$CO_3$when stored at 4 °C.

Layered double hydroxides of uncalcined- and calcined-Mg/Al- CO_3 were used successfully as supports for the immobilization of acid phosphatase by the means of

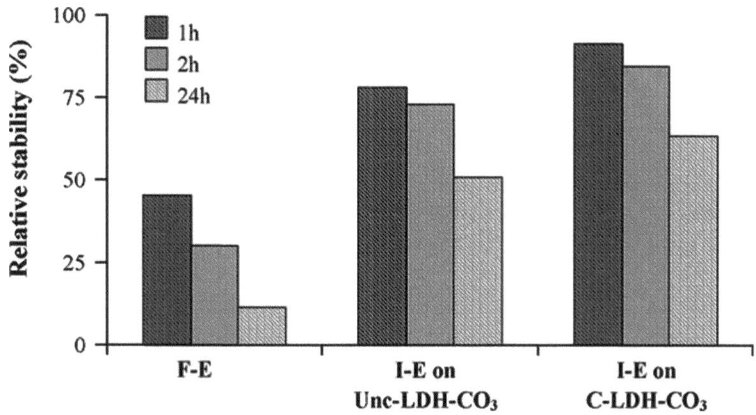

Fig. 14.11 Hydrolytic stability of free (F-E) and immobilized acid phosphatase on LDHs (I-E on Unc-LDH-CO_3 and I-E on C-LDH-CO_3)

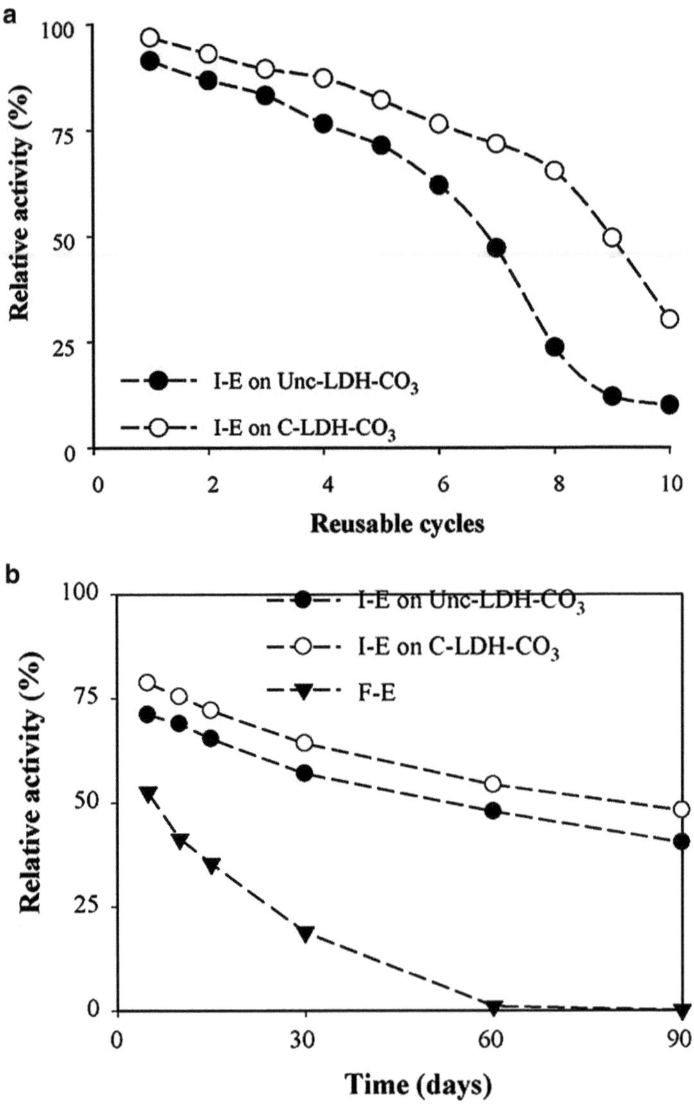

Fig. 14.12 (**a**) Reusability of immobilized acid phosphatase on LDHs (I-E on Unc-LDH-CO_3 and I-E on C-LDH-CO_3). (**b**) Storage stability of free (F-E) and immobilized acid phosphatase on LDHs (I-E on Unc-LDH-CO_3 and I-E on C-LDH-CO_3)

direct adsorption. Little activity loss, excellent thermal stability, hydrolytic stability, reusability and storage stability of the immobilized acid phosphatases revealed their promising potentials for practical application in the fields such as agricultural production and environmental remediation.

14.6 Nanobiosensors for in Changing Climate

Now-a-days, nanotechnology and nanomaterials are intertwined in the construction of almost every bioelectronics or biosensor devices due to their exceptional physicochemical, electrical, optical and mechanical properties. Nanomaterials have overcome the obstacles of low sensitivity, selectivity and analytical interference of old methods by controlling the size, shape and composite (Marx et al. 2004).

14.6.1 Real Time Urease Activity Monitoring

Nanobiocomposite Sensors:

Biosensors are reported for different enzymatic activity detection. Presence of urea can be detected based on the detection of Urease (Ur) and Glutamate dehydrogenase (GLDH). Ur catalyzes decomposition of urea into hydrogen bicarbonate and ammonium ions (NH_4^+). NH_4^+ ions are known to be unstable and easily disperse in the environment. GLDH immediately catalyzes the reaction between NH_4^+, α-ketoglutarate (α-KG) and nicotinamide adenine di nucleotide (NADH) to produce NAD^+ and α-glutamate. Immobilization of Ur onto a suitable matrix is crucial for the development of an electrochemical urea sensor. In this context, metal oxide nanoparticles-chitosan (CH) based hybrid composites have attracted much interest for the development of a desired biosensor.

Metal oxide nanoparticles such as iron oxide (Fe_3O_4), zinc oxide (ZnO), cerium oxide (CeO_2) etc. have been suggested as promising matrices for the immobilization of desired biomolecules. These nanomaterials exhibit large surface to volume ratio, high surface reaction activity, high catalytic efficiency and strong adsorption ability that can be helpful to obtain improved stability and sensitivity of a biosensor. Moreover, nanoparticles have a unique ability to promote fast electron transfer between electrode and the active site of an enzyme. Among various metal oxide nanoparticles, Fe_3O_4 nanoparticles due to biocompatibility, strong superparamagnetic behaviour and low toxicity have been considered as interesting for immobilization of desired biomolecules.

The proposed mechanism relating to the preparation of CH-Fe_3O_4 nanobiocomposite film and immobilization of Ur-GLDH onto CH-Fe_3O_4 nanobiocomposite film is shown in Fig. 14.13. It can be seen that surface charged Fe3O4 nanoparticles interact with cationic biopolymer matrix of CH via electrostatic interactions and hydrogen bonding with NH_2/OH group to form hybrid nanobiocomposite. The Ur-GLDH molecules exist in anionic form at pH 7 because pH of solution is above isoelectric point (5.5) of Ur-GLDH molecules that facilitate interactions with positively charged CH of nanobiocomposite via electrostatic interactions (Fig. 14.13). In the nanobiocomposite, presence of Fe_3O_4 nanoparticles results in increased electroactive surface area of CH for loading of the enzymes due to affinity of the Fe_3O_4 nanoparticles towards oxygen atoms of enzymes. This

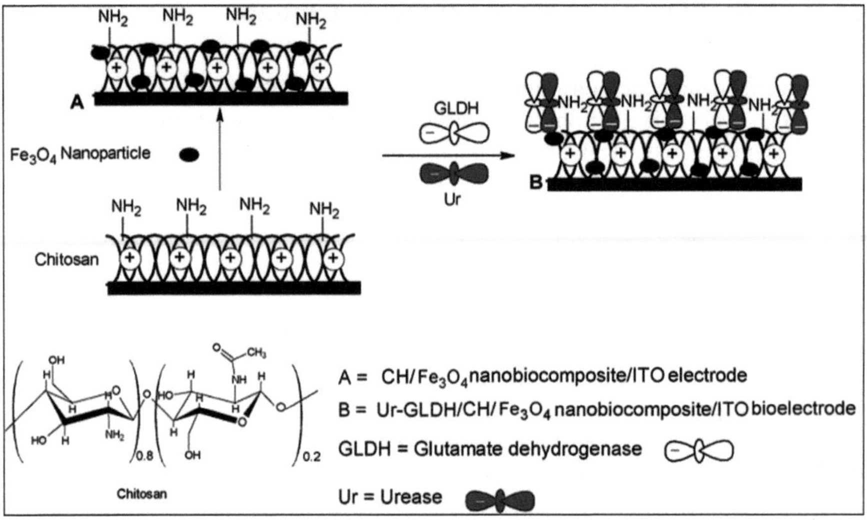

Fig. 14.13 Proposed mechanism for preparation of CH-Fe_3O_4 nanobiocomposite and immobilization of Ur-GLDH onto CH-Fe_3O_4 nanobiocomposite film

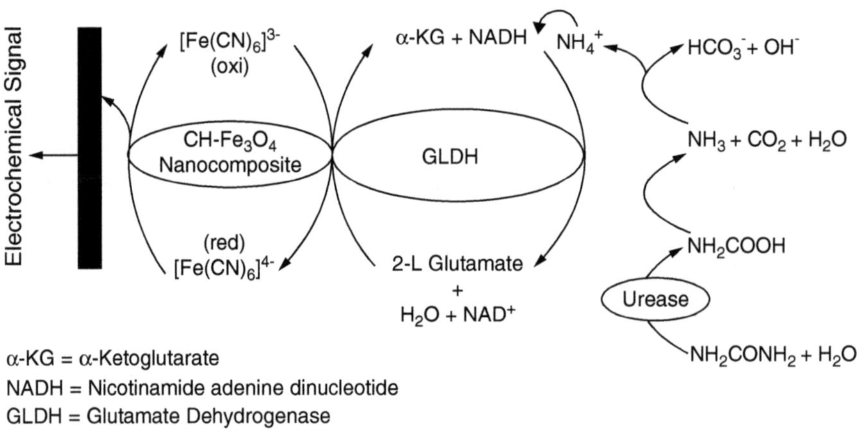

Fig. 14.14 Biochemical reaction during electrctrochemical detection of urea using Ur-GLDH/CH-Fe_3O_4 nanobiocomposite

suggests that Ur-GLDH molecules easily bind with charged CH-Fe_3O_4 hybrid nanobiocomposite matrix via electrostatic interactions.

The proposed biochemical reaction during the urea detection is shown in Fig. 14.14. Ur catalyzes hydrolysis of urea to carbamine acid that gets hydrolyzed to ammonia (NH_3) and carbon dioxide (CO_2). GLDH catalyzes the reversible reaction between α-KG and NH_3 to NAD^+ and linked oxidative deamination of

L-glutamate in two steps. The first step involves a Schiff base intermediate being formed between NH3 and α-KG. The second step involves the Schiff base intermediate being protonated due to the transfer of the hydride ion from NADH resulting in L-glutamate. NAD^+ is utilized in the forward reaction of α-KG and free NH_3 that are converted to L-glutamate via hydride transfer from NADH to glutamate. NAD^+ is utilized in the reverse reaction, involving L-glutamate being converted to α-KG and free NH_3 via oxidative deamination reaction. The electrons generated from the biochemical reactions are transferred to the CH -Fe_3O_4/ITO electrode through the Fe(III)/Fe(IV) couples that help in amplifying the electrochemical signal resulting in increased sensitivity of the sensor (Kaushik et al. 2009).

14.6.2 Pesticide Detection

Along with the existing practices, nanomaterials including carbon nanotubes (CNTs) (multi-wall CNTs and single-wall CNTs), metal nanoparticles (MNPs) [gold (Au) and silver (Ag)], graphene and graphene-based nanocomposites, nanocrystal coordination polymers, quantum dots (QDs) and magnetic NPs have had the opportunity to be used in various OPs bio-detection systems (Laschi et al. 2008).

14.7 Conclusion

Nanotechnology seems to play important role in development of intelligent delivery of agrochemicals, development of noel superabsorbent polymers, real time monitoring of agrochemicals *via* nanobiosensors, increased resilience and resilience of microbes under heat stress which are very relevant in changing climate scenario.

14.8 Future Prospects

Nanotechnological developments in agricultural sciences are still in infancy. Long term evaluation of nanoproducts (fertilizers, pesticides, nanogels, biosensors and heat stable enzymes) under various agroclimatic conditions, diverse soil types and crop ecologies would give specific solutions for tacking ever changing agricultural production constraints.

References

IPCC. (2014). Climate Change 2014: Synthesis Report. Contribution of Working Groups I, II and III to the Fifth Assessment Report of the Intergovernmental Panel on Climate Change [Core Writing Team, R.K. Pachauri and L.A. Meyer (eds.)]. IPCC, Geneva, Switzerland, p 151.

Kaushik, A., Solanki, P. R., Ansari, A. A., Sumana, G., & Ahmad, S. (2009). Iron oxide-chitosan nanobiocomposite for urea sensor. *Sensors and Actuators B Chemical, 138*, 572–580.

Kumar, A. (2016). Microbial resistance and resilience of balanced fertililized and nanomaterials added soils under heat stress. M.Sc thesis submitted to Department of Soil Science and Agricultural Chemistry, Bihar Agricultural University, Sabour, Bhagalpur, Bihar, India.

Laschi, S., Bulukin, E., Palchetti, I., Cristea, C., & Mascini, M. (2008). Disposable electrodes modified with multi-wall carbon nanotubes for biosensor applications. *Irbm, 29*, 202–207.

Mandal, N., Datta, S. C., & Manjaiah, K. M. (2015). Synthesis, characterization and controlled release study of Zn from Zincated nanoclay polymer composites (ZNCPCs) in relation to equilibrium water absorbency under Zn deficient Typic Haplustept. *Annals of Plant and Soil Research, 17*, 187–195.

Mandal, N., Datta, S. C., Manjaiah, K. M., Dwivedi, B. S., Kumar, R., & Aggarwal, P. (2018). Zincated Nanoclay polymer composites (ZNCPCs): Synthesis, characterization, biodegradation and controlled release behaviour in soil. *Polymer-Plastics Technology and Engineering*, 1. https://doi.org/10.1080/03602559.2017.1422268.

Marx, S., Zaltsman, A., Turyan, I., & Mandler, D. (2004). Parathion sensor based on molecularly imprinted sol-gel films. *Annals of Chemistry, 76*, 120–126.

Sarkar, S., Datta, S. C., & Biswas, D. R. (2014). Synthesis and characterization of Nanoclay–polymer composites from soil clay with respect to their water-holding capacities and nutrient-release behavior. *Journal of Applied Polymer Science, 131*. https://doi.org/10.1002/app.39951.

Singh, A., Sarkar, D. J., Singh, A. K., Parsad, R., Kumar, A., Parmar, B. S., & Singh, B. S. (2011). Studies on novel nanosuperabsorbent composites: Swelling behavior in different environments and effect on water absorption and retention properties of sandy loam soil and soil-less medium. *Journal of Applied Polymer Science, 120*, 1448–1458.

Zhang, B., Horn, R., & Hallett, P. D. (2005). Mechanical resilience of degraded soil amended with organic matter. *Soil Science Society of America Journal, 69*, 864–871.

Zhu, J., Huang, Q., Pigna, M., & Violante, A. (2010). Immobilization of acid phosphatase on uncalcined and calcined mg/Al-CO_3 layered double hydroxides. *Colloids and Surfaces B: Biointerfaces, 77*, 166–173.

Chapter 15
Harnessing Under-utilized Crop Species-A Promising way towards Sustainability

Madhumita, S. Sheraz Mahdi, Suborna Roy Coudhary, and Aziz Mujtaba Aezum

Abstract Agriculture is reeling under intense pressure to constantly produce increased quantities of food, feed and biofuel out of limited land resources. Present over-reliance on a handful of major staple crops has inherent agronomic, ecological, nutritional and economic risks and is probably unsustainable in the long run. Modern agricultural systems that promote cultivation of a very limited number of crop species have downgraded indigenous crops to the status of neglected and under-utilized crop species (NUCS). NUCS are indispensable in reducing food and nutrition insecurity, owing to their wider resilience to climate variability and inherent nutritional composition. Currently underutilized food sources ranging from minor grains and pulses, root and tuber crops and fruits and vegetables to non-timber forest products have the potential to make a substantial contribution to food and nutrition security, to protect against internal and external market disruptions and climate uncertainties, and lead to better ecosystem functions and services, thus enhancing sustainability. The integration of these species diversifies agricultural system and makes it much more resilient as well as strengthens its adaptation, mitigation and coping mechanisms. Most of the these crops do not require high inputs and can be successfully grown in marginal, degraded and wastelands with minimal inputs and at the same time can contribute to increased agricultural production, enhanced crop diversification and improved environment and have the potential to contribute useful genes to breed better varieties capable of withstanding and sustain the climate change scenario. However, what is required to promote NUCS is scientific research including agronomy, breeding, post-harvest handling and value addition, and linking farmers to markets. The paper largely emphasizes on –the potential

Madhumita (✉)
Department of Extension, Bihar Agricultural University, Sabour, Bhagalpur, Bihar, India

S. Sheraz Mahdi
Mountain Research Centre for Field Crops, Sher-e-Kashmir University of Agricultural Sciences and Technology of Kashmir, Khudwani, Anantnag, Kashmir, J & K, India

S. R. Coudhary
Department of Agronomy, Bihar Agricultural University, Sabour, Bhagalpur, Bihar, India

A. M. Aezum
Department of Soil Science, SKUAST-K, Shalimar, Srinagar, India

S. Sheraz Mahdi (ed.), *Climate Change and Agriculture in India: Impact and Adaptation*, https://doi.org/10.1007/978-3-319-90086-5_15

of neglected and under-utilized crops in the present context owing to global menace of climate change and raised concerns of food and nutritional security for growing population, viable solutions and recommendations to promote its conservation as well as effective use in mainstream agriculture.

Keywords Neglected and under-utilized crops (NUCS) · Agro-diversity · Sustainability · Resilience · Food and nutritional security

15.1 The Growing Need

The unprecedented rise in population has engendered various challenges, most pertinent among them being food security and climate change. Agriculture is reeling under intense pressure to produce greater quantities of food, feed and biofuel on limited land resources for the projected nine billion people on the planet by 2050. It is envisioned that agricultural production has to increase by 70% by 2050 to cope with an estimated 40% increase in world population (Ebert 2014).

While about 7000 plant species are found useful in agriculture, only about 150 species among them are largely used and less than 30 plant species meet about 90 per cent of world's food requirement. The more recent intensification of agricultural research, production and associated policy support at the national and global levels had been narrowing the species base with emphasis only on a few of them belonging to cereal and other crop groups, while many species are left out of priority. Such shrinking species content in the food basket is a matter of major concern (Ravi et al. 2010a, b). Staple crops have to face the pressing need of producing sufficient food for the increasing population. It is thus strongly demanded that the diverse agricultural resources are tapped of their potential and so the burden on major crops is released off. This diversification away from over-dependency on staple crops will be significant as part of the progress towards the goal of achieving security of food production (Thakur 2014). Dependence on a handful of major crops has inherent agronomic, ecological, nutritional and economic risks and is probably unsustainable in the long run, especially in view of global climate change. It is now generally accepted that climate change will have a major impact on both biotic and abiotic stresses in agricultural production systems and threaten yield and crop sustainability. Greater diversity, which builds spatial and temporal heterogeneity into the cropping system, will enhance resilience to abiotic and biotic stresses (Ebert 2014).

The underutilized plant species of economic importance are the key to sustainable agriculture in most of the developing countries facing resource constraints as well as rapid depletion of natural resources due to ever-increasing population pressure. From past UUC's have continuously contributed for the subsistence and economy of poor people throughout the developing countries. Despite their potential for dietary diversification and the provision of micro-nutrients such as vitamins and minerals, they still continue to attract little research and development attention (Thakur 2014).

15.2 Underutilized Crops/ Neglected and Underutilized Crop Species

Underutilized species, refers to lesser known species in terms of marketing and research but do have ability to survive in marginal or stress conditions. They can be defined as crops that have not been previously classified as major crops, have previously been under-researched, currently occupy low levels of utilization and are mainly confined to smallholder farming areas (Chivenge et al. 2015). These crops belonging to categories such as cereals and pseudo cereals, legumes, vegetables, oilseeds, roots and tubers, aromatic and medicinal plants, fruits and nuts, have earned collective names such as 'neglected and underutilized' or 'forgotten', 'orphan', 'minor' crops (Padulosi and Hoeschle-Zeledon 2008).These species hold the potential to improve people's livelihoods, as well as food security, but their potential remains largely unrealized or unrecognized due to their limited competitiveness with commodity crops in mainstream agriculture. While they face under realization of potential on a national level, but are of significant importance locally, being highly adapted to marginal, complex, and difficult environments and contributing significantly to diversification and resilience of agro ecosystems. This manifests their significance in future adaptation of agriculture to climate change (Padulosi et al. 2011).

Underutilized species include not just food plants but also many other species—wild or cultivated—used as sources of oil, fuel, fiber, fodder, beverages, stimulants, narcotics, ornamental, aromatic compounds, and medicine. To be considered as an 'underutilized food crop', a plant must have the following features:

- Crop must have a scientific or ethno botanical proof of food value.
- Crop must have been cultivated, either in the past, or only being cultivated in a specific geographical area,
- It must be currently cultivated less than other conventional crops,
- Crop must have weak or no formal seed supply system,
- Crops are recognized to have indigenous uses in localized areas,
- Received little attention from research, extension services, farmers, policy and decision makers and technology providers,
- May be highly nutritious and/or have therapeutic medicinal or therapeutic properties or other multiple uses (Thakur 2014).

Neglected or underutilized crops have the potential to play a number of roles in the improvement of food security in India that include being:

- Part of a focused effort to help the poor for subsistence and income,
- A way to reduce the risk of over-dependency on very limited numbers of major staple food crops,
- A way to increase sustainability of agriculture through a reduction in inputs,
- Increase the food quality;
- A way to preserve and celebrate cultural and dietary diversity,

- A way to use marginal and wastelands for agricultural purposes to meet the ever increasing food demand.

In addition, under- utilized crops are also seen as offering economic advantages due to their uniqueness, suitability to environments in which they are grown and low input requirements (Mabhaudhi et al. 2016).

Most of the under- utilized crops have numerous potentialities within them which could be significantly useful to mankind. Uses of few of those have been done in Table 15.1.

15.3 Processed Products from Under-utilized Crops

Various processed products such as canned jackfruit bulbs in syrup, squash, raw jack pickle, roasted jack seeds, jack seed flour and candied jackfruit have been prepared from jack fruit (Chadha and Pareek 1988; Chandra and Prakash 2009). Various processed products such as nectar, suash, slab, toffee powder, etc. can be made with Bael pulp. Ber can be processed to prepare murabba, candy, dehydrated ber, pulp, jam and ready to serve beverage (Khurdiya 1980; Pareek 2001). Jamun fruits can be processed to prepare excellent quality fermented and non- fermented beverages. Besides that good quality jelly, jam, leather can be prepared. The seeds can be processed into powder which is very useful to cure diabetes (Khurdiya 2001a, b) (Fig. 15.1).

15.4 Their Inherent Potential Owing to the Global Menace –Climate Change

There is a large number of plant resources which holds promise to humanity in terms of nutrition or agricultural yield even in harsh or adverse conditions to which main or commercial crops succumb. Amaranth, cucurbits, and water spinach (Ipomoea aquatica) are some of the few crop choices under such extreme conditions (Kuo et al. 1992, Wang et al. 2012). Water spinach proved to be heat tolerant and amaranth moderately heat tolerant, whereas majority of vegetable crops are either heat sensitive or only slightly heat tolerant (Kuo et al. 1992). As a C4-cycle plant, amaranth can sustain high photosynthetic activity and water use efficiency under high temperatures and high radiation intensity, making it an ideal crop for abiotic stress conditions under changing climates (Wang et al. 2012). Amaranth is a very nutritious leafy vegetable, both in raw and cooked form. The nutritional value of this crop is comparable to spinach, but much higher than cabbage and Chinese cabbage. Amaranth is increasingly gaining importance both for household consumption and commercial production in Africa and Asia. There is a good market potential for this crop, both in the high-price and low-price segments.

Table 15.1 Some Under-utilized crops and their uses

S. No	Name of the crop	Family	Common Names	Uses	References
1.	Bael	Rutaceae	Bel, Bael, belli, wood apple, golden apple	Pulp used in diarrhea, dysentery and other stomach ailments; marmelosin extracted from fruits have therapeutic properties, trifoliate leaves used in puja/prayer of Lord Shiva, treatment of digestive and gastrointestinal disorders, digestion, respiratory infections, scurvy, curing peptic ulcer, diabetes, chronic inflammation, snake bites.	Chadha and Pareek 1988;Ved 1991; Patnaik et al. 1996; Mazumdar 2004; Bael fruit 2011; Kumari et al. 2011
2.	Artocarpus heterophyllus	Moraceae	Jackfruit, Kathal	Fruit contains isoflavones, antioxidants and phytonutrients all of which are credited for their cancer-fighting properties, anti-ulcer properties, and is also good for those suffering from indigestion; anti-ageing properties, treatment of a number of skin problems.	Chadha and Pareek 1988; Parimala 2007; Patti 2010
3.	Averrhoa carambola	Oxiladaceae	Carambola, star fruit	Rich in anti-oxidants, potassium and vitamin C; low in sugar, sodium and acid.It is a potent source of both primary and secondary polyphenolic antioxidants.	Ved 1991
4.	Carissa spp.	Apocynaceae	Karonda, Karmada, Karvanda	Curing anemia and as astringent, anti-scorbutic and as a remedy for biliousness; anticonvulsant; cardiotonic; antioxidant, hepatoprotective; antiviral and antibacterial	Vohra and De, 1963; Jigna et al. 2005; Devmurari et al. 2009; Kumari et al. 2011.
5.	Grewia subequinalis	Tiliaceae	Phalsa	Unripe fruits are said to remove vata, kapha and biliousness; astringent properties and used for several stomach ailments	Chadha and Pareek 1988; Ali and Rab 2000
6.	Millets (Pennisetum, Eleusine, Setaria, Panicum, Paspalum)	Poaceae	Pearl, Thinai, Varagu, finger, sorghum and Jowar etc.	These tiny grain is gluten- free and packed with vitamins and minerals; act as prebiotic, rich in Ca, P, Mg, Mn, tryptophan, fibre, vitamin B group; antioxidant, antidiabetic	Ravi 2004; Gruere et al. 2007; Upadhyay 2009; Ravi et al. 2010a, b

(continued)

Table 15.1 (continued)

S. No	Name of the crop	Family	Common Names	Uses	References
7.	Simmondsiachinensis schneider	Simmondsiaceae	Jojoba	Cosmetics purposes, treat sores, cure stomach problems and restore hair	Bhatnagar et al. 1991
8.	Zizyphus mauritiana	Rhamnaceae	Ber, Indian jujube, Indian plum, desert apple	Rich source of calcium, phosphorus, protein, minerals, vitamin C and A. Seeds and bark cure for dysentery and boils and fruit as laxative and aphrodisiac; fruits are applied on cuts and ulcers; are employed in pulmonary ailments and fevers; and, mixed with salt and peppers, are given in indigestion and biliousness.	Jawanda and Bal 1978; Chadha and Pareek 1988; FACT 1998; Ved 1991; Kumari et al. 2011.
9.	Syzium cumini	Myrtaceae	Jamun, jambula, black plum	Antioxidant activity, stomachic, carminative, antiscorbutic and diuretic, antimicrobial properties.	Chadha and Pareek 1988;Ved 1991; Luximon- Koley et al. 2011
10.	Tamarindus indica	Fabaceae	Tamarind	Culinary use, antimicrobial, antidiabetic	Chadha and Pareek 1988;Ved 1991; Ali and Rab 2000; Maiti et al. 2004; Doughari 2006
11.	Zizyphus mauritiana	Rhamnaceae	Ber, Indian jujube, Indian plum, desert apple	Rich source of calcium, phosphorus, protein, minerals, vitamin C and A. Seeds and bark cure for dysentery and boils and fruit as laxative and aphrodisiac; fruits are applied on cuts and ulcers; are employed in pulmonary ailments and fevers; and, mixed with salt and peppers, are given in indigestion and biliousness.	Jawanda and Bal 1978; Chadha and Pareek 1988; FACT 1998; Ved 1991;

Source (Thakur 2014)

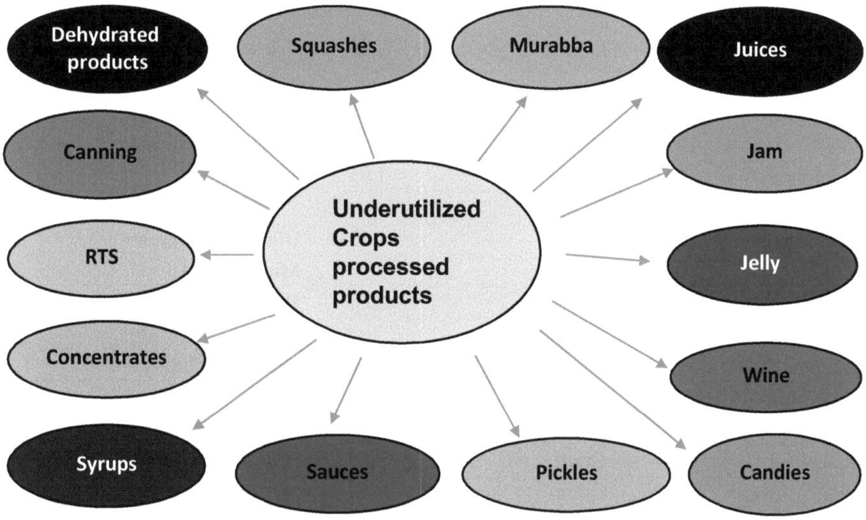

Fig. 15.1 Various processed food products from underutilized fruits (Source (Thakur 2014))

15.5 NUCS "Wonder Plants": Indian Perspective

Southern part of Rajasthan is predominantly a tribal dominated area having harsh climate, hence, only indigenous vegetables (IVs) which are hardy, drought resistant and have short duration grow well. Some of them namely kachari (Cucumis melo var. agrestis), snap melon (Cucumis melo var. momordica Duthie and Fuller), spine gourd (Momordica dioica Roxb. Ex Wild.), bitter melon (M. balsamina L.) and hill colocynth [Cucumis hardwickii (Royle) Gabaev, grow naturally during rainy season and generate good source of income for the tribals. These vegetables possess very good nutritive and medicinal value with resistance to biotic and abiotic stresses but till now no systematic efforts have been made to improve the existing land races of these IVs (Maurya et al. 2006).

India is endowed with a number of excellent crops. First crop in this list is a hardy tree- moringa (*Moringa oleifera*), the "wonder tree," which as well as its drought-resistance trait also has leaves of high nutritional content. Likewise, species from India including custard apple (*Annona squamosa*), Indian gooseberry (*Emblica officinalis*), ber (*Zizyphus mauritania*), tamarind (*Tamarindus indica*), and neem (*Azadirachta indica*) are also well recognized for their drought tolerance and ability to thrive in poor soils and marginal lands (Hegde 2009). A globally renowned hardy and multipurpose tree species known for its drought resistance is prosopis (Prosopis spp.), a reliable crop for both human consumption and animal feed in difficult areas (Pasiecznik et al. 2001). A good example is that of bambara groundnut (*Vigna subterranea*), a nutritious legume originating from west Africa and cultivated throughout sub-Saharan Africa (Heller et al. 1997). This legume, known for its drought tolerance (Andika et al. 2008), is found growing in harsh climates and

Table 15.2 List of improved varieties released in different underutilized plants in India

Crop/Variety	Year	Economic product	Yield	Recommended areas/regions
Amaranth				
Annapurna	1986	Grain	2.25	Northwest hills
GA-1	1991	Grain	2.50	Gujarat, Maharashtra
Suvarna	1994	Grain	1.95	Karnataka
Buckwheat				
Himpriya	1991	Grain	1.50	High-altitude region
VL-7	1992	Grain	1.00	Mid-hills of UP
PRB-1	1997	Grain	2.00	Hills
Winged bean				
AKWB-1	1991	Green pods	10.50	All winged bean areas
Rice-bean				
RBL-1	1987	Grain	1.50	Punjab state
RBL-6	1991	Grain	1.80	NW and NE regions
PRR-2	1997	Grain	1.50	North-west hills
Faba- bean				
VH 82–1	1994	Grain	4.20	Northern plains
Jojoba				
EC 33198	1986	Oil	4.20	Arid regions and coastal areas
Guayule				
Arizona-2	1986	Rubber	1.35	Arid and semi-arid regions
HG-8	1991	Rubber	1.50	Arid and semi-arid regions

Source (Joshi et al. 2002)

marginal soils (Heller et al. 1997); but in spite of these traits the crop still suffers from a status of neglect because of its unpredictability in yields, long cooking time, and negative social image (Mayes et al. 2009). Other underutilized crops known for their drought tolerance are the minor millets, a category of several "coarse" cereals used particularly in South Asia whose drought-resistant traits coupled with an excellent nutritious profile offer major opportunities for the development of areas increasingly affected by water shortages such as those in the marginal hills of Tamil Nadu or Karnataka States of India (Bala Ravi 2004; Padulosi et al. 2009).

Among perennial species, a good example is that of the sea buckthorn (*Hippophae rhamnoides*), a species naturally distributed from Europe to Central Asia and China, which has been found to be more tolerant to abiotic stresses than apple and pear—tolerance which seems associated with its high levels in ascorbic acid and myo-inositol (Kamayama et al. 2009).

Also, there area number of improved varieties released in underutilized crops like amaranth, buckwheat, winged bean, rice bean, faba bean, jojoba and guayule which have been described in the Table 15.2.

15.6 Advantages of Crop Diversity

Pest Suppression
It is a persistent challenge with the farmers which consumes a voluminous amount of cost of cultivation. This very challenge will further aggravate due to impacts of climate change. The raised temperatures, higher humidity due to heat and other phonological changes have conducive effect on pest proliferation. This abundance will be accompanied by higher rates of population development, growth, migration, and overwintering. Diversity in crops helps break the pest cycle or pest –crop association that becomes integral in monocropping type situations. Also, with greater plant species richness and diversity in spatial and temporal distribution of crops, diversified agro ecosystems mimic more natural systems and are therefore able to maintain a greater diversity of animal species, many of which are natural enemies of crop pests (Lin 2011).

Disease Suppression
Increasing diversification of cereal cropping systems by alternating crops, such as oilseed, pulse, and forage crops, is another option for managing plant disease risk. Disease cycles get interrupted through crop rotation by interchanging cereal crops with broadleaf crops that are not susceptible to the same diseases. Variety in plants as well as greater temporal and spatial diversity in agricultural systems hinders the disease infestation Reduced tillage could enhance soil biodiversity, leading to greater disease suppression, and stand densities could be adjusted to allow for better microclimatic adjustments to disease growth.

Climate Variability Buffering and Mitigation
Agricultural vulnerabilities have been found in a number of important crop species. Temperature maximums and minimums, as well as seasonal shifts, can have large effects on crop growth and production. Research has shown that crop yields are quite susceptible to changes in temperature and precipitation, especially during flower and fruit development stages. Here, the importance of diversified agro ecosystems comes into picture, as complex systems help mitigate the effects of such fluctuations on crop production.

15.7 Their Relevance in the Present Context-

Under the overarching goals of food security, poverty elimination and environmental sustainability, underutilized species should be selected on the basis of their capacity to best address such challenges:

Food Security: Attention should be paid to both quantity and quality of food.

Underutilized species offer untapped potentials to contribute to fight malnutrition. Their enhanced use can bring about better nutrition (vitamin C in the fruit of the Barbadoscherry -Malpighia glabra- is more than ten times higher than in the kiwi fruit–notably very rich in this micro nutrient; nutritional value of the Himalayan chenopod grains, Chenopodium spp., is superior to that of most major cereals). Emphasis should thus be given to those species having comparative advantages in providing better food, being affordable by the poor and more available both in time and space (www.fao.org/docs/eims/upload/207051/gfar0089.pdf).

Nutrient Security: Food and Agriculture Organization (FAO) statistics reveal (Swaminathan 1999) that while about 800 million children, women and men are currently suffering from protein-calorie under nutrition over 2 billion suffer from hidden hunger and there is a high frequency of low birth-weight children caused by the deficiency of micronutrients in the diet, particularly iron. Such micronutrients are in plenty in Panicum miliaceum (proso millet), Paspalum scrobiculatum (kodo millet), Chenopodium (chenopod), Amaranthus (amaranth), Fagopyrum (buckwheat) and so on. These underutilized plants can help to make diets more balanced and hence can play an important role in combating silent hunger (Joshi et al. 2002). Many vegetables – indigenous vegetables in particular – have high levels of micronutrients and could significantly contribute to nutritional security if eaten as part of the daily diet.

Poverty Elimination: Multiple uses offer greater opportunities to raise income of local people by diversifying valuable plant products. The greater the number of uses, the greater the chances to strengthen local markets and contribute to improve well- being of people. In terms of numbers, the recorded 3000 vascular species of economic importance are part of a much larger diversity basket, largely unexplored by R&D. As for figures on income generation, it is estimated that the use of minor forest products in India employs as a whole more than 10 million people per year.

Environmental Sustainability: Underutilized species have recognized ability to grow in marginal areas. Selection criteria should thus take into consideration their comparative advantages in halting soil erosion, contribute to land rehabilitation, ability to withstand difficult soils (excess of salt, lack of water, etc..), contribute to maintain balanced ecosystems and ability to tolerate heat, cold and other abiotic stresses (www.fao.org/docs/eims/upload/207051/gfar0089.pdf).

Acting as Crop Wild Relatives (CWR): Crop Wild relatives (CWR) as gene donors for plant breeding have been a major contributor to economic development and food security. With the accelerated rate of change predicted for future climate and recognition of the need to find quick solutions to expect increases in abiotic and biotic stresses, it is expected that the demand for such genetic traits will also rise significantly.

Resilience to Climatic Variability's: Resilience has been defined as the capacity of a system to absorb shock while maintaining function (Resilience Alliance 2008). Thus, a resilient agro ecosystem will continue to provide a vital service such as food production if challenged by severe drought or by a large reduction in rainfall. In agricultural systems, crop biodiversity may provide the link between stress and resilience because a diversity of organisms is required for ecosystems to function and provide services (Heal 2000).

The other way in which underutilized species help agriculture to adapt to climate change is through their contribution in enhancing the diversification and resilience of agro ecosystems in order to withstand the impacts of climate change scenarios (e.g., drought and increased frequency and intensity of extreme weather events such as cyclones and hurricanes) (Padulosi et al. 2011).

15.8 Advantages of UUC's

The benefits of these underutilized plant species are manifold:

- They have potential to contribute to poverty elimination through employment opportunities and income generation and also through improved efficiency and profitability of farm household labour use in both rural and urban environments.
- With the use of underutilized crops, there is a way to reduce the risk of over-reliance on very limited number of major crops.
- They can contribute to sustainable livelihoods through household food security as they can widen the food edibility options.
- They add nutrients to the diet and are sometimes convenience food for low income urban people. They are adapted to fragile environments and can contribute to the stability of agro ecosystems, particularly in the arid, semi-arid lands, mountains, steppes and tropical forests.
- They provide a broad spectrum of crops to improve productivity and global food security and to meet new market demands.
- They assist development of rural community through small-scale investment.
- They have a strong cultural and sacred identify and are associated with traditional customs and beliefs. Therefore, a best way to preserve and celebrate cultural and dietary diversity. (Thakur 2014)

15.9 Constraints in Utilization and Marketing of UUC's

Overall, the slow progress and poor popularity in the effective development and utilization of underutilized crops results from a number of constraints which are summarized below:

- Lack of information on production, nutritional quality, consumption and utilization of many of the underutilized plant products which are unpopular compared to major fruits.
- Lack of awareness on economic benefits and market opportunities.
- Lack of technology for value addition through village level food processing.
- Lack of improved quality planting material.
- Lack of technology to reduce the gestation period and enhance the fruit production.
- Lack of interest by researchers, agriculturists and extension workers.
- Lack of producer interest.
- Low yield.
- Post-harvest and transport losses.
- Non-existence of marketing network and infrastructure facility for underutilized fruits.
- Lack of national policy.
- Lack of credit and investment.
- Non-availability of scientific resources for testing, valuation and post- harvest management of different underutilized fruits.
- Disorganized communities.

15.10 Global Initiatives/Organizations towards their Conservation

AVRDC (Asian Vegetable Research and Development Center (AVRDC)
The World Vegetable Center, previously called Asian Vegetable Research and Development Center (AVRDC): Its gene bank is a major source of germplasm for breeding abiotic stress tolerant vegetable crops. Heat-tolerant tomato, pepper, Chinese cabbage, and mungbean varieties developed by AVRDC have enabled increased production of these crops in tropical lowlands. Indigenous vegetables tolerant to degraded, drought-prone, or saline areas provide great potential to combat malnutrition and mitigate the risks that climate change poses to farmers in developing countries.

International Centre for Underutilized Crops (ICUC)
This is a research, development and training organization. It provides expertise and acts as a knowledge hub and supported research on national priorities for germplasm collections, agronomy and post- harvest methodology of underutilized species and associated scientific conferences and training events.

Global Facilitation Unit (GFU)
The GFU is a multi-institutional initiative that acts globally to promote a wider use of underutilized plant species through supporting and facilitating the work of other stakeholders.

Crops for the Future (CFF)
has been an independent, international organization that works with its partners and has a mandate to promote and facilitate the greater use of neglected and underutilized crops to advocate research, policies, and capacity building on underutilized crops for the diversification of agricultural systems and diets (Thakur 2014).

15.11 Conclusion

Owing to the present as well as the future food and other agricultural demands combined with the alarming menace of climate change to global agriculture, it is high time that we realize the importance of under-utilized crops. There is an urgent need to promote/revive indigenous crop varieties and reverse the loss of agro-biodiversity caused due to market drivers. The benefits or uses they offer is no less than wonder or treasure for nations. Many of these wonder plants were once more widely grown but are today falling into disuse for a variety of agronomic, genetic, economic and cultural factors. Farmers and consumers are using these crops less because they are in some way not competitive with other crop species in the same agricultural environment. The general decline of these crops may erode the genetic base and prevent the use of distinctive useful traits in crop adaptation and improvement. So, the factors or reasons that limit their full use must be identified as well as addressed adequately. This will actualize when there will be optimum research as well as promotion of these "jewel" crops as they truly hold greater promise for entire humanity and are awaiting to be explored.

References

Ali, R. & Rab, F. (2000). Research needs and new products development from under-utilized tropical crops. Acta Hort. (ISHS) 518:241–248. Retrieved from http;//www.actahort.org/books/518/518_33.htm. Accessed on 10 April 2014.

Andika, D. O., Onyango, M. O. A., & Onyango, J. C. (2008). Role of Bambara groundnut (*Vigna subterranea*) in cropping systems in western Kenya. In J. Smartt & N. Haq (Eds.), *New crops and uses: Their role in a rapidly changing world, Centre for Underutilized Crops*. Southampton: University of Southampton.

Bael Fruit. (2011). Bael Fruit –Medicinal properties and health benefits. Retrieved from blog.onlineherbs.com/bael-fruit-medicinal- properties-an…-United States. Accessed on 11 March 2014.

Bhatnagar, N., Bhandari, D. C., Dwivedi, N. K., & Rana, R. S. (1991). Performance and potential of jojoba in the Indian arid regions. *Indian Journal of Plant Genetic Resources, 4*(2), 57–66.

Chadha, K. L., & Pareek, O. P. (1988). Genetic Resources of Fruit Crops: Achievements and Gaps. *Indian Journal of Plant Genetic Resources, 1*(1and2), 43–48.

Chandra, D.S. & Prakash, J. (2009). Minor fruits: a livelihood opportunity for the tribal peoples of Tripura. *IInd International Symposium on pomegranate and minor including Mediterranean fruits*, ISPMMF 2009.

Chivenge, P., Mabhaudhi, T., Modi, A. T., & Mafongoya, P. (2015). The potential role of neglected and underutilized crop species as future crops under water scarce conditions in sub-Saharan Africa. *International Journal of Environmental Research and Public Health, 12*(6), 5685–5711.

Devmurari, V., Shivanand, P., Goyani, M. B., Vaghani, S., & Jivani, N. P. (2009). A review: Carissa congesta: Phytochemical constituents, traditional use and pharmacological properties. *Pharmacognosy Reviews, 3*(6), 375.

Doughari, J. H. (2006). Antimicrobial activity of Tamarindus indica Linn. *Tropical Journal of Pharmaceutical Research, 5*(2), 597–603.

Ebert, A. W. (2014). Potential of underutilized traditional vegetables and legume crops to contribute to food and nutritional security, income and more sustainable production systems. *Sustainability, 6*(1), 319–335.

FACT. (1998). Ziziphus mauritiana – a valuable tree for arid and semi-arid lands. Retrieved from: http://www.winrock.org/fnrm/factnet/factpub/FACTSH/ziziphus.htm. Accessed on 29 June 2012.

Gruere, G.P., Nagarajan, L., King, E.D.I. & Oliver. (2007). Collective action and marketing of underutilized plant species. International Food Policy Research Institute (IFPRI). Series No. 69. Retrieved from: http://www.ifpri.org/publication/collective-action-and-marketing-underutilized-plant-species. Accessed on 11 April 2014.

Heal, G. (2000). *Nature and the marketplace: Capturing the value of ecosystem services*. Washington, DC: Island Press.

Hegde, N. G. (2009). Promotion of underutilized crops for income generation and environmental sustainability. *Acta Horticulturae, 806*, 563–577. ISHS.

Heller, J., Begemann, F. L. & Mushonga, J. (1997). Promotion, conservation and use of underutilized neglected crops. Bambara groundnut. *Proceedings of the workshop on conservation and improvement of Bambara groundnut,* November 14–16, 1995.

Jawanda, J. S., & Bal, J. S. (1978). Ber-highly paying and rich in food value. *Indian Horti., 23*, 19–21.

Jigna, P., Rathish, N., & Sumitra, C. (2005). Preliminary screening of some folklore medicinal plants from preliminary screening of some folklore medicinal plants from western India for potential antimicrobial activity eastern India for potential antimicrobial activity. *The Indian Journal of Pharmacology, 37*(6), 408–409.

Joshi, V., Gautam, P. L., Mal, B., Sharma, G.D., & Kochhar, S. (2002). 33 Conservation and use of underutilized crops: An Indian perspective.

Kamayama W Ohkawa, Chiba E, Sato K et al. (2009) Nutritional component and nitrogen fixation in seabuckthorn (Hippophae rhamnoides L.). Acta Horticulturae 806: 309–322. ISHS.

Khurdiya, D. S. (1980). New beverage from dried ber (Zizyphus mauritiana Lam). *Journal of Food Science and Technology, 17*, 158.

Khurdiya, D. S. (2001a). Post harvest management of underutilized for fresh marketing. In *Winter school on exploitation of underutilized fruits* (pp. 266–274).

Khurdiya, D. S. (2001b). Post harvest management of underutilized fruits for processed products. In *Winter school on exploitation of underutilized fruits* (pp. 291–298).

Koley, T. K., Barman, K., & Asrey, R. (2011). Nutraceutical properties of jamun (Syzygium cumini L.) and its processed products. *Indian Food Industries, 30*(4), 34–37.

Kumari, P., Joshi, G. C., & Tewari, L. M. (2011). Diversity and status of ethno-medicinal plants of Almora district in Uttarakhand, India. *International Journal of Biodiversity and Conservation, 3*(7), 298–326.

Kuo, C. G., Chen, H. M., & Sun, H. C. (1992). Membrane thermostability and heat tolerance of vegetable leaves. *Adaptation of food crops to temperature and water stress*, 160–168.

Lin, B. B. (2011). Resilience in agriculture through crop diversification: Adaptive management for environmental change. *Bioscience, 61*(3), 183–193.

Mabhaudhi, T., O'Reilly, P., Walker, S., & Mwale, S. (2016). Opportunities for underutilized crops in southern Africa's post–2015 development agenda. *Sustainability, 8*(4), 302.

Maiti, R., Jana, D., Das, U. K., & Ghosh, D. (2004). Antidiabetic effect of aqueous extract of seed of Tamarindus indica in streptozotocin-induced diabetic rats. *Journal of Ethnopharmacology, 92*(1), 85–91.

Maurya, I. B., Arvindakshan, K., Sharma, S. K., & Jalwania, R. (2006, December). Status of indigenous vegetables in southern part of Rajasthan. In *I international conference on indigenous vegetables and legumes. Prospectus for fighting poverty, hunger and malnutrition* (Vol. 752, pp. 193–196).

Mayes, S., Basu, S., Murchie, E. et al. (2009). BAMLINK. Across disciplinary programme to enhance the role of Bambara groundnut (Vigna subterranea L. Verdc.) for food security in Africa and India. *Acta Horticulturae 806*: 39–47. ISHS.

Mazumdar, B. C. (2004). Minor fruit crops of India: Tropical and subtropical. Daya Books.

Padulosi, S. & Hoeschle-Zeledon, I. (2008). Crops for the future: Paths out of poverty. Strategic Plan 2009-2013, Bioversity International Regional Office for Asia, the Pacific and Oceania, Selangor, Malaysia. 16 p.

Padulosi, S., Mal, B., Bala Ravi, S., Gowda, J., Gowda, K. T. K., Shanthakumar, G., & Dutta, M. (2009). Food security and climate change: Role of plant genetic resources of minor millets. *Indian Journal of Plant Genetic Resources, 22*(1), 1.

Padulosi, S., Heywood, V., Hunter, D., & Jarvis, A. (2011). Underutilized species and climate change: current status and outlook. In *Crop adaptation to climate change* (1st ed., pp. 507–521). New York: Wiley.

Pareek, O.P. (2001). Ber. International Centre for Crops. Southampton (U.K.).

Parimala. (2007). .Medicianal uses of jack fruit. Retrieved from http://jaspari.info/2007/03/medicial-uses-of-jackfruit.html. Accessed on 26 July 2013.

Pasiecznik, N. M., Felker, P., Harris, P. J., Harsh, L., Cruz, G., Tewari, J. C., & Maldonado, L. J. (2001). *The'Prosopis Juliflora'-'Prosopis Pallida'Complex: A monograph* (Vol. 172). Coventry: HDRA.

Patti, A.K. (2010). Jackfruit (Artocarpus heterophylla). By Abhay Kumar Patti, Odisha, India, Retrieved from: http://prlog.org/books/518_31htm. Acta Hort., (ISHS) 518:233–236.

Pattnaik, S., Subramanyam, V. R., Bapaji, M., & Kole, C. R. (1996). Antibacterial and antifungal activity of aromatic constituents of essential oils. *Microbios, 89*(358), 39–46.

Ravi, B. S. (2004). Neglected millets that save the poor from starvation. *LEISA India, 6*(1), 1–8.

Ravi, S. B., Hrideek, T. K., Kumar, A. K., Prabhakaran, T. R., Mal, B., & Padulosi, S. (2010). Mobilizing neglected and underutilized crops to strengthen food security and alleviate poverty in India.

Resilience Alliance. (2008). Website: http://www.resalliance.org

Swaminathan MS. (1999). Enlarging the basis of food security: role of underutilized species. In: *Proceedings of the International Consultation organized by the Genetic Resources Policy Committee (GRPC) of the CGIAR, M.S. Swaminathan Research Foundation, Chennai, India,* 17–19 February 1999, p. 22.

Thakur, M. (2014). Underutilized food crops: Treasure for the future India. *Food Science Research Journal, 5*(2), 174–183.

Upadhyay, H.D. (2009). Sustainable conservation and utilization of genetic resources of two underutilized crops-finger millet and foxtail millet- to enhance productivity, nutrition and income in Africa and Asia. Monograph. .Retrieved from http://oar.icrisat.org/id/eprint/5199. Accessed on 12 May 2014.

Ved, P. (1991). In S. S. Samant, U. Dhar, & P. LMS (Eds.), *Indian medicinal plant: Current status in Himalayan medicinal plants: Potential and prospects* (pp. 45–63). Nainital: Gramodaya Prakashan.

Vohra, M. M., & De, N. N. (1963). Comparative cardiotonic activity of Carissa carandas L. and Carissa spinarum A. DC. *The Indian Journal of Medical Research, 51*, 937–940.

Wang, S. T., & Ebert, A. W. (2013). Breeding of leafy amaranth for adaptation to climate change. In R. Holmer, G. Linwattana, P. Nath, & J. D. H. Keatinge (Eds.), *High value vegetables in Southeast Asia: Production, supply and demand; Proceedings of the SEAVEG 2012. Regional Symposium* (pp. 36–43). Tainan/Taiwan: AVRDC – The World Vegetable Center.

Chapter 16
Weather Based Information on Risk Management in Agriculture

K. K. Singh, A. K. Baxla, Priyanka Singh, and P. K. Singh

Abstract Weather and climate information plays a major role in the entire crop cycle right from selecting the most suitable crop/variety/ field preparation up to post harvest operations and marketing; and if provided in advance can be helpful in inspiring the farmer to organize and activate their own resources in order to reap the benefits by judicious application of costly inputs. It becomes more and more important to supply meteorological information blended with weather sensitive management operations before the start of cropping season in order to adapt the agricultural system to increased weather variability. India Meteorological Department (IMD), Ministry of Earth Sciences (MoES) in collaboration with Indian Council of Agricultural Research (ICAR) and State Agriculture Universities (SAUs) is rendering weather forecast based District level Agro meteorological Advisory Services (AAS) to the farmers in the country under the scheme "Gramin Krishi Mausam Sewa (GKMS)" since monsoon 2008. AAS provides advance weather information along with crop specific agromet advisories to the farming community by using state of the art instruments and technology through efficient delivering mechanism of the information which ultimately enables farmers to take appropriate actions at farm level. This present system of delivering the services at district level is underway to extend up to sub-district/ block level with dissemination up to village level to meet the end users' requirements in both the irrigated and rainfed systems and facilitate the agriculture risk management effectively.

Keywords Weather Forecast · Agromet advisory services · Gramin Krishi Mausam Sewa · Rainfed and irrigated system

K. K. Singh (✉) · A. K. Baxla · P. Singh · P. K. Singh
Agromet Advisory Services Division, India Meteorological Department, New Delhi, India

S. Sheraz Mahdi (ed.), *Climate Change and Agriculture in India: Impact and Adaptation*, https://doi.org/10.1007/978-3-319-90086-5_16

16.1 Introduction

Agriculture production is governed by various factors out of which weather is the only factor over which human has no control and hence it has an overwhelming dominance over the success or failure of agricultural enterprise. It is an accepted fact that food production is inextricably linked with climate and weather. It is also reported that weather induced variability of food production is more than 10 per cent. This variability can be as high as 50 per cent of the normal production in respect of smaller areas situated in arid and semi-arid regions. In order to reduce risks of loss in food production due to the vagaries of weather, weather per se, should be taken into account as one of the major inputs in agricultural planning. Hence forecast of weather parameters play a vital role in agricultural production. It also aids in minimizing crop losses to a considerable extent. Thus development and refinement of the art of weather prediction has been essential since time immemorial. Therefore prediction of weather systems in different spatial and temporal scale over the Indian region assumes considerable importance. The advent of new meteorological modeling capabilities provides opportunities for the meteorological community to develop better products and information to decision-makers in the various user sectors (i.e. energy, health, agriculture, water). Both post-processed General Circulation Model (GCM) output and forecasts made by Weather Research and Forecasting (WRF) models can be used in agricultural decision making. In the recent past, IMD has made enormous improvement in the accuracy and lead time of forecasts for various usage including tactical decisions at field level agricultural applications based on medium range forecast (e.g. frost protection, irrigation and fertilization) & for strategic decisions based on longer time scales, ranging from several weeks to months. This type of climate information is provided by GCMs and seasonal ensemble predictions. The potential value for agriculture of an accurate medium range forecast is enormous. If rainfall behavior were predicted with sufficient lead time and with a high degree of confidence, farmers could for example respond to forecasts by changing crop varieties, changing crop species, implementing soil and water conservation techniques, increasing or decreasing area planted, adjusting timing of land preparation, increasing or decreasing soil inputs and selling or purchasing livestock herds (Motha 2007).Weather aberrations may be nullified to a large extent by timely communication of adoptive measures disseminated through Agromet Advisory Services (AAS) to the farming community (Chattopadhyay and Lal 2007; Rathore et al. 2009; Chattopadhyay and Rathore 2013; Rathore et al. 2013). In the present paper it has been showed that how the National Meteorological Services like India Meteorological Department in collaboration with other organizations has geared up its activities to face the challenges of such weather aberrations, particularly in providing the risk management solution to the farmers of the country.

16.2 Weather Forecast and Agromet Information

India experiences large spectra of weather events having spatial scale of less than 1 km to more than 1000 km and temporal range of less than an hour to more than a week. Different parts of the country experience different kinds of weather conditions such as Winter season (Jan-Feb) is characterized by Western Disturbances, Cold Wave, Fog; Pre-Monsoon (Mar-May) by Cyclonic Disturbances, Heat Wave, Thunder Storms, Squalls, Hail Storm, Tornado; Monsoon (Jun-Sep) by Southwest Monsoon Circulation, Monsoon Disturbances; and Post-Monsoon (Oct-Dec) by Northeast Monsoon, Cyclonic Disturbances. All these weather systems individually or together affect the crop physiology and growth severely by means of inducing stress. Growth and development of crop depends upon all the weather variables therefore prediction of these weather systems in different spatial and temporal scale is of considerable importance to predict the weather induced stresses in crops.

IMD having mandate to issue weather forecast for different time scale in advance, it provides opportunity to efficiently minimize the loss from adverse weather and took the benefit from benevolent weather.

16.2.1 Nowcast and Special Weather Forecast for Extreme Events

Nowcast having temporal resolution of 3 h to 6 h, derived products of Doppler weather Radar form a very important guiding tool for improving the nowcasting system. IMD has implemented nowcasting of thunderstorms, squalls and hailstorms for the areas covered by DWRs. Agriculture sector are also benefitted by nowcast/forecast of severe weather as time of rain fall occurrence and quantum of rain may enable farmers to plan the agriculture activities which in turn may reduce/protect from the loss of inputs, enhance its use efficiency like pesticide spray, fertilizer application and thus yield more production. Special weather forecast for agriculture provides the necessary meteorological input to assist farmers in making decisions. The requirements for these special forecasts will vary from season to season and crop to crop. Special forecast issued are as follows:

- Tropical storms (cyclones, hurricanes, typhoons, etc.) associated with high winds, flooding and storm surges.
- Floods (other than those related to tropical storms) heavy rains due to monsoons, water logging and landslides.
- Severe thunderstorms, hail storms, tornadoes and squalls.
- Drought and heat waves.
- Cold spells, low temperature, frost, snow and ice-storms.
- Dust storms and sand storms.

16.2.2 Medium Range Weather Forecast

IMD is issuing quantitative district level (646 districts) weather forecast up to 5 days and the products comprise of quantitative forecasts for 8 weather parameters viz., rainfall, maximum temperature, minimum temperatures, wind speed, wind direction, relative humidity I,II and cloudiness. This weather forecast is generally valid for a period of 5 days and prepared using the GFS-1534 at 12.5 km spatial resolution. This Medium Range Weather Forecast has been generated on Every Tuesday and Friday by NWP division of IMD and sent to RMC/MC for value addition in forecast by local expertise. Value added districtwise forecast further disseminated to 130 AgoMeteorological Field Units of IMD for Agroadvisory generation. Group of experts in agriculture discipline issues the advisory for next five days based on the forecast. Medium Range Weather forecast is of prime importance for farmers in order to take the tactical decisions. Farmers are using these advisories for sowing and transplantation of crops, fertilizer application, predictions regarding pests and diseases and measures to control them, weeding/thinning, irrigation (quantities and timing), and harvest of crops.

16.2.3 Extended Range Forecast

Long breaks in critical growth periods of agricultural crops lead to substantially reduced yield. Thus, the forecast of this active/break cycle of monsoon, commonly known as the Extended Range Forecasts (ERF) is very useful. The forecasts of precipitation on this intermediate timescale are critical for the optimization of planting and harvesting. Prediction of monsoon break 2 to 4 weeks in advance, therefore, is of great importance for agricultural planning (sowing, harvesting, etc.) and yield forecasting, which can enable tactical adjustments to the strategic decisions that are made based on the longer-lead seasonal forecasts, and also will help in timely review of the ongoing monsoon conditions for providing outlooks to farmers.

IMD has been issuing experimental extended range forecast since 2009 using available products from statistical as well as multi-model ensemble (MME) based on outputs available from dynamical models (NCEP_CFS, IITM_CFS, JMA, ECMWF etc) from various centers in India and abroad. The MME forecast is being prepared once in a week with the validity for subsequent four weeks. However model runs is made for 45 days every week. The latest generation coupled models are found to be very useful in providing skillful guidance on extended range forecast. The performance of extended range forecasts for the southwest monsoon seasons clearly captured the delay/early onset of monsoon over Kerala, active/break spells of monsoon and also withdrawal of monsoon in the real time in providing guidance for various applications. On experimental basis the MME forecast on meteorological subdivision level up to two weeks are also being used in providing the agromet advisory for farming community. During the other season the MME based ERF also provides encouraging results in case of northeast monsoon rainfall over southern peninsula and tropical cyclo-genesis over the north Indian Ocean during the post monsoon season from October to December (OND). In addition, the MME based ERF forecast also provides useful guidance pertaining to rainfall associated with Western Disturbances (WD) over northwest India during winter. The ERF forecast for minimum and maximum temperatures during winter and summer seasons are also found to be very useful.

16.2.4 Long Range Forecast

IMD has been issuing long-range forecasts (LRF/Seasonal forecast) for monsoon based on statistical methods for the southwest monsoon rainfall over India (ISMR) for more than 100 years. The forecast for the South-West monsoon rainfall is issued in two stages. The first stage forecast for the seasonal (June to September) rainfall over the country as a whole is issued in April and the update of the April forecast is issued in June. Along with the update forecast, forecast for seasonal rainfall over four broad geographical regions of India and July rainfall over country as a whole are also issued. Rainfall induced stress associated with amount and date of occurrence viz. early, mid and late drought is predicted by long range forecast. Long range forecast provides lead time for starategic planning in agriculture.

16.2.5 Customized Agromet Products for Advisory Preparation

Various products has been derived from weather forecast, satellite observation, remote sensing data, gridded product of IMD for monitoring and forecasting of soil moisture, drought and crop health. Aridity anomaly maps gives information about the moisture stress experienced by growing plant. The crop water stress condition during the monsoon anomalies are helpful for early warning of crop stress occurrence. Standard precipitation Index (SPI) (Guhathakurta et al. 2011) used for monitoring rainfall departure status from the normal and helps in monitoring of rainfall status. NOAA/AVHRR Satellite derived state wise weekly Normalized Difference Vegetation Index (NDVI) monitors the crop condition. These entire derived products guide in irrigation give advisory for irrigation and other farm activities such as mulching (Fig. 16.1).

16.2.6 Gramin Krishi Mausam Sewa

Weather and climate information plays a vital role in agriculture production to render the need of weather information to farming community efforts has been made since 1945 with the initiation of Farmers Weather Bulletin. Since then with the

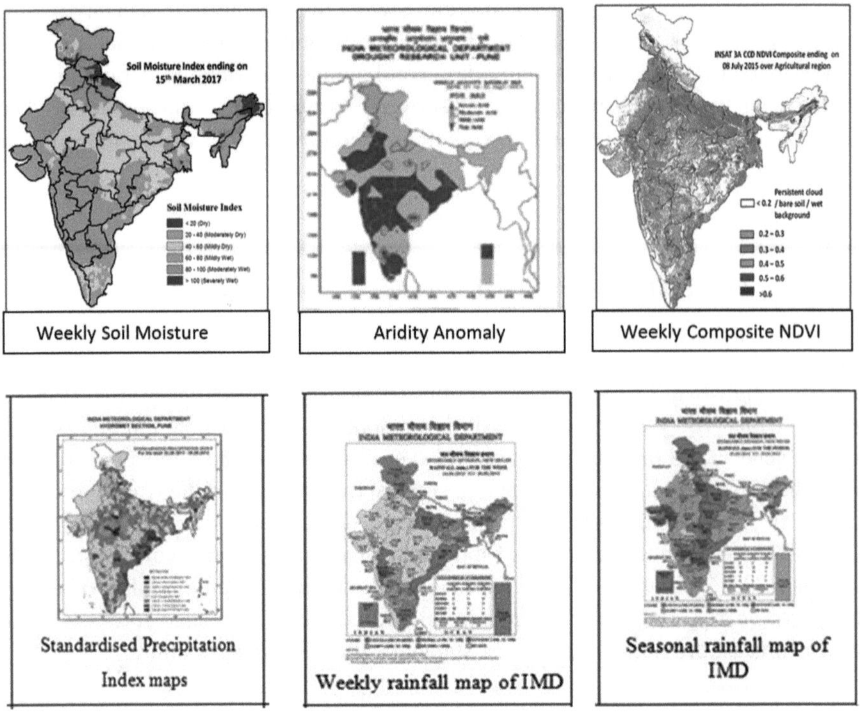

Fig. 16.1 Derived Agromet product facilitate in Agro Advisory preperation

advent of technology, extension of weather observation network and weather forecast the customized weather information in the form of advisory also follows era of evolution. In the year 1976 Agro advisory Services initiated at state level followed by Agroclimatic Zone level AAS using medium range weather forecast in the year 1991. In the year 2008 District Level agro advisories has been initiated by IMD, MoES in collaboration with ICAR and State Agriculture Universities. The main emphasis of the existing AAS system is to collect and organize climate/weather, soil and crop information, and to amalgamate them with weather forecast to assist farmers in taking management decisions. This has helped to develop and apply operational tools to manage weather related uncertainties through agro-meteorological applications for efficient agriculture in rapidly changing environments. Being a multidisciplinary and multi-institutional project, AAS project is being implemented through tiered structure to set up different components of the service spectrum. It include meteorological (weather observation & forecasting), agricultural (identifying weather sensitive stress & preparing suitable advisory using weather forecast), extension (two way communication with user viz. farmers and planners) and information dissemination (Media, Information Technology, Telecom) agencies. Under GKMS scheme weather based crop and locale-specific agro-advisories for 633 rural districts are prepared and disseminated to farmers deploying various modes of information dissemination e.g. radio, television, print media, internet, Kisan Call Centres and mobile phones. Presently 20.6 million farmers receive abridged advisories through SMS on their mobile phone.

The services at its current spatial resolution made significant contribution to reduce risk and improve agricultural productivity farm income, despite local climate variations. It also focuses on environment friendly integrated solutions that are within the farmers' capabilities. It was observed that there has been substantial increase in productivity for cereals, oilseeds and vegetable. National Council of Applied Economic Research (NCAER), an independent agency, during 2015 estimated that the economic benefit from the use of weather information as the product of the percentage of farmers receiving information, scenario-wise, times the percentage of farmers benefiting from the information times average profit, crop-wise, attributable to weather information times the total national production of crops. Conversion factors, crop-wise, were used to convert farmers' financial profits to economic profits (NCAER 2015). At present only 24 percent of the farmers are benefitting from the SMS services. The economic profit estimates Rs. 42,000 crore. Service has the potential of generating net economic benefit up to Rs. 3.3 lakh crores on the 22-principal crops when AAS is utilized by 95.4 million farming households.

Further to improve the relevance of this service at block level, high –resolution weather forecast will be utilized to develop the services. As a part of Gramin Krishi Mausam Sewa it is planned to establish 660 District Agromet Units (DAMUs) in the premises of Farm Science Centre, called Krishi Vigyan Kendra (KVKs), in each district in a phase manner. Efforts are being made to automize the process of farm advisory preparation and dissemination through Kisan portal. Service delivery at village level will be established using all the dissemination channels including DD Kisan, Kisan portal, the Ministry of IT and the Department of Electronics and IT (DeitY), the Department of Post, CSC etc. and other initiatives under Digital India Movement.

16.3 Structuring a Weather Risk Management Tools/ Solution

The emerging weather and climate risk clearly offers new risk management tools and opportunities for agriculture. Identifying the location wise risk to weather, time period during which risk is prevalent and further quantifying and designing a weather risk management strategy based on an index is more pertinent to neutralize the risk in agriculture. Following are the ingredients of a typical Agromet Advisory Bulletin to reap benefits of benevolent weather and minimize or mitigate the impacts of adverse weather;

(i) District specific weather forecast, in quantitative terms, for next 5 days for rainfall, cloud, max/min temperature, wind speed/direction and relative humidity, including forewarning of hazardous weather event likely to cause stress on standing crop and suggestions to protect the crop from them.

(ii) Weather forecast based information on soil moisture status and guidance for application of irrigation, fertilizer and herbicides etc.

(iii) The advisories on dates of sowing/planting and suitability of carrying out intercultural operations covering the entire crop spectrum from pre-sowing to post harvest to guide farmer in his day–today cultural operations.

(iv) Weather forecast based forewarning system for major pests and diseases of principal crops and advises on plant protection measures.

(v) Advisory for extreme events suggest the measures under Drought/Dry spell, extreme cold, Heat wave, hailstorm, cyclone, intense rainfall and flood.

(vi) Propagation of techniques for manipulation of crop's microclimate e.g. shading, mulching, other surface modification, shelter belt, frost protection etc. to protect crops under stressed conditions.

(vii) Advisory on contingency plan under extreme weather situations.

(viii) Reducing contribution of agricultural production system to global warming and environment degradation through judicious management of land, water and farm inputs, particularly pesticides, herbicides and fertilizers.

(ix) Advisory for livestock on health, shelter and nutrition.

Under GKMS Scheme IMD in collaboration with ICAR, State Department of Agriculture and other agencies is planning to establish an automated system of agro

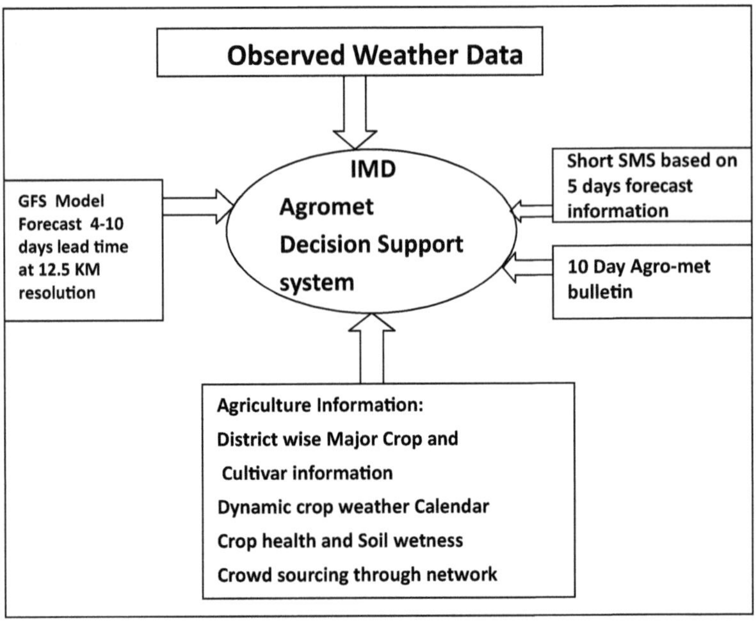

Fig. 16.2 System Overview of Agromet-Decision Support System

advisory generation at district level. An advanced system with detailed location specific information of agriculture, a dynamic crop weather calendar for major crop of the area, remote sensing and ground information based crop health and water balance based soil moisture information along with the high resolution weather forecast will make the system exhaustive and capable to generate the customized advisory for rainfed and irrigated farming system. Knowledge of field experts and the State agriculture officers will train the system for automation (Fig. 16.2).

16.4 Dissemination of Information

Dissemination of information under the aberrant weather condition plays a vital role in minimizing crop loss. The information Dissemination Process are broadly divided into three channels viz. Mass and electronic media, Group methods and Individual contacts. Presently AAS disseminated to the farmers regularly and also in case of extreme weather through Short Message Service (SMS) messaging and Interactive Voice Response Technology (IVR) in regional/English language. In a public-private partnership (PPP) arrangement, AMFUs are preparing and sending district AAS bulletins twice weekly to private companies including IFFCO Kisan Sanchar Limited (IKSL), NOKIA, Reuter Market Light (RML), Handygo, Mahindra Samriddhi comprising of weather forecasts and advisories on extreme events along with crop, pest, disease, seed and fertilizer information all over the country. AMFUs are also uploading the agromet advisories in the form of SMS in regional languages in Kisan Portal started by Ministry of Agriculture, Govt. of India and their respective web sites of Universities or National Institutes.

In general, the use of more than one channel gives a greater chance of reaching the client or user. The Agromet Advisory Services provided by IMD/MoES through various channels have resulted in significant increases in farm productivity, resulting in increased availability of food and higher income generation. There is a need for dissemination of AAS information to farmers on a wider scale and convincing them about its positive impacts on a sustainable basis.

16.5 Conclusion

The risks confronting with agriculture production are very high due to weather variability. Weather forecast translated in agro advisory helps in taking the decision at field level in advance and avoids the associated risk. Increasing number of observatories, use of advance tools and techniques, expansion of existing Agromet Field Unit network, customized Decision Support System for advisory generation and advance dissemination tools & techniques are highly efficient to support the weather smart agriculture under existing scenario. Though useful services are being provided to the farmers through the Agromet Advisory Services by using the latest state of art technology, there is an urgent need for a better understanding of the changing climate patterns and how they affect agriculture, better weather forecasting at small scale and awareness of information provided to the farmers.

References

Chattopadhyay, N., & Lal, B. (2007). *Agrometeorological risk and coping strategies-perspective from Indian subcontinent: Managing weather and climate risks in agriculture* (pp. 83–97). New York: Springer Publication.

Chattopadhyay, N., & Rathore, L. S. (2013). Extreme Events: Weather service for Indian Agriculture. *Geography and You, 13*(79), 12–16.

Guhathakurta, P., Koppar, A.L., Krishnan, U., Menon, P. (2011). New rainfall series forthe district, meteorological sub-divisions and country as whole of India, National Climate Center Research Report No. 2/2011 (www.imdpune.gov.in).

Motha, R. P. (2007). Implications of climate change on long-lead forecasting and global agriculture. *Australian Journal of Agricultural Research, 58*(10), 939–944

National Council of Applied Economic Research (NCAER). (2015). *Impact assessment and economic benefits of weather and marine services*. New Delhi: Supported by Ministry of Earth Sciences

Rathore, L. S., Chattopadhyay, N., & Singh, K. K. (2013). Reaching farming communities in India through farmer awareness programmes. In *the book of "climate exchange, World Meteorological Organisation (WMO)"* (pp. 20–23). United Kingdom, (UK): Tudor Rose Publication

Rathore, L. S., Roy Bhowmik, S. K., & Chattopadhyay, N. (2009). Integrated Agromet Advisory Services in India. In *Challenges and Opportunities in Agrometeorology*. New York: Springer Publication.

Chapter 17
Adaptation and Mitigation of Climate Change by Improving Agriculture in India

Rattan Lal

Abstract Between 1800 and 2050, the population of India would increase from 255 million to 1.71 billion, by a factor of 7, with a strong environmental impact. Rapid urbanization and its encroachment on agricultural land is a consequence of increase in population. Between 1950 and 2025, the population (106) would increase from 1.4 to 28.6 (20.4 times) of New Delhi, 4.5 to 20.1 (4.5 times) of Kolkata, 2.9 to 25.8 (8.9 times) of Mumbai, 0.6 to 6.6 (11.0 times) of Pune, 1.1 to 8.9 (8.1 times) of Hyderabad, 0.7 to 9.5 (13.6 times) of Bengaluru, and 1.5 to 9.6 (6.4 times) of Chennai. The city of Mumbai generates 11 thousand Mg of waste per day or 4 million Mg per year, which if recycled effectively, can improve urban and peri-urban agriculture. It takes about 40,000 ha of land to provide accommodation and infrastructure to 1 million people. An annual increase of 11.5 million people in India encroaches upon 0.5 million hectare (Mha) of agricultural land. Thus, there is a strong need to protect prime agricultural land against other uses. By 2025, India will have 7 cities of >10 million people, and a city of 10 million consumes 6000 Mg of food per day. Thus, nutrients brought into the city must be returned to the land by recycling waste as compost and for producing energy. Climate change, with increase in frequency of extreme events, is exacerbating vulnerability of agricultural soils to degradation processes. Land area (Mha) in India already affected by degradation includes 93.7 by water erosion, 9.5 by wind erosion, 14.3 by waterlogging, 5.9 by salinity/alkalinity, 16.0 by soil acidity and 7.4 by complex problems. In addition to the impacts of changing and uncertain climate, soil degradation is exacerbated by burning of crop residues, use of cow dung for household cooking rather than as manure, uncontrolled grazing, unbalanced use of fertilizers, and other extractive farming practices. The drought-flood syndrome, caused by water misuse and mismanagement, adversely affects agronomic productivity and wellbeing of millions of people despite the fact that India receives 4000 km^3 of annual precipitation.

A systematic understanding is needed of the coupled cycling of water, carbon, nitrogen, phosphorus and sulfur at ecoregions and watershed scale to enhance provisioning of essential ecosystem services from agroecosystem (e.g., food feed, fiber,

---

R. Lal (✉)
Carbon Sequestration Center, The Ohio State University, Columbus, OH, USA
e-mail: lal.1@osu.edu

S. Sheraz Mahdi (ed.), *Climate Change and Agriculture in India: Impact and Adaptation*, https://doi.org/10.1007/978-3-319-90086-5_17

fuel, water, biodiversity). In addition to the drought-flood syndrome, other ramifications of the mismanagement of coupled cycling include emission of greenhouse gases from agroecosytems, especially of CH_4 and N_2O with global warming potential of 21 and 310, respectively. Adaptation and mitigation of agroecosystems to climate change necessitate adoption of the strategies of sustainable intensification. The latter implies "producing more from less": more agronomic yield per unit of land area, and input of water, energy, fertilizers, pesticides and gaseous emissions. The large yield gap, difference in agronomic yield of research plots and the national average yield, can be abridged by adoption of the best management practices (BMPs). Thus, soil health must be restored by increasing soil organic carbon (SOC) concentration to the threshold level of ~1.5–2.0% in the root zone (0–40 cm depth). Soils of agroecosystems in India, similar to those of other countries in South Asia and Sub-Saharan Africa, are severely depleted of their SOC stocks. The magnitude of depletion is high in soils prone to accelerated erosion by water and wind, and other degradation processes. The SOC stock can be restored by adaptation of BMPs which control erosion and create a positive soil/ecosystem C budget. Important among these are afforestation of degraded and marginal soils, restoration and management of wetlands, use of conservation agriculture in conjunction with mulch farming/cover cropping, integrated nutrient management, and establishment of biofuel plantations on degraded lands. The SOC restored must also be stabilized/ protected to prolong its mean residence time to centennial/millennial scale. There is no one-size-fit-all BMP, and site-specific adaptation/fine-tuning is essential with due consideration of the biophysical, socio-economic and cultural (the human dimensions) factors. In addition to adaptation and mitigation of climate change, restoration of degraded soils is also essential to local, national, regional and global peace and harmony. Fertile soils of good health, rich in SOC stock, and teaming with biodiversity of intense activity are essential to advancing food and nutritional security, improving water quality and renewability and adapting and mitigating climate change. Healthy soils are the engine of economic development especially under changing and uncertain climate.

Keywords Climate change · Soil carbon · Carbon sequestration and soil health

17.1 Introduction

The population of India has and is growing at a fast rate, which (million) was 255 in 1800, 295 in 1900, and 1014 in 2000. It increased by only 40 million during the nineteenth century but by 719 million during the twentieth century. It is projected to increase by another 645 million to a total of 1.66 billion by 2050. The present population of India is 1.34 billion, and will increase to 1.51 billion by 2030, 1.66 billion by 2050, but decreases to 1.52 billion by 2100 (UN, 2017). The present population of India of 1.34 billion is 17.7% of the world population of 7.55 billion or one in 6 persons is living in India on merely 2.4% of the world's total land area.

Population of major cities (in million) of India is projected to increase between 1950 and 2025 by a factor of 20.4 for New Delhi (1.4 to 28.6), 4.5 for Kolkata (4.5 to 20.1), 8.9 for Mumbai (2.9 to 25.8), 11.0 for Pune (0.6 to 6.6), 8.1 for Hyderabad (1.1 to 8.9), 13.6 for Bangleru (0.7 to 9.5), and 6.4 for Chenai (1.5 to 9.6). An annual increase of India's population by 11.5 million, equivalent to adding the entire population of Cuba or Tunisia to that of India, may require an additional land area of 0.5 Mha to provide for infrastructure and accommodation. Therefore, the net land area available for agricultural production is shrinking because of the competing non-agricultural uses.

Agricultural land area in India is also shrinking because of soil degradation. Vulnerability to degradation is caused by land misuse and soil mismanagement. The widespread use of extractive farming practices, based on plowing and residues removal along with uncontrolled grazing, leads to the severe problems of soil degradation. Land area affected by a range of soil degradation processes in India is estimated at 146.8 Mha (Bhattacharyya et al. 2015), or 44% of the total land surface of 329 Mha. Severe problems of soil and environmental degradation, driven by anthropogenic factors, are also intricately interconnected to the changing and uncertain climate characterized by extreme events, drought-flood syndrome, and heat waves. Accelerated erosion and other degradation processes aggravate the emission of greenhouse gases (GHGs) from soils.

Notable among these gases are carbon dioxide (CO_2), methane (CH_4) and nitrous oxide (N_2O). Degraded soils are a major source of these gases because of the nitrification/de-nitrification, methanogenesis and mineralization processes. The objective of this article is to deliberate the importance of adopting best management practices (BMPs) to restore soil health, create a positive soil/ecosystem carbon (C) budget, sequester atmospheric CO_2 in soils of agro ecosystems as soil organic matter (SOM), enhance soil resilience to climate change, and improve agronomic productivity and use efficiency of inputs.

17.2 Extractive Farming Practices

The downward spiral of soil and environmental degradation is set-in-motion by the widespread use of extractive farming practices (Table 17.1). Notable among these, with adverse impacts on soil health, are excessive ploughing, residue removal and in-field burning, uncontrolled and free grazing, flood and excessive irrigation, unbalanced use of fertilizers (more N and less P and K), using animal dung for cooking rather than as manure, and scalping topsoil for brick making. Total number of bricks produced in India is 250 billion per annum, which decapitates a large area of agricultural land.

As a consequence of extractive farming, there is a severe problem of soil degradation or decline in soil health and its functionality. Notable among soil degradation processes are: (i) physical degradation such as accelerated erosion by water and wind, structural decline causing crusting and compaction, and waterlogging, (ii)

Table 17.1 Extractive farming practices which degrade soil and environment quality

	Traditional practice	Degradation processes
1.	Excessive ploughing	Erosion by water and wind, decline of SOM content, reduction in aggregation.
2.	Removal of crop residues for competing uses (e.g., fodder, fuel, construction)	Reduction in SOM content, decline in soil biodiversity, decrease in aggregation, vulnerability to crusting and compaction, nutrient mining.
3.	In-field burning of crop residues	Emission of GHGs (CO_2, N_2O, CH_4) and other compounds (CO_2, NO_x, SO_2, SO_x, NH_3, particulate matter, soot of black C (BC), reduction in SOM.
4.	Uncontrolled and free grazing	Compaction, crusting, decline in SOM.
5.	Flood irrigation with brackish water	Secondary salinization, decline in water table, solublization of arsenic in alluvial soils
6.	Using animal dung for household fuel rather than manure	Decline in SOM content, depletion of soil fertility, emission of obnoxious gases (CO_2, SO_2, NH_4, PM, soot BC), pulmonary diseases.
7.	Scalping topsoil for brick making	Depletion of soil fertility, deficiency of micronutrients in soil, food of poor nutritional quality, low use efficiency of inputs.
8.	Imbalanced use of fertilizer (more N but less P and K)	Elemental imbalance in soil, low crop yield, nutrient depletion.

chemical degradation including salinity, alkalinity, acidification, nutrient depletion, and elemental imbalance (toxicity, deficiency), (iii) biological degradation characterized by loss of soil biodiversity, decrease in microbial biomass C, depletion of SOM content to below the critical level in the root zone, and (iv) ecological degradation comprising of disruption in biogeochemical and bio geophysical cycling, and perturbation of water and energy balance etc. (Fig. 17.1) Although estimates of soil degradation in India are tentative and incomplete, as much as 146.8 Mha of land area (45% of the total area of the country) is affected by soil degradation (Bhattacharyya et al. 2015).

The biophysical processes of soil degradation are driven by social, economic and cultural factors. These interactive effects (biophysical × socioeconomic) are exacerbated by causes related to land misuse and soil mismanagement (Fig. 17.2). Therefore, risks of soil degradation by diverse processes can be minimized by judicious management of factors and causes through adoption of BMPs.

17.3 Depletion of Soil Organic Carbon and Environmental Pollution by Soil Degradation

Soils of agroecosystems of India are strongly depleted of their soil organic carbon (SOC) concentration and stocks. Whereas the threshold level of SOC concentration in the root zone ranges from 1.1 to 1.5%, the actual levels in soils of the Indo-Gangetic Plains (IGPs) may be as low as 0.1% or less. Consequently, crop yields are

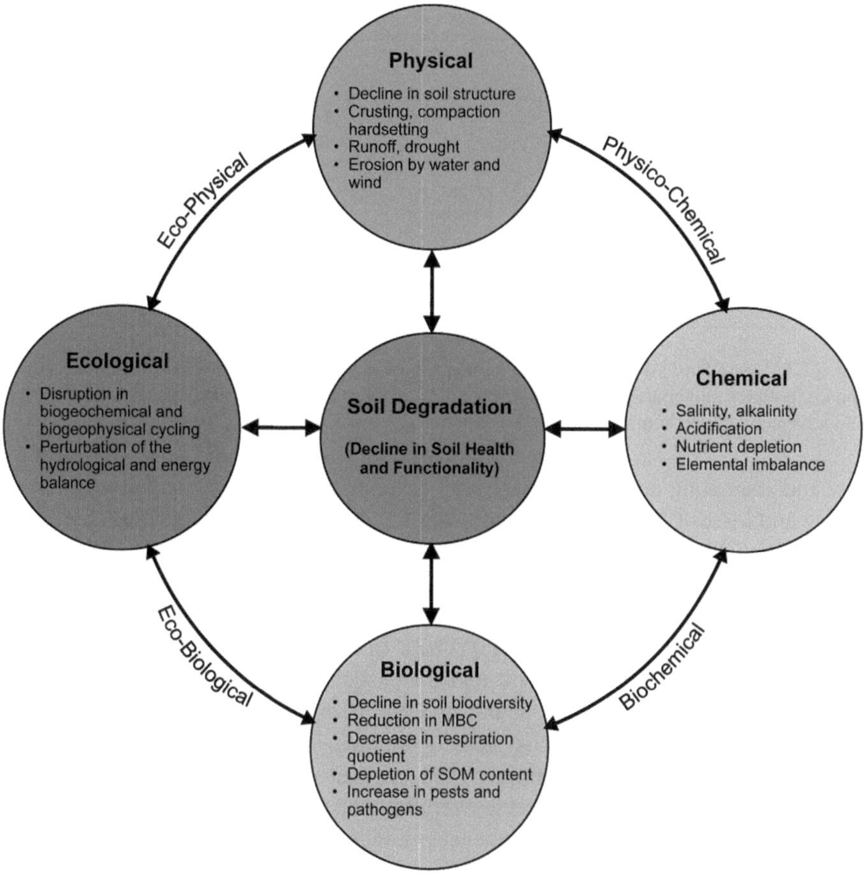

Fig. 17.1 Processes of soil degradation driven by biophysical, social, economic and cultural factors

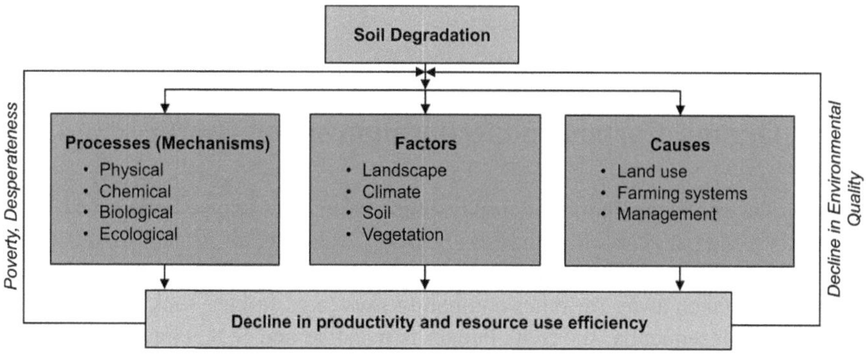

Fig. 17.2 Processes, factors and causes of soil degradation

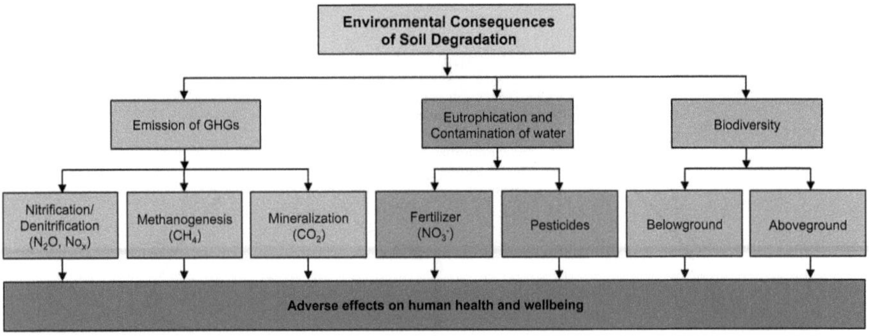

Fig. 17.3 Degradation of soil exacerbate emission of greenhouse gases from soil into the atmosphere, pollutes and contaminates water resources, reduces above and below-ground biodiversity, and adversely affects human health and wellbeing

low and stagnating and use efficiency of inputs (e.g., fertilizers, irrigation, energy) is low and losses (by erosion, decomposition, volatilization and leaching) are high. Furthermore, soil degradation (e.g., accelerated erosion) increases the emission of GHGs from soil especially that of N_2O by nitrification/de-nitrification and CH_4 by methanogenesis. Soil degradation also leads to pollution and contamination of water and decline of biodiversity (Fig.17.3). In addition to the decline in quantity and quality of food produced, soil and the associated environmental degradation also adversely impacts human health and wellbeing. Indeed, the health of soils, plants, animals, human and ecosystem is one and indivisible. Everything is connected to everything else (Commoner 1971), because when we try to pick one thing in nature, we find that it is hitched to everything else (Muir 1911). Soil degradation has severe and long lasting consequences. Landscape is a mirror image of people who live on it. Landscape with degraded and depleted soils supports meager living and impoverished lifestyle. Some options of BMPs in to minimize risks of soil degradation and restore degraded soils of India are outlined in Table 17.2 and briefly discussed below.

17.4 Crop Residue Management for Sequestration of Soil Organic Carbon and Restoration of Soil Health

India produces a large amount of crop residues estimated to range between 510 and 836 million Mg per year (Hiloidhari et al. 2014; Cardoen et al. 2015a; Ravindranath et al., 2005). However, most of the crop residues are either burnt in-field (~100 million Mg/yr), taken away for other competing uses (e.g., fodder, fuel, construction) or grazed. Consequently, the SOC budget is negative and SOC concentration and stocks are perpetually depleting. Hilodhari et al. (2014) estimated that 234 million Mg of crop residues are surplus, after the fodder and other needs are met. Other

Table 17.2 Basic principals of soil management and the associated best management practices

Basic principle		Associated Best Management Practices
1.	Creating a positive soil C budget and managing soil organic matter	Mulch farming, forages/cover cropping, conservation agriculture, agroforestry, and increasing input of biomass-C.
2.	Soil fertility improvement	Integrated nutrient management (INM), biological nitrogen fixation (BNF), mycorrhizal inoculation, balanced application of fertilizers, replacing whatever nutrients are removed, soil-test based fertilizer use, yield-based fertilizer input.
3.	Soil water management	Increasing soils available water capacity by increasing infiltration and reducing run-off and evaporation, increasing SOM content, improving root growth in the sub-soil, using drip sub-irrigation (rather than flood irrigation), direct-seeded aerobic rice.
4.	Improving soil structure and tilth	Conservation agriculture, mulch farming, manuring, use of gypsum in acidic soils.
5.	Managing soil compaction	Controlled grazing, minimizing vehicular traffic, growing deep-rooted crops (pigeon pea) in the rotation cycle, increasing activity of soil biota (earthworms), adopting conservation agriculture.
6.	Creating disease-suppressive soils	Improving soil biodiversity by increasing input of biomass-C, using compost and manure, enhancing microbial biomass-C.
7.	Integrated farming systems	Adopting complex cop rotations, growing pulses and cover crops, forages, agroforestry.

Table 17.3 Land use in South Asia and India

	Land area (10^6 ha)	
Parameter	South Asia	India
Total land area	688.0	358.7
Land area	640.0	297.3
Agricultural area	316.2	179.6
Arable land	22.4	156.4
Permanent crops	17.8	13.0
Permanent meadows and pastures	78.0	10.2
Forest	94.0	70.5
Other land	229.9	47.2
Inland water	48.0	31.4
Area equipped for irrigation	110.9	70.4

estimates of surplus amounts of residues include 279 million Mg and 383 million Mg. Assuming average surplus of crop residues of ~300 Mg/yr. and with C concentration of 40%, 120 Mg of biomass-C can be recycled back to the soil. With humification efficiency of 20%, 24 TgC can be sequestered through recycling of the surplus amount of crop residues. With arable land area of 156 Mha (Table 17.3), this would be equivalent to SOC sequestration rate of 150 kg/ha yr. With societal value of SOC of $120/Mg C (Lal 2014), farmers can be compensated at an average rate of $18/ha yr. through payments for ecosystem services.

Table 17.4 Nutrient Concentration of crop residues and animal dung (Recalculated from Cardoen et al. 2015a)

	Concentration (%)		
Residues	C	N	P
Cereals	41.1	0.49	0.08
Legumes	38.5	0.87	0.21
Dung	39.1	0.21	–

Table 17.5 Grain production in India (2014–2015)

Crop	Production (10^6 Mg)
Rice	104.8
Wheat	88.9
Maize	23.7
Millet	9.7
Sorghum	5.7
Coarse cereals	41.7
Barely	1.6
Small millets	0.4

FAO STATS 2017

In addition to crop residues, application of cattle dung as manure can also lead to improvements in soil health though SOC sequestration. There are 500 million livestock in India (FAO 2017), which produce 500 to 700 million Mg of fresh dung (average of 562 million Mg/yr). The present use of dung involves 60% as manure, 37% for fuel and 3% for other uses (Dikshit and Birthal 2010; Lohan et al. 2015). With availability of clean cooking fuel, 40% of the dung (225 million Mg/yr) can be used as manure. With 40% C and 20% humification efficiency, additional soil application of 225 million Mg of dung as manure would also lead to SOC sequestration of (225×10^6 Mg $\times$ 0.4 $\times$ 0.2) of 18 Tg C/yr., at an average annual rate of 115 kg/ha yr. over 156 M ha of cropland. Inputs of crop residues and manure also add to plant nutrients (Table 17.4), especially N.P and S. Thus, recycling of nutrients can reduce the input of chemical fertilizers and enhance use efficiency of inputs. Recycling of biomass canalso improve the nutrient budget in agriculture (Pathak et al. 2010).

17.5 Agronomic Productivity

Food grain production in India is second only to that of China (Tables 17.5 and 17.6). Total food grain production in 2016–17 is estimated at 274.4 million Mg (Ray 2017), which is 8.7% more than that of the last year. Food grain production in India has increased from merely 50 million Mg in 1947 when the population of India was

Table 17.6 Pulse production in India (2014–2015)

Pulse	Production (10^6 Mg)
Pigeon pea	2.8
Chickpea	7.2
Black gram (urd)	1.9
Green gram (moong)	4.5
Other pulses	3.9
Groundnut	6.6
Soybean	10.5

FAO STATS 2017

344 million (105 kg/person) to 274.4million Mg in 2017 when the population is 1342 million (203 kg/person). Thus, the rate of food production has outpaced that of population growth. Nonetheless, there is no cause for complacency and even the bigger challenges lie ahead.

While food production must be increased from the soil and water resources already appropriated, the rate of population growth (1.18%/yr) must also be decreased to the replacement level or lower. Thus, a double-prong strategy is essential.

17.6 Management of Water Resources

Similar to soil, India is also endowed with a large amount of renewable water resources. However, the large water footprint of agriculture must be reduced. This would require use of drip sub-irrigation rather than flood irrigation, and use of conservation-effective measures to reduce losses of water by runoff and evaporation, and storage of water in the soil. Soil water storage can also be enhanced by increasing SOC concentration through retention of crop residue mulch and application of animal manure. India's water footprint can be reduced by conversion of blue water (runoff, deep seepage) into green water (plant-available water) (Hoekstra and Mekonnen 2012). In this regard, the importance of the judicious use of gray/black water (urban wastewater) cannot be over-emphasized. Following proper treatment to reduce the risks of pathogens, gray/black water can be used for irrigation and for recycling water and nutrients. Mega cities (>10 million population) consume ~6000 Mg of food per day. Thus, through prudent management, city waste, a rich source of nutrients, can be an asset rather than a liability. Nutrient contained in the municipal solid waste must also be recycled (Fiksel and Lal 2017).

17.7 Sustainable Intensification of Agroecosystems

India is endowed with vast amounts of soil and water resources distributed over a wide range of biomes (climates) and eco-regions. Yet, increase in population at 1.18% with annual addition of 11.5 million new mouths to feed and accommodate is a major challenge which must be critically addressed. In addition to restoring degraded soils (146.8 Mha), urban encroachment and brick making must also be controlled through appropriate policy interventions. The prime agricultural land must be protected from other competing non-agricultural uses (e.g., urbanization, infrastructure development, recreational facilities).

Agronomic yields in India are lower than the potential attainable yields. This is true both for irrigated and rainfed agro ecosystems. The yield gap can be abridged through adoption of BMPs (Table 18.2) and adopting the principles of sustainable intensification (SI). The latter implies producing more per unit of land area and input of fertilizers, pesticides, irrigation and energy. The goal is to reduce losses and increase the use efficiency of inputs. Thus, India can meet the needs of present and the future population from lesser land and water resources than those allocated at present. The land and water thus saved can be used for nature conservancy. With the average per capita food grain need of 250 kg/yr., India's population of 1.7 billion will need a total production of 425 million Mg year. With production of 273 million Mg in 2017, the annual rate of growth between 2017 and 2050 (33 yr) must be 4.6 million Mg/yr. This rate of growth is achievable, especially with increase in productivity with increase in SOC stock in the root zone (Lal 2006), and with adoption of system-based CA system (Lal 1995, 2015).

17.8 Conclusion

India's population of 1342 million in 2017 is projected to increase to 1513 million by 2030, 1659 million by 2050 and decrease to 1517 million by 2100. Rate of increase of food grain production of 4.6 million Mg/yr. between 2017 and 2050 (from 274 million Mg in 2016–17 to 425 million Mg by 2050) must be obtained from lesser land and water resources than those being appropriated at present. The strategy is to abridge the yield gap through SI of agroecosystems by "producing more from less" through:

1. Restoring degraded soils (146.8 M ha),
2. Preventing any new soil degradation,
3. Banning use of topsoil for brick making,
4. Prohibiting in-field burning of crop residues, and using conservation agriculture,
5. Using dung as compost and providing clean cooking fuel to rural communities,

6. Sequestering SOC by recycling of surplus crop residues and animal dung into agroecosystems,
7. Enhancing water use efficiency by using drip sub-irrigation, prohibiting flood irrigation and reducing losses by runoff and evaporation,
8. Increasing use efficiency of nutrients by restoring soil health and using improved formulations of fertilizers,
9. Incentivizing farmers through payments for ecosystem services (C and water credits), and
10. Promoting soil-centric technologies.

References

Bhattacharyya, R., Ghosh, B. N., Mishra, P. K., et al. (2015). Soil degradation in India: Challenges and potential solutions. *Sustainability., 7*, 3528–3570.

Cardoen, D., Joshi, P., Diels, L., Sarma, P. M., & Pant, D. (2015). Agriculture biomass in India: Part 2. Post harvest losses, cost and environmental impacts. *Resources, Conservation and Recycling, 101*, 143–153.

Commoner, B. (1971). *The closing circle:Nature, man and technology*. New York: Knopf.

Dikshit, A., & Birthal, P. (2010). Environmental value of dung in mixed crop-livestock systems. *Indian Journal of Animal Sciences, 80*(7), 679–682.

FAO. (2017). FAO, Rome, Italy. http://www.fap/prg/faostat/en/#data/RL.

Fiksel, J. and R. Lal. (2017). Transforming waste into resources for the Indian economy. Current Science (Submitted).

Hiloidhari, M., Das, D., & Baruah, D. C. (2014). Bioenergy potential from crop residue biomass in India. *Renewable and Sustainable Energy Reviews, 32*, 504–512.

Hoekstra, A. Y., & Mekonnen, M. M. (2012). The water footprint of humanity. *PNAS, 109*, 3232–3237.

Lal, R. (1995). The role of residue management in sustainable agricultural systems. *Journal of Sustainable Agriculture, 5*, 51–78.

Lal, R. (2006). Enhancing crop yields in the developing countries through restoration of soil organic carbon pool in agricultural lands. *Land Degradation& Development, 17*, 197–209.

Lal, R. (2014). Societal value of soil carbon. *Journal of Soil and Water Conservation, 69*, 186A–192A.

Lal, R. (2015). A system approach to conservation agriculture. *Journal of Soil and Water Conservation, 70*(4), 82A–88A.

Lohan, S. K., Dixit, J., Kumar, R., Pandey, Y., Khan, J., Isaq, M., Modasir, S., & Kumar, D. (2015). Biogas: A boon for sustainable energy development in India's cold climate. *Renewable and Sustainable Energy Reviews, 43*, 95–101.

Muir, J. (1911). *My first summer in the sierra*. Boston: The Riverside Press Cambridge.

Pathak, H., Mohanty, S., Jain, N., & Bhatia, A. (2010). Nitrogen, phosphorus, and potassium budgets in Indian agriculture. *Nutrient, Cycling, Agroecosystems, 86*, 287–299.

Ravindranath, N. H., Somasekhar, H. I., Nagraja, M. S., et al. (2005). *Biomass and Bioenergy, 29*, 178–190.

Ray, R.K. (2017). Indian foodgrain output up 8.7% at a record 273.38 MT in 2016–17. Hindustantimes, 8th July 2017.

U.N. (2017). World Population Prospects: Key Findings and Advance Tables. 2017 Revision. Division of Economic and Social Affairs. U.N., New York, p 46.

Chapter 18
Scenario of Crop Production in Temperate and Cold Arid Region with Respect to Climate Change

Latief Ahmad, Sabah Parvaze, S. Sheraz Mahdi, R. H. Kanth, and B. S. Dhekale

Abstract Agriculture and forestry will be particularly sensitive to changes in mean climate and climate variability. Climate change will definitely impact agriculture and agricultural productivity especially with the introduction of new crop species and varieties, higher crop production and expansion of suitable areas for crop cultivation. However this may be accompanied by a rise in need for plant protection, risk of nutrient leaching and accelerated breakdown of soil organic matter. The rise in carbon-dioxide level may result in increased water use efficiency, which will recompense for some of the adverse effects of increasing water constraint and extreme weather events. This on the other hand will also lead to low harvest yields, higher variation in yield and reduction in areas suitable for growing traditional crops. In Kashmir Valley, the increased precipitation is expected to be large enough to compensate for the increased evapotranspiration. On the other hand, increased precipitation, cloudiness and rain days and the reduced duration of snow cover and soil frost may negatively affect the agricultural production.

Keywords Crop production · Temperate region · Cold arid region · Climate change · Ladakh · Kashmir

---

The original version of this chapter was revised: The author groups of this chapter have now been restored as in the original submission. The correction to this chapter is available at https://doi.org/10.1007/978-3-319-90086-5_21

L. Ahmad (✉) · S. Parvaze · R. H. Kanth
Division of Agronomy, SKUAST-Kashmir, Shalimar, Srinagar, Kashmir, India

S. Sheraz Mahdi
Mountain Research Centre for Field Crops, Sher-e-Kashmir University of Agricultural Sciences and Technology of Kashmir, Khudwani, Anantnag, Kashmir, J & K, India

B. S. Dhekale
Division of Agri. Statistics, SKAUST, Kashmir, Shalimar, India

S. Sheraz Mahdi (ed.), *Climate Change and Agriculture in India: Impact and Adaptation*, https://doi.org/10.1007/978-3-319-90086-5_18

18.1 Introduction

Agriculture is very intensely influenced by weather and climate. While farmers are often compliant in managing weather and annual variability, there is however a high degree of adaptation to the local climate. This may be in the form of conventional infrastructure, native agricultural practices or personal experience. Climate change an in this way be relied upon to affect the agriculture, possibly threatening established characteristics of farming systems and provide chances for improvements. The role of climate change in influencing human activities and the natural environment is all around acknowledged. This may go from an extreme event enduring just a couple of hours (i.e. .e. hurricane, thunderstorm and showers) through to the multi-year dry spells (Maracchi et al. 2005).

Increase in concentration of greenhouse gases in the atmosphere further contribute to climate change. This results in a strong impact on human activities and natural environments. Essential areas, for example agriculture and forestry, will be more susceptible than auxiliary and tertiary segments, for example manufacturing and retailing (Parry et al. 1988).

Agriculture is the fundamental activity by which people live and survive on the earth. Evaluating the impacts of climate change on agriculture is a vital job. In developed as well as developing nations, the influence of climate on crops and domesticated animals regardless of water system, enhanced plant and creature cross breeds and the developing utilization of synthetic manures. The continuous dependency of agricultural production on light, heat, water and other climatic factors, the reliance of a great part of the total populace on farming exercises, and the noteworthy extent and quick rates of conceivable atmosphere changes all consolidate to make the requirement for an exhaustive thought of the potential effects of atmosphere on worldwide agriculture (Rosenzweig and Liverman 1992).

Climate change impacts on agriculture differ regionally. Climate change impacts on subtropical and tropical areas will be predominately negative, particularly where agriculture is right now minimal as for high- temperature and moisture-deficit conditions. The most susceptible agricultural systems are present in arid, semi-arid, and dry sub-humid regions in the developing world, home to half of the world's malnourished populace, where high precipitation variability and recurring droughts and floods regularly disrupt food production and where poverty is unescapable. Agriculture in cold-limited (high-latitude and high-altitude) areas could benefit from humble levels of warming that adequately increment the length of the growing season (Padgham 2009).

18.2 Climate Change and Agriculture

In light of a portion of the past encounters, effect of climate change on agriculture will be one of the real central components impacting the future sustenance security of humanity on the earth. Agriculture isn't just sensitive to climate change yet additionally one of the significant drivers for climate change. Understanding the weather changes over some stretch of time and modifying the management practices towards accomplishing better harvest are snags to the development of agricultural sector in general. The sensitivity of agriculture to climate is questionable, as there is territorial variability in precipitation, temperature, crops and cropping systems, soils and management practices. The inter-annual variations in temperature and precipitation have been much higher than the predicted changes in temperature and precipitation. The crop losses may increase if the predicted climate change increases the climate variability.

Different crops differ in their response to global warming will have a complex impact. The tropics are more reliant on farming as 75% of total populace lives in tropics and 66% of these individuals' fundamental occupation is farming.

With low levels of innovation, extensive variety of pests, diseases and weeds, land degradation, unequal land distribution and rapid population growth, any effect on tropical agriculture will influence their business. Rice, wheat, maize, sorghum, soybean and barley are the six chief crops in the world grown in 40% cropped area, and comprise 55% of non-meat calories and over 70% of animal feed (FAO 1990). Subsequently, any impact on these crops would adversely disturb the food security.

18.3 Study Area

The State of Jammu and Kashmir lies between 32°17' and of 37°5' North and 73°26′ and 80°30' East. The State is located amidst three climatic regimes of Asia n its south fringe lies the frail rainstorm zone of Punjab. On the north-east the State is bounded by the vast arid plateau of Tibet while the North-west border areas confront the eastern furthest reaches of Mediterranean climatic district. This topographical position, combined with the differed physiography, gives the State a wide climatic variety.

The State has been alienated into four broad macro-climatic zones (i) sub-tropical (ii) valley temperate (iii) intermediate, and (iv) cold-arid. The State has generally a mountainous area and occupies a central position in the Asian continent. The state comprises nearly two-thirds of the mountainous area of India. 2.3 million ha out of 3.5 million ha of mountainous area of India belong to Jammu and Kashmir. The State is limited on the north by Chinese and Russian regions, on the east by Tibet, on the south by Punjab (India) and on the West by Pakistan (Fig. 18.1).

The topography of the state is highly uneven comprising of many snow-covered peaks having altitude range of 554 to 7077 m amsl in North and North-west, which are followed towards the South by lower range of hills. Total geographical area of the State is 2, 22, 236 km^2 including 78,114 km^2 (35.15%) under Pakistan, and 42,735 km^2 (19.23%) under China. Ladakh is the largest hilly arid zone which involves 58,321 km^2 (42.00%).

Kashmir area includes the temperate zone of the state of Jammu and Kashmir. his zone encounters wet and regularly extreme winters with frost, snow and rain and comparatively dry and warm summer. Snowfall, a critical type of precipitation,

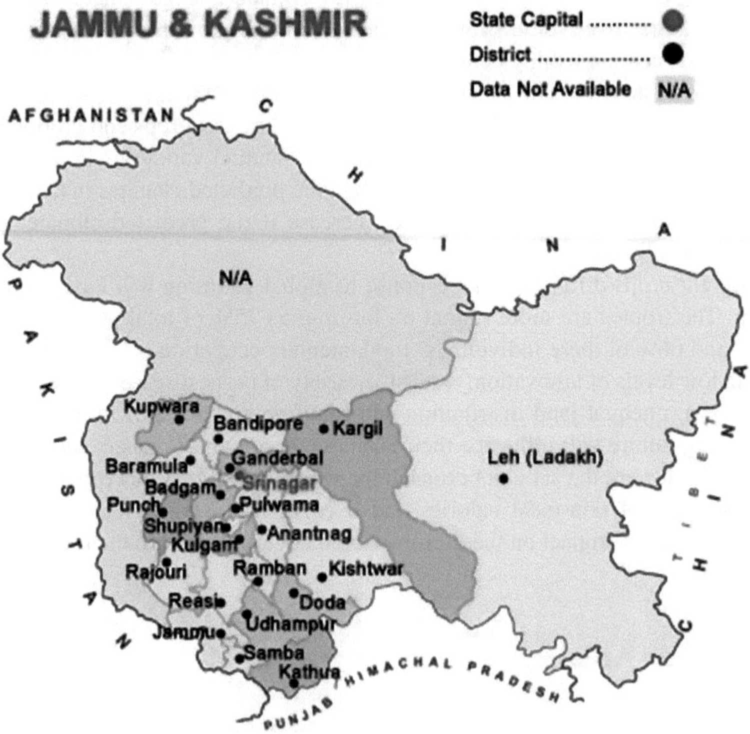

Fig. 18.1 Map showing different districts of J&K State and its neighboring areas

keeps up sufficient moisture supply amid summer when precipitation is sparse. The valley temperate zone involves the zones of varied relief. The plain valleys have an elevation of 1560 m amsl, which ascends to 1950 m in low altitude Karewas in mid belts, 2400–3000 m in the upper belts and to 4200 m in snow bound ranges. The soils of Kashmir valley are alluvial in nature having 62% of area under irrigation. The prominent meteorological features of temperate zone demonstrate that the zone receives annual precipitation of around 680 mm, of which nearly 70 per cent is received during winter and spring seasons (from December to May). The overall average temperature in different months varies from 1.2 °C to 24.5 °C with cold thermal index and humid hydric index.

In India, arid zone encompasses 3, 87, 390 km 2 area of which 1, 07, 545 km 2 lie in the cold arid region of Western Himalayas. The cold arid region of Western Himalaya primarily contains Ladakh area of Jammu and Kashmir State and certain sections of Lahul-Spiti sub-division in Himachal Pradesh. The region in J&K lies in the northern most tip of Asian sub-continent between Karakoram and greater Himalayan ranges and is entwined with bare and rough mountains.

18.4 Climate Change in Kashmir and Ladakh

With regards to India, particularly Jammu and Kashmir, which settles in delicate Himalayan Ecosystem; there markers of climate change are obvious at this point. Climate change represents a genuine risk to the agriculture, horticulture, water resources, tourism, species diversity, habitats, forests, wildlife, and livelihood in the region. As per United Nations Environment Programme (UNEP) report some parts of the State are moderate to highly vulnerable. As per Indian Network on Climate Change Assessment (Parvaze et al. 2017) evaluation the number of rainy days in the Himalayan region in 2030s may increase by 5–10 days on an average, with an increase by more than 15 days in the eastern part of the Jammu and Kashmir region. The intensity of rain fall is probably going to increment by 1–2 mm/day. This is probably going to affect a portion of the horticultural crops. The rate of subsidence of ice sheets are allegedly differing which is being credited to winter precipitation, atmosphere warming and anthropogenic components. Temperature, precipitation and cold wave are well on the way to fundamentally affect the agricultures sector. Scarcity in food production is growing in Jammu & Kashmir. With the diminishment in precipitation, the rain-fed agriculture will endure the most. Horticultural crops like apple are likewise demonstrating decrease underway especially because of decrease in snowfall. Around 34% and 39% of the forested frameworks are probably going to experience shifts in vegetation type with a trend towards increased occurrence of the wetter forest types.

The climate of Kashmir and Ladakh Valley has witnessed a different change in climate and climatological variables as compared to the other parts of India. During last few decades the maximum and minimum temperatures of the region have shown no significant increasing or decreasing trends. The amount of rainfall has however shown an increase in quantity during this period (Parvaze et al. 2016). Compared to other parts of the country which are witnessing externalities of climate change, long term annual rainfall in the state of Jammu and Kashmir showed positive growth momentum with annual growth rate of 1.39 percent during last 28 years. Overall, annual maximum and minimum temperature for the State almost exhibited a uniform trend during last 40 years (Jammu and Kashmir Agricultural Production Department).

According to a report by Jammu and Kashmir Environmental Information System (J&KENVIS), "the climate change projections predict an increase from 0.9±0.6 °C to 2.6±0.7 °C in the 2030s. The net increase in temperature ranges from 1.7 °C to 2.2°C with respect to the1970s. Seasonal air temperatures also show a rise in all seasons. The annual rainfall in the Himalayan region is likely to vary between 1268±225.2 and 1604±175.2 mm in 2030s. The projected precipitation is likely to increase by 5% to 13% in 2030s with respect to 1970s. The intensity of rain fall is likely to increase by 1-2mm/ day. The water yield in the Himalayan region, mainly covered by the river Indus, is likely to increase by 5%–20% in most of the areas, with some areas of Jammu and Kashmir showing an increase of up to 50% with respect to the1970s" (J&KENVIS 2015).

18.5 Vulnerability of Agriculture and Allied Activities Due to Climate Sensitivity

Temperature, precipitation and cold wave significantly impacts the agricultures sector and enhances its vulnerability. This happens due to the early onset of rain and number of dry days. In western Himalaya, the traditional farming operation cultivating operation is an unpredictable result of yield farming crop husbandry, animal husbandry and forest resources constituting interlinked diversified production systems. Nonetheless, with changing land use, the area under cultivation of many traditional crops has been reduced and some others are at the brink of extinction. Deficit in food production is growing in recent times in Jammu & Kashmir. With the reduction in rainfall, the rain-fed agriculture will suffer the most. Plant crops like apple are additionally indicating decrease underway and a genuine scope especially because of decrease in snowfall.

18.6 Impact of Climate Change in Temperate Kashmir Valley

Crop yields have been reliably observed to be higher in temperate regions than in the tropics (FAO 1990). Numerous factors contribute to this result. Soils in the humid tropics have a tendency to be profoundly drained of nutrients and are consequently unproductive because of high temperatures, intense rainfall, and erosion. Temperate soils are generally viewed as more favorable to agriculture than tropical soils because of higher nutrient levels. However, there are exceptions such as poorly developed and infertile soils in temperate regions. Solar radiation, which is basic to plant development, and whose intensity is controlled by the angle of the sun, day length, and cloudiness, is lower in winter and higher in summer in temperate zones. In the tropics, sunlight based radiation is regularly restricted by cloudiness amid the rainy seasons.

18.7 Impact on Forests

Vegetation designs (circulation, structure and nature of woods) crosswise over globe are controlled for the most part by the climate. Even with global warming of 1–2 °C, significantly less than the latest projections of warming amid this century, most biological systems and landscapes will be affected through changes in species composition, productivity and biodiversity. It has been exhibited that upward movement of plants will occur in the warming scene.

Because of increment in temperatures, change in vegetation, fast deforestation and shortage of drinking water, natural habitat devastation and corridor fragmentation may prompt an awesome risk to annihilation of wild flora and fauna. It is normal that with the expansion in dry spell cycles and corresponding increment in forest fires some pine species will infringe upon alternate species and will assume an essential part in diminishing the yield of non-timber forest products (NTFP) from these forests. The ecosystem services created would likewise modify. Unavoidably, any adjustment in the forest (distribution, density and species composition) under CC would massively impact economies like forestry, agriculture, livestock husbandry, NTFPs and medicinal plants based livelihoods.

Some impacts of climate change on forests are listed below (J&KENVIS 2015):

- Marked extension (11%) in Temperate deciduous, cool mixed and conifer forests at the cost of alpine pastures which are probably going to shrivel.
- The uncommon patterns of winter movement of birds in the wetlands of Jammu, Kashmir and Ladakh
- Decline financially vital species like Deodar, Fir and spruce and increase in Blue Pine in Kashmir Valley and Chir Pine in Jammu.
- The spread of invasive alien species like Parthenium, Lantana, Ageratum, Xanthium, Anthemis, etc.,
- Decreasing tree density and forest fragmentation
- Increased net primary productivity.
- Increased incidences of forest fires.

18.8 Impact on Agriculture

Agriculture is highly dependent on weather and changes in weather cycle have a major effect on crop yield and food supply. Mountain agriculture is mostly rainfed and driven by biomass energy of surrounding forests and confined to terraces carved out of hill slopes. Climate change may result in the following major impacts on agriculture in Kashmir valley (J&KENVIS 2015):

- Irrigated rice, wheat and mustard productions might be decreased by 6%, 4% and 4%, individually.
- The shortfall in food production in Kashmir region has touched 40%, while the deficiency is 30% in vegetable production and 69% in oilseed production, putting food security at a more serious hazard.
- Invasion of weeds in the croplands and those are consistently removed by the farmers (e.g., Lantana camara, Parthenium sp.
- Increased rate of recurrence of insect-pest attacks
- Decline in crop yield.
- More and more paddy land being changed over to rainfed orchard or dry land. The huge chunk of paddy land has been converted into rain-fed dry land in many regions of the valley as of late.
- Area under apple cultivation increased however yield per hectare has - fundamentally declined amid the previous decades..

These factors have led to loss in agri-diversity and change in crops and cropping patterns. Conventional farming in the Himalayan mountains has been a rich storehouse of agro-biodiversity and strong to crop maladies.

18.9 Impact on Agriculture in Ladakh

Agriculture is one of most delicate areas to effects of climate change due its high reliance on atmosphere and climate. Over hundreds of years, agriculturists in Ladakh have developed self-supported cultivating frameworks not withstanding in the midst of a climatically difficult condition. In fact, it has been the pillar of economy and gives nourishment and occupation security to the general population of Ladakh.

However, recent environmental changes are harmful for the fabric of sustainability for the general population of Ladakh. The effect of environmental change in the course of the most recent couple of years has unfavorably influenced farming in this district in various ways. Snow and water from glaciers is the chief source of irrigation for nine-tenths of the farming community. Thus, glacial melt determines the productivity in the region to a large extent. Besides glaciers are the primary sources of water for the farmers in the region. Excessive glacial melt has now become a cause of floods in the region, putting the lives of more than 80% farmers to risk.

Continuous retreating of glaciers in the region will have an adverse effect on t agriculture. The economy and food security of the region will be threatened due to changes in agricultural produce. It is supposed that increase in temperature could modify agricultural produce. The change may be either positive or negative. The impact on agriculture and its outcomes on the general public are probably going to differ locally contingent upon the kind of environmental change that has occurred here and the choices accessible to farmers. It could cut down the yield of wheat.

However, higher temperature may enhance the productivity of wheat in the high altitude areas. Warmer temperatures ay influence many crops to develop all the more rapidly, however hotter temperatures could likewise diminish yields. Crops have a tendency to grow quickly in warmer conditions. Nonetheless, for a few crops, (for example, grains), speedier development decreases the measure of time that seeds need for growth and maturity. Throughout the years, agriculturists have brought new plants into their farmlands.

Crops tend to grow faster in warmer conditions. However, for some crops (such as grains), faster growth reduces the amount of time that seeds have to grow and mature. Over the years, farmers have introduced plants into their farmlands. Looking at the brighter side of the impact of climate change, it merits saying that farmers in Ladakh have profited from the current global warming as they can develop number of new vegetable products.

Climate change is likewise connected with an increase in the incidence of extreme climatic events, for example drought, cloudburst and floods. The nomads known as the Changpas living in the Changthang region intensely feel environmental change. These individuals are reliant on the raising of animals, the pashmina goats for their fleece. As of late, the migration paths of the Changpas have changed because of decline in pasture land. Further, indeterminate snowfall has led loss of numerous domesticated animals in the district. In future, environmental change in Ladakh is probably going to influence agriculture and increment the danger of water for irrigation. The villagers could confront serious water shortage which is credited to overwhelming runoff from rapid snow melt. Besides, an expansion in the occurrence of pest attacks and weeds could also make destruction in the district.

References

Food and Agriculture Organization. (1990). FAO Yearbook. Production. Vol. 44. Food and Agriculture Organization of the United Nations. Rome, 1991.

Maracchi, G., Sirotenko, O., & Bindi, M. (2005). Impacts of present and future climate variability on agriculture and forestry in the temperate regions: Europe. In *Increasing climate variability and change* (pp. 117–135).

Padgham, J. (2009). Agricultural development under a changing climate: Opportunities and challenges for adaptation. World Bank.

Parry, M. L., Carter, T. R. and Knoijin, N. T. (eds.). (1988). The impact of climatic variations on agriculture, Kluwer, Dordrecht, 876.

Parvaze, S., Parvaze, S., Haroon, S., Khurshid, N., & Khan, J. N. (2016). Projected change in climate under A2 scenario in dal lake catchment area of Srinagar city in Jammu and Kashmir. *Current World Environment, 11*(2), 429.

Parvaze, S., Ahmad, L., Parvaze, S., & Kanth, R. H. (2017). Climate change projection in Kashmir Valley (J and K). *Current World Environment, 12*(1), 107–115.

Rosenzweig, C., & Liverman, D. (1992). Predicted effects of climate change on agriculture: A comparison of temperate and tropical regions. In D. S. K. Majumdar (Ed.), *Dalam Global Climate Change: Implications, Challenges, and Mitigation Measures* (pp. 342–361). Pennsylvania: The Pennsylvania Academy of Sciences.

J&KEVIS. (2015). ENVIS newsletter October – December, 2015 Climate Change & Concerns of J&K, J&KENVIS Centre Department of Ecology, Environment & Remote Sensing Jammu & Kashmir.

Chapter 19
Climate Scenario in Cold Arid Region and its Future Prediction

Latief Ahmad, Sabah Parvaze, S. Sheraz Mahdi, and R. H. Kanth

Abstract Ladakh region of Jammu and Kashmir State is characterized by excessive aridity and severe moisture deficit throughout the year. On one side of the region are the Karakoram ranges while on the other side lie the mighty Greater Himalayas and Zanskar ranges. Together these impart a rain shadow effect to the region resulting in very low annual precipitation. With scarce water resources, such regions show high sensitivity and vulnerability to the change in climate and need urgent attention. The objective of this study is to understand the climate scenario in the cold arid region of Ladakh region and to characterize its changing climate. The climate over Leh has shown a warming trend with reduced precipitation. The reduced average seasonal precipitation may be associated with some indications of reducing number of days with higher precipitation amounts over the region.

Keywords Climate · Cold-arid · Climate change · Ladakh

19.1 Introduction

Ladakh is a cold-arid region which lies between 32°15 'to 36°0 'N and 75°15 'to 80°15′ E. It is the northern-most territory of Jammu and Kashmir province of India. Ladakh lies in the Trans-Himalayan Area and is limited in the north-upper east by Tibet, in the northwest by Baltistan and in the west by Himachal Pradesh. Ladakh is

The original version of this chapter was revised: The author groups of this chapter have now been restored as in the original submission. The correction to this chapter is available at https://doi.org/10.1007/978-3-319-90086-5_21

L. Ahmad (✉) · S. Parvaze · R. H. Kanth
Division of Agronomy, SKUAST-Kashmir, Shalimar, Srinagar, Kashmir, India

S. Sheraz Mahdi
Mountain Research Centre for Field Crops, Sher-e-Kashmir University of Agricultural Sciences and Technology of Kashmir, Khudwani, Anantnag, Kashmir, J & K, India

S. Sheraz Mahdi (ed.), *Climate Change and Agriculture in India: Impact and Adaptation*, https://doi.org/10.1007/978-3-319-90086-5_19

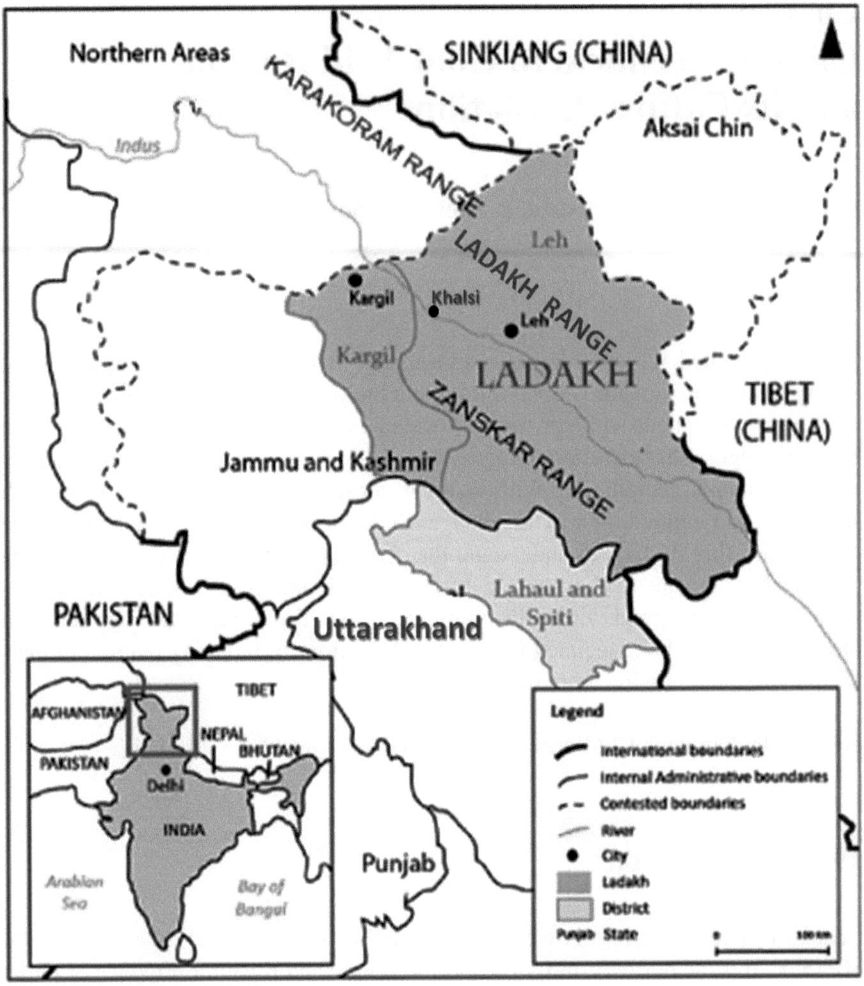

Fig. 19.1 Map of Jammu and Kashmir showing the Ladakh region (Le Masson and Nair 2012)

the biggest zone of Jammu and Kashmir state and involves a territory of 58,321 km^2 which is almost 42% of the aggregate region of the state. The height ranges from more than 2400 m with peaks extending from 7200 to 8400 m. The zone is one of the most astounding occupied places on the planet. The map of Jammu Kashmir State demonstrating Ladakh area is given in Fig. 19.1.

Ladakh consists of two districts Leh and Kargil. Leh with an area of 45,110 km^2 makes it largest district in the country in terms of area. The topography is mostly mountainous and bared. Except for few places, the area is mostly devoid of vegetation. The cold arid zone experiences severe cold and dry winter and moderately hot and dry summer. The zone receives about 80–90 mm rainfalls in Leh to about

Table 19.1 Land Utilization Pattern of Leh (Area in Hectares)

Year		2003–04	2014–15
Reporting area		45,167	45,167
Forest area		–	0
Area not available for cultivation	Land put to non-agricultural uses	2908	2978
	Barren and uncultivable land	25,163	27,474
Other uncultivable excluding fallow	Permanent pastures and other grazing lands	1092	NA
	Land under misc. tree, crop, grooves not including net area sown	1148	1732
	Culturable wasteland	4412	2567
Fallow land	Fallow other than current fallows	143	37
	Current fallows	190	376
	Net area sown	10,111	9982
Area sown more than once		313	632
Total cropped area		10,424	10,614

300 mm in Kargil. Severe aridity with very cold thermal index is a characteristic of the zone. Soils of this zone have a high permeability and low water holding capacity.

The state of Jammu and Kashmir has been divided into three main regions on the basis of physiography. The Outer Himalayas consisting of Jammu province, the Lesser Himalayas consisting of Kashmir province and the Inner Himalayas consisting of Ladakh province. Ladakh lies in the cold-arid agro-climatic zone of Jammu and Kashmir and is divided into three agro-ecological zones (Ahmad and Kanth 2014):

1. Upper Agricultural Zone (UAZ) 3597–4420 m. (a.m.s.l)
2. Central Agricultural Zone (CAZ) 3048–3597 m. (a.m.s.l)
3. Lower Agricultural Zone (LAZ) < 3048 m. (a.m.s.l)

The physiography of the region is divided into hilly and valley lands. About 85% of the area comes under hills and only 15% area is under valley. The population density of the region is very low being only 4.6 per km^2. The land use statistics of Leh district is presented in Table 19.1 (LAHDC 2015). The region has no forest cover Most of the land in the region is barren or uncovered. The total cultivated land forms less than 25% of the total area of Ladakh.

The district Leh has a location of 34° 17′ N and 77° 58′ E. Besides being the district head quarter, it is also the capital of Ladakh Autonomous Hill Council in the state of Jammu and Kashmir, India. The city is located in the foothills of the Ladakh range.it lies in the Indus River catchment with an average altitude of 3500 m amsl. The Indus valley is bound by the Ladakh range in the north and Zanskar range in the south.

19.2 Agriculture

Ladakh region possesses a wide range of flora and fauna. The forest vegetation mostly comprises of few trees species and few shrubs like popular, willow, sea-buck thorn etc. The major crops grown in the region are barley and wheat. Grasses like alfalfa are also grown in the region. Horticulture is a major part of livelihood of people. Varieties of Grapes (Resin type), prunes and drying varieties of apricots, sea-buck thorn are grown widely in the region. Cultivation of apple, walnut and currants is also practiced in certain areas of the region. Ladakh region is the major producer of apricot crop in Jammu and Kashmir. The major share of the fruit (almost 40%) alone comes from Kargil district of the Ladakh region.

Agriculture in Ladakh is mainly subsistent in nature yet one of a kind and a typical of Tibetan plateau farming system. The people of Ladakh until recently followed an integrated farming system. They produced and consumed their own grains, cereals and vegetables. The manure, seeds and other agricultural inputs were produced by the people themselves. They also raised their own animals and arranged their own farms in an integrated and adjusted way in order to confront the harsh climate and hostile environment of Ladakh.

Nonetheless, many factors like cropping intensity at a low level, lesser productivity and a brief agriculture season has brought about the reliance of its developing populace on import of food grains, vegetables and spices. Moreover, as Ladakh opens to the world, its conventional agricultural framework and produces confront serious interruptions. Agriculture has now occupied a secondary position to the fast development of tourism in the region. Nevertheless, the potential for agricultural development in the region is still immense n spite of its high elevation and arid climate. Both agricultural and horticultural crops are grown in Ladakh especially niche or specialty crops. These crops if properly marketed after value addition can bring about its very own green in this region.

Besides, as Ladakh opens to the world, its traditional system and crops face severe disruptions. Nonetheless, there is a huge back seat to the rapid growth of tourism in the region. Nonetheless, there is a huge potential for agriculture in Ladakh, despite its high altitude and arid climate. Ladakh grows niche crops, both agricultural and horticultural, which with an appropriate value added and proper marketing can herald a green revolution of its own sort in this region. Aiming at production of vegetables in winters, market in the prolonged winter season can prove to be very beneficial to local farmers. The area under cultivation of different crops is given in Table 19.2 (LAHDC 2015).

Table 19.2 Area (in hectares) sown under different crops in Ladakh, J&K

Year	Rice	Maize	Wheat	Grim	Other Millets	Pulses	Fruits	Vegetables	Spices	Other food crops	Total food crops	Oil seeds	Fodder	Total non-food crops	Total Area sown
1	2	3	4	5	6	7	8	9	10	11	12	13	14	15	16
2001–02	–	–	2604	4734	436	270	98	240	–	–	8382	73	2068	2141	10,523
2002–03	–	–	2653	4702	377	272	85	246	–	–	8335	73	2070	2143	10,478
2003–04	–	–	2894	4504	349	349	85	245	–	–	8327	73	2024	2097	10,424
2004–05	–	–	2894	4480	384	249	35	299	–	–	8341	67	2020	2087	10,428
2005–06	–	–	2973	4463	375	272	38	310	–	–	8431	65	2089	2154	10,585
2006–07	–	–	2973	4463	375	272	38	310	–	–	8431	65	2089	2154	10,585
2007–08	–	–	2968	4452	359	286	39	313	–	–	8436	74	2089	2163	10,599
2008–09	–	–	2634	4639	379	306	91	348	–	–	8402	86	2028	2114	10,516
2009–10	–	–	2690	4646	342	297	90	358	–	–	8427	86	2095	2181	10,608

(continued)

Table 19.2 (continued)

Year	Rice	Maize	Wheat	Grim	Other Millets	Pulses	Fruits	Vegetables	Spices	Other food crops	Total food crops	Oil seeds	Fodder	Total non-food crops	Total Area sown
2010–11	–	–	2579	4421	–	192	131	229	–	–	7552	86	1947	–	11,692
2011–12	–	–	2595	4488	97	243	40	280	–	–	7743	90	1991	2081	9824
2012–13	0	0	1092	2869	556	258	88	321	3	1	5188	86	2093	2179	7367
2013–14	0	0	1077	3934	622	251	134	393	3	1	6415	80	2100	2180	8595
2014–15	0	0	2776	4288	564	258	104	388	5	1	8364	89	2161	2250	10,614

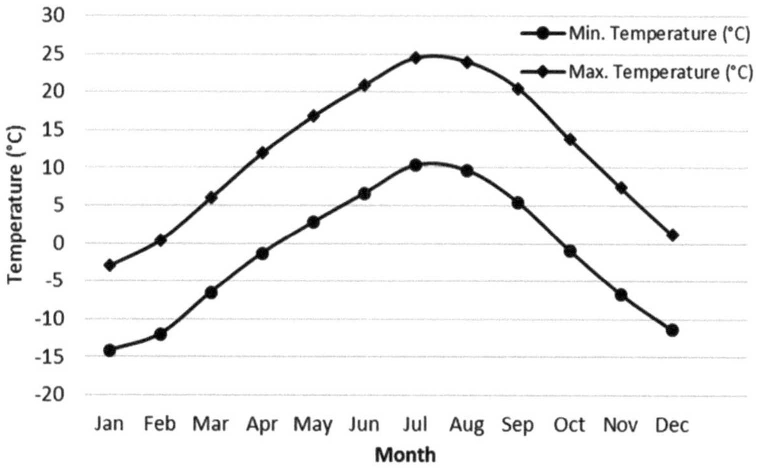

Fig. 19.2 Monthly variation on Maximum and minimum temperatures in Ladakh

19.3 Climate Scenario

19.3.1 Temperature

Ladakh is located on the rain shadow region of the Himalayas. The monsoon winds after losing their moisture in the Himalayan Mountain and the plains make their way to Leh. The district brings together the elements of both arctic and desert climate, thus attributing it the title of "Cold Desert. It has unique attributes as wide diurnal and seasonal fluctuations in temperature with −40 °C in winter and 35 °C in summer. The average monthly variation of Maximum and minimum temperatures in Ladakh (2000–2016) is presented in Fig. 19.2.

January, February and December are the coldest months of the year. The temperature seldom goes above the subzero condition beginning from December. The night-time is chilling and temperatures float around −20 °C. Days are warmer with 2 °C as average temperature. January also witnesses maximum snowfall and thick ice sheet formations over rivers and lakes are common. Manali - Leh and Srinagar - Leh Highway are closed and Ladakh remains practically inaccessible by road during these months. Frostbites are common and most locals develop scales and cracks on their skins.

March and April are warmer than January, February and December. During March, temperatures hover between 6 °C in the day and – 5 °C in the night. April is still better and day temperatures go up as much as 12 °C. Occasional snowfall during these months can result in lower temperatures.

During May to August, life begins to move towards normal in Ladakh. The Highways open up the last week of April or the first week of May. The weather

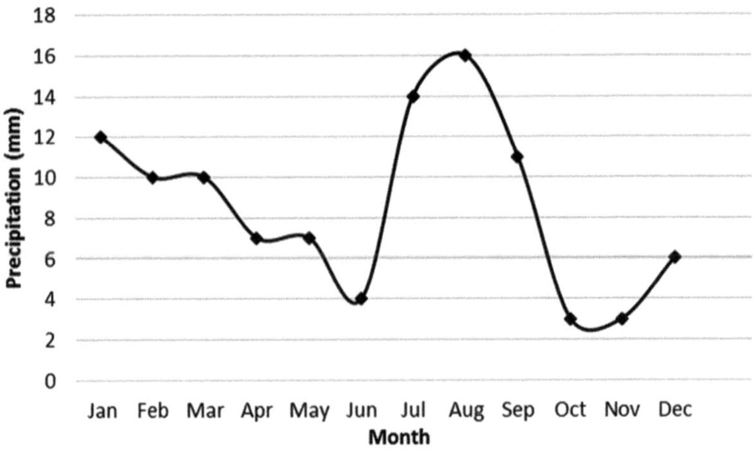

Fig. 19.3 Monthly variation of precipitation in Ladakh region

starts clearing during May. Temperatures are a pleasant 16 °C during the day and a barely manageable 3 °C in the night. Even during May, the breeze carries the winter chill. June is when the mercury really kicks up and temperatures soar up to 21 °C (day). The night temperatures are still low being about 7 °C. July is arguably the warmest month of the year. The day and night temperatures are 25 °C and 10 °C respectively. This is when Ladakh receives the odd rain, otherwise, the weather remains pleasant and inviting all through the month. The weather starts to get chilly once again in August. The breeze is a lot gustier and carries a lot of bite.

The weather is again cold during September to November. All through September, the weather is somewhat similar to that in May. The days are a lovely 20–22 °C. The nights are back to being bitterly cold. However, it's amid the periods of October and November that Ladakh backpedals to its miserable self. Since temperatures begin dipping under 0 °C on significantly more standard premise by and by, all interstate roads and passes are closed again for 5–6 months.

19.3.2 Precipitation

Precipitation is very low with annual precipitation of 10 cm mainly in the form of snow. Roughly one-third of the total annual precipitation is associated with westerly disturbances and occurs between December and February. Another third falls from July to August, so the summer monsoon is only of minor importance in Ladakh. Air is extremely dry and relative humidity ranges from 6–24%. Because of high elevation and low humidity the radiation level is very high. The value of global solar radiation is one of the highest in the world being 6–7 KW-hmm^{-1}. It is very usual to experience dust storms in the afternoon. The soil in the region is thin, sandy and porous. The entire area completely deprived of any natural vegetation. The water for irrigation is chiefly obtained from the glacier-melted snow.

The region has inherent constraints like pro-longed winter, scanty rainfall, rugged terrain's and consequently limited availability of productive land and there exists possibility for expansion of economic activity through various scientific methods.

The average monthly variation of precipitation (2000–2016) in Ladakh region is shown in Fig. 19.3 The monthly analysis of rainfall shows that summer months of have shown an average precipitation of nearly 40 mm with August (16 mm) being the rainiest month followed by July and then June. The precipitation in the summer is usually in the form of rainfall.

During winter, January is the most precipitous month with average 11 mm precipitation followed by February, December and November. Precipitation during winter is mostly in the form of snow.

19.4 Climate Change and Future Trends

Climate change has turned out to be one of the essential worldwide issues of discussion on the globe today. Additionally, it is presently broadly perceived and acknowledged by established researchers that there are different types of environmental change in various parts of the world. A consensus has emerged that the mountain regions are more vulnerable to the adverse impacts from climate change due to its fragile environment. Ladakh is one such area which has witnessed changes in the trends of extreme weather and climatic events. Climate change in the region has also been indicated by recent research and historical data analysis. There has also been a significant change in the pattern of rainfall and snowfall in the region.

An overall significant warming trend over the last 100 years has been observed in Ladakh (Chevuturi et al. 2016). It has been observed that from 1901 to 1979 there was a warm period over Leh followed by lowering temperature during 1979–1991, after which from 1991 onwards, the temperature again increased rapidly. The precipitation records show that after a low precipitation period before 1970, there was a period of increasing precipitation trend from1970 to 1995, but after 1995 onwards, there is again a decreasing trend in the precipitation.

In addition to this, the variation in temperature and precipitation trends in Leh indicate that is to some degree an inverse relationship between temperature and precipitation. A warm period was observed in Leh before the 1970s, which was accompanied by lower precipitation over the region. The temperature was low between the 1970s to mid-1990s s, when the region received higher precipitation. During the last two decades, the precipitation has witnessed a decreasing trend while the temperature is on a rising trend. Moreover some sign of diminishing number of days having high precipitation has also been observed.

The winter precipitation has been on a slight rise during last century (1901–2000). The rise is of the order of nearly 0.04 mm per year which accounts to 0.40 mm per decade (Shafiq et al. 2016).

The precipitation in the summer is usually in the form of rainfall with linear trend and has shown an overall decline during past century (1901–2000) with recent decades showing some positive trend in summer season rainfall. A reduction of 0.127 mm per year in the summer precipitation has been observed over the time period 1901–2000 in Leh.

In the past 20 to 25 years there has been a trend towards greater precipitation, but the pattern remains highly irregular. For instance, in 2002, a large amount of snow was recorded but the rainfall recorded for the region was very low. However in 2006 out of the total precipitation recorded, 150 mm was rain and 232 mm was snow. This was much more than the average over 1995 to 2007 of being 32 mm rain and 215 mm of snow (Paul 2010).

The climate change projection models have indicated an increase in the maximum and minimum temperature of Ladakh region. Precipitation would increase but with an uneven distribution. Furthermore, the changes in maximum and minimum temperature during the next few decades would not be significant whereas precipitation variation would be quite significant.

19.5 Impact of Climate Change on Agriculture

Agriculture is one of most delicate areas to effects of climate change due its high reliance on atmosphere and climate. Over hundreds of years, agriculturists in Ladakh have developed self-supported cultivating frameworks not withstanding in the midst of a climatically difficult condition. In fact, it has been the pillar of economy and gives nourishment and occupation security to the general population of Ladakh.

However, recent environmental changes are harmful for the fabric of sustainability for the general population of Ladakh. The effect of environmental change in the course of the most recent couple of years has unfavorably influenced farming in this district in various ways. Snow and water from glaciers is the chief source of irrigation for nine-tenths of the farming community. Thus, glacial melt determines the productivity in the region to a large extent. Besides glaciers are the primary sources of water for the farmers in the region. Excessive glacial melt has now become a cause of floods in the region, putting the lives of more than 80% farmers to risk.

Continuous retreating of glaciers in the region will have an adverse effect on t agriculture. The economy and food security of the region will be threatened due to changes in agricultural produce. It is supposed that increase in temperature could modify agricultural produce. The change may be either positive or negative. The impact on agriculture and its outcomes on the general public are probably going to differ locally contingent upon the kind of environmental change that has occurred here and the choices accessible to farmers. It could cut down the yield of wheat.

However, higher temperature may enhance the productivity of wheat in the high altitude areas. Warmer temperatures ay influence many crops to develop all the more rapidly, however hotter temperatures could likewise diminish yields. Crops have a tendency to grow quickly in warmer conditions. Nonetheless, for a few crops, (for example, grains), speedier development decreases the measure of time that seeds need for growth and maturity. Throughout the years, agriculturists have brought new plants into their farmlands.

Crops tend to grow faster in warmer conditions. However, for some crops (such as grains), faster growth reduces the amount of time that seeds have to grow and mature. Over the years, farmers have introduced plants into their farmlands. Looking at the brighter side of the impact of climate change, it merits saying that farmers in Ladakh have profited from the current global warming as they can develop number of new vegetable products.

Climate change is likewise connected with an increase in the incidence of extreme climatic events, for example drought, cloudburst and floods. The nomads known as the Changpas living in the Changthang region intensely feel environmental change. These individuals are reliant on the raising of animals, the pashmina goats for their fleece. As of late, the migration paths of the Changpas have changed because of decline in pasture land. Further, indeterminate snowfall has led loss of numerous domesticated animals in the district. In future, environmental change in Ladakh is probably going to influence agriculture and increment the danger of water for irrigation. The villagers could confront serious water shortage which is credited to overwhelming runoff from rapid snow melt. Besides, an expansion in the occurrence of pest attacks and weeds could also make destruction in the district.

19.6 Leh Cloudburst 2010

A cloudburst occurred near Leh in Jammu and Kashmir on August 6, 2010 at around 0130–0200 hours IST (Indian Standard Time) as per the report given by India Meteorological Department (IMD 2010). The National Disaster Management (NDM) cell under the Ministry of Home Affairs (MHA) gave the report on the situation dated 6th August 2010. According to the report, this cloudburst hit the Leh region at around 0100–0200 hrs after torrential rains in the interceding night of 5 and 6 August, 2010. The report also notes that this cloudburst triggered flash floods in Choglumsar and Pathar Sahib Villages and the surrounding areas of Leh Township. As of 25th Aug 2010 the figure of human deaths reached 196 (including six foreigners) and the normalcy in basic civic amenities, like water supply, electricity, communication, health, transport, aviation and tourism was restored only by end of August.

According to the India Meteorological Department (IMD), "a cloudburst features very heavy rainfall over a localized area at a very high rate of the order of 100mm per hour featuring strong winds and lightning. It is a remarkably localized phenomenon affecting an area not exceeding 20–30 km^2. A cloudburst occurs during monsoon season over the regions dominated by orography like Himalayan region, northeastern states and the Western Ghats".

The IMD report following the Leh cloudburst, in light of the satellite images presented that "a convective system developed in the easterly current associated with monsoon conditions over the region. The convective cloud band extending from southeast to northwest developed over Nepal and adjoining India in the afternoon of 5th August. It gradually intensified and moved west-northwestward towards Jammu & Kashmir. An intense convective cloud cluster developed to the east of Leh by 2130 hours IST of 5th August which resulted in a cloudburst event between 0130-0200 IST of 6th Aug 2010".

Ladakh experienced a substantial and ill-timed rainfall, followed by the terrific flash floods resulted from cloudburst in August 2010. This resulted in flooding of agricultural fields along with mudslides consisting of rock and sand. This resulted in further deterioration of agricultural fields and reduction of area available for cultivation. Also numerous agricultural fields and roads were washed away due to consistently happening floods in the course of the most recent couple of years, prompting decrease in net farming area.

References

Ahmad, L., & Kanth, R. H. (2014). Characterization of climate of Leh district of cold-arid Himalaya. *Journal of Agrometeorology, 16*(2), 214.

Chevuturi, A., Dimri, A. P., & Thayyen, R. J. (2016). Climate change over Leh (Ladakh), India. *Theoretical and Applied Climatology*, 1–15.

LAHDC. (2015). District statistical handbook 2014–2015, Ladakh Autonomous Hill Development Council.

Le Masson, V., & Nair, K. (2012). Chapter 5 Does Climate Modeling Help when Studying Adaptation to Environmental Changes? The Case of Ladakh, India. In *Climate change modeling for local adaptation in the Hindu Kush-Himalayan region* (pp. 75–94). Bingley: Emerald Group Publishing Limited.

Shafiq, M. U., Bhat, M. S., Rasool, R., Ahmed, P., Singh, H., & Hassan, H. (2016). Variability of precipitation regime in Ladakh region of India from 1901-2000. *J Climatol Weather Forecasting, 4*(165), 2.

Paul, S. (2010). Climate change and the challenges of conservation in Ladakh in International Association for Ladakh Studies Ladakh Studies NR.

IMD. (2010). India meteorological department annual report (Ministry of Earth Sciences, Govt. of India).

Chapter 20
Impact of Climate Change on Temperate Fruit Production in Kashmir Valley, North Western Himalayan Region of India – Challenges, Opportunities and Way Forward

Nazeer Ahmed, F. A. Lone, K. Hussain, Raihana H. Kanth, and S. Sheraz Mahdi

Abstract Over the past few decades, anthropogenic changes in the climate of the earth have become the focus of scientific and social attention. The entire temperate western Himalayan region extending from Jammu and Kashmir, Himachal Pradesh and Uttrakhand has a unique and fragile eco-system, where the very sustenance and livelihood of more than 75 percent of people is dependent on agriculture and draws about 60 percent of its Gross Domestic Product (GDP) from the surrounding ecological resources. The temperate climate in this area makes this region unique and offers tremendous opportunities to produce high quality fruits like apple, peach, plum, almonds, apricot, walnut etc. and of high value off season vegetables and ornamental crops. But in the recent decades, the annual mean temperature of Kashmir valley has increased significantly. Accelerated warming has been observed during 1980–2014, with intense warming in the recent years (2001–2014). During the period 1980–2014, steeper increase in annual mean maximum temperature than annual mean minimum temperature has been observed. In addition, mean maximum temperature in plain regions has shown higher rate of increase when compared with mountainous areas. In case of mean minimum temperature, mountainous regions have shown higher rate of increase. Analysis of precipitation data for the same period shows a decreasing trend with mountainous regions having the highest rate of decrease which can be quite hazardous for the fragile mountain environment of the Kashmir valley housing a large number of glaciers. Increased temperature is

N. Ahmed (✉) · F. A. Lone · K. Hussain
Sher-e-Kashmir University of Agricultural Sciences and Technology of Kashmir, Srinagar, J & K, India
e-mail: vc@skuastkashmir.ac.in

R. H. Kanth
Division of Agronomy, SKUAST-Kashmir, Shalimar, Srinagar, Kashmir, India

S. Sheraz Mahdi
Mountain Research Centre for Field Crops, Sher-e-Kashmir University of Agricultural Sciences and Technology of Kashmir, Khudwani, Anantnag, Kashmir, J & K, India

S. Sheraz Mahdi (ed.), *Climate Change and Agriculture in India: Impact and Adaptation*, https://doi.org/10.1007/978-3-319-90086-5_20

affecting vernalization of these high chill requiring fruits like apple, pear, walnut apricot almond and cherries leading to slow growth in production and productivity especially in rain fed areas. In addition,with due to erratic and extreme weather conditions, all kinds of pome and stone fruits are getting heavily damaged and quite often there is a coincidence of snowfall and flowering in most of the fruit and nut crops resulting in severe frost injury and in some cases the higher average temperature during winter inducing early bloom and maturity. The impact of fluctuation in temperature change is so much that most apple and almond trees sprout 2–3 weeks earlier instead of their normal sprout in mid March and April respectively. Frequency of natural disasters like drought, floods and strong winds in some areas has also increased. Cherries are also fast disappearing from their traditional growing areas of Kashmir valley. The diseases like *Alternaria* leaf spot and scab in apple, gummosis in stone fruits and nuts have become severe. The aphid attack is occurring approximately two weeks earlier under increased temperature. The red mite, white grub and scale insects have emerged seriously in almost all crops impacting productivity and quality of the produce. In the light of changing climate, there is a need to develop climate resilient varieties where crop architecture and physiology may be genetically altered to adapt to changing environmental conditions. At the same time, measures need to be taken for mitigation of the climate change both at local, regional and state level with equal and proactive participation of all stake holders including, farmers, scientific community and administrators of the line/ field department. This chapter aims to discuss about the recent trends of climate change and its impact on temperate fruit crops. Some of adaptive and mitigation strategies to combat the ill effects of climate change on temperate fruit crops have also been discussed.

Keywords Climate change · Temperate fruit crops · Impact · Adaptive · Mitigation measures

20.1 Introduction

The warming of climate system is showing its impacts on almost all spheres of life with predominant effects on the environment, water resources, industrial production, agricultural activities, and human lives (Shi and Xu 2008; Yao and Chen 2015). In a recent report, IPCC (2014) ascertains that the global climate has shown a warming trend of about 0.87 °C for the period spanning from 1901 to 2013 and mainly attributed the anthropogenic factors as the primary cause. Annual surface air temperatures over India also have shown increasing trends of similar magnitude during the period 1901–2013 (Annual Climate Summary 2014). Generally, climate studies have been performed on global scales (Pielke et al. 2000), but in order to understand the complexities, the scale needs to be narrowed down to the local level. The complexities become increasingly important over mountainous regions, which display great climatic variability taking topographical factors into consideration, because of their unparalleled impact on the local climatic conditions. Himalayas is one of the world's most sensitive hotspots to global climate change, with impacts manifesting at a particularly rapid rate. Numerous studies suggest an increased rate of warming, particularly in the minimum temperatures over most of the mountainous regions which among others include Himalayas and Tibetan Plateau as compared to the nearby low land areas (Rangwala and Miller 2012). The rate of warming around the globe has been increasing steadily but seems to have heightened since the last five decades, and highest rates of increase were recorded in the last two decades (Bhutiyani et al. 2010; Qiang et al. 2004). Conforming to the global trends in the rate of warming, high elevated areas particularly Alps, Rockies and Himalayas have shown that these highlands have also warmed significantly, as revealed by variousstudies done in these areas (Bhutiyani et al. 2010; Dimri and Dash 2012). The ongoing debate about the current and future climate change impacts have gained impetus due to the fragile ecosystems of the Himalayan region which is also extremely vulnerable to natural hazards (Cruz et al. 2007). The concerns of the changing climate arise from the fact that Himalayas encompass the potential of floods, droughts, and landslides (Barnett et al. 2005), simultaneously affecting the human health, biodiversity, agricultural production, sustenance of livelihood, and food security (Xu et al. 2009; Rasool et al. 2016). Although some studies suggest that pre-monsoon or spring (March to May) and summer cooling have been reported in some portions of Himalayan and Indus basins (Yadav et al. 2004; Fowler and Archer 2006), but most of the studies suggest significant increasing annual temperatures in the last century (Shafiq et al. 2016; Sharma et al. 2000; Bhutiyani et al. 2010). Variation in the topography and vast spatial expanse of the Himalayas, its sub-regions respond differently to the small variations in the climate system (Chevuturi et al. 2016). Any small variation in the temperature and precipitation can prove detrimental to the already fragile environs of Kashmir Himalayas (main Himalayan range), which is an amalgamation of many glaciers, water resources, and forests. Increasing temperature due to climate change is likely to affect future winter chill of temperate fruit crops and could have other major impact on the fruit and nut industry. This study provides the information on climate change, trends and its impacts on temperate fruit crops production of Kashmir Valley and reviews evidence of warming as well as variability in precipitation and extreme events. Understanding and anticipating the impacts of climate change on horticulture sector particularly fruit production and the services they provide to people are critical. Efforts to develop and implement effective policies and management strategies for climate change mitigation and adaptation requires particular new research initiatives under these rugged topographical landscapes.

20.2 Climate Change Trends Over Kashmir Valley

Jammu and Kashmir, state of India is located in the northern part of the Indian subcontinent in the vicinity of the Karakoram and westernmost Himalayan mountain ranges. Kashmir Valley is a mesogeographical region with an area of around 15, 948 km^2. Topographically, the valley depictsan elliptical bowl-shaped character, encapsulated between mighty Pir-Panjal range in its south and southwest and the great Himalayan range in the north and east. The mountain ranges rise to the height of about 5550 m in the northeast and dip down to a height of 2770 m in south. The valley of Kashmir stretches between 32° 22′ to 34° 43′ N latitude and 73° 52′ to 75° 42′ E longitudes (Hussain 1987). The climate of Kashmir varies greatly owing to its rugged topography.The recent analysis of the monthly and annual trends in temperature conducted by Shafiq et al. (2018) for the period 1980–2014 revealed a significantly increasing trend in the annual mean maximum temperature in the whole Kashmir valley during the 35-year period ($p < 0.01$) at a rate of 0.03 °C/year. The results further suggested that except summer, all other seasons are showing statistically significant increasing trend. The increasing trend in winter can prove to be detrimental for the agriculture and glacial land forms of the valley. Corresponding trends in annual mean minimum temperature reveal that the same have been increasing at a rate of about 0.02 °C/year, with the Mann-Kendall test showing a statisticallyincreasing trend ($p < 0.01$). Further, with mean minimum temperature only, in spring and autumn seasons are showing statistically significant increasing trends. These results show that at the regional level, climate has slowly but surely shown a consistently increasing trend particularly in case of temperatures.

With regard to precipitation, the results showed a consistent decrease in the annual precipitation at an alarming rate of about −7.9 mm/year which was statistically significant. The rainfall data, when analyzed, seasonally shows that it is decreasing at the rates of −9.12, −5.58, and − 1.09 mm/year for the winter, spring, and summer seasons, respectively. Only the autumn rainfall follows an almost consistent uniform increasing trend at the rate of 0.5 mm/year.

20.3 Extremes in Weather Conditions During Critical Periods of Growth

Due to climate change there is now a general observation of the extreme events viz; heavy rains/cold weather in critical period of growth viz., during blooming period thus off setting pollination and subsequently fruit set, strong winds resulting in the fruit fall, hail storms and the emergence of insect and fungal diseases. In some areas less rainfall particularly in karewas leads to water scarcity thereby severely affecting the quality of the fruits. Due to unusual hailstorms and windstorms in summers, fruits like cherry, apple, plum, peach and appricot are getting damaged severely (Choudhary et al. 2015). There have been several incidences of agriculture drought in past thirty years. The longest spell of drought for 8 weeks has been reported in 1987 (Aug-Sept) followed by 1997 (6-weeks in June–July, 2000 (6-weeks in Sept-Oct), 2004 (6 weeks in Sept-Oct), 1993 (5 weeks in June–July) and 1997 (5 weeks in July–August). In some areas due to excessive water logging the fruit trees develop many diseases thus severely affecting the fruit set. With regard to projections for extreme weather over Valey, Gujree et al. (2017) has reported that the futuristic frequency and intensity in the extreme temperature events is showing a significant increase in plain regions, whereas, mountain fronts shall have more extremes in precipiation and minimum temperatures.

20.4 Fruit Production in North Western Himalayas Region

The north western temperate Himalayan region extending from Jammu and Kashmir, Himachal Pradesh to Uttrakhand has a unique and fragile eco-system. The temperate climate in this area makes this region unique and offers tremendous opportunities to produce high quality fruits like apple, peach, plum, almonds, apricot, walnut etc. and of high value off season vegetables and ornamental crops. Presently the temperate fruits and nuts in theses three Himalayan states are grown in more than 7.65 lakh hectares with a production of about 33.17 lakh metric tonnes but the productivity has remained low (3.35 in HP, 3.83 in Uttrahkhand and 5.2 mt/ha in J & K) compared to 17.17 mt/ha on all India basis. The total area under fruit cultivation, production and productivity in J&K, Himachal Pradesh, and Uttrakhand according to 2014–15 estimates is given in

Table 20.1 Data on the area, production and productivity of temperate fruits in North Western Himalayas (2014–15)

	Fruits		
Regions	Area (000 ha)	Production (000MT)	Productivity (MT/ha)
Jammu and Kashmir	336.45	1779.44	5.2
Himachal Pradesh	224.35	751.94	3.35
Uttranchal	204.96	785.97	3.83
Total	765.76	3317.35	–
All India basis	6109.67	86601.68	17.17

Source: Database of National Horticulture Board, Ministry of Agriculture

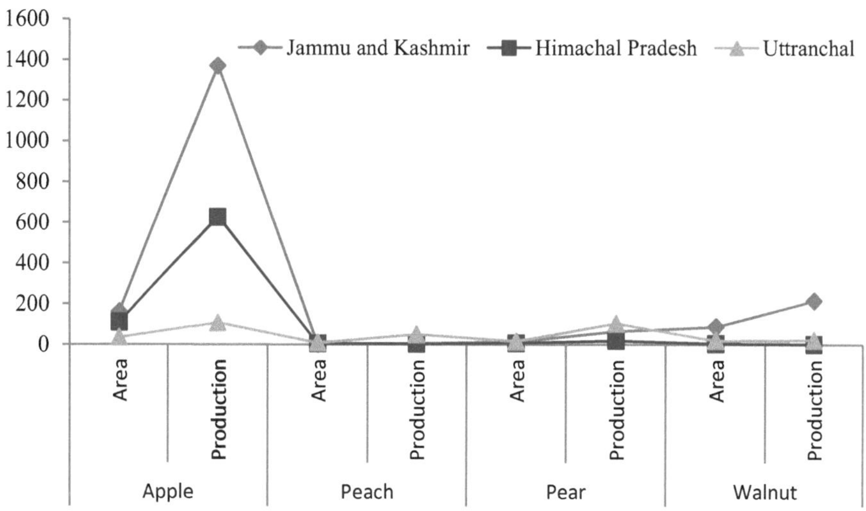

Fig. 20.1 Data (area and production) of the major temperate fruits in three north western Himalayan states of India (2014–15)

Table 20.1. Amongst these three states J&K state leads in apple and walnut production (13.68 and 2.13 lac mt respectively). Whereas, Uttaranchal leads in the production of peach and pear producing 4.98 and 10.4 lac mt respectively (Fig. 20.1). In J&K State in spite of increase in temperature there is an increase in the area and productivity in most of the temperate fruit crops since 1974–75 except almond where a decline in the area has been noticed after 1990's mostly due to various climatic factors. The comparative area, production and productivity of various temperate fruits (apple, pear, almond, walnut from 1974–75, 1990–91 and 2014–15 in J & K state is given in Table 20.2.

Table 20.2 Variation in area under cultivation of different fruit crops in J&K from 1974–75 to 2014–15

Crop		Year		
		1974–75	1990–91	2014–15
Apple	Area (000 ha)	46.2	68.7	160.25
	Production (000 mt)	190.5	658.2	1368.63
	Productivity (mt/ha)	4.1	9.6	8.84
Pear	Area (000 ha)	2.31	8.1	13.67
	Production (000mt)	7.7	16.7	65.41
	Productivity (mt/ha)	3.3	2.1	4.78
Walnut	Area (000 ha)	13.2	19.2	86.10
	Production (000mt)	10.1	2.2	213.86
	Productivity (mt/ha)	0.8	0.1	2.51
Almond	Area (000 ha)	9.4	19.2	15.72
	Production (000mt)	1.5	2.2	8.98
	Productivity (mt/ha)	0.2	0.1	1.75
Total	Area (000 ha)	82.5	176.3	336.45
	Production (000mt)	216.2	769.9	1779.44
	Productivity (mt/ha)	2.6	4.4	5.29

Table 20.3 Area, Production & Productivity of apple during 2001–02 and 2014–15 in north western Himalayan states

State		2001–2002	2014–2015
Jammu & Kashmir	Area ('000 ha.)	90.1	160.25
	Production ('000 mt)	909.6	1368.63
	Productivity (mt/ha.)	10.1	8.54
Himachal Pradesh	Area ('000 ha.)	92.8	109.55
	Production ('000 mt)	180.6	625.20
	Productivity (mt/ha.)	1.9	5.70
Uttarakhand	Area ('000 ha.)	51.8	34.69
	Production ('000 mt)	59.3	106.10
	Productivity (mt/ha.)	1.1	3.05
All India	Area ('000 ha.)	241.6	319.20
	Production ('000 mt)	1158.3	2133.84
	Productivity (mt/ha.)	4.8	6.68

Source: Database of National Horticulture Board, Ministry of Agriculture

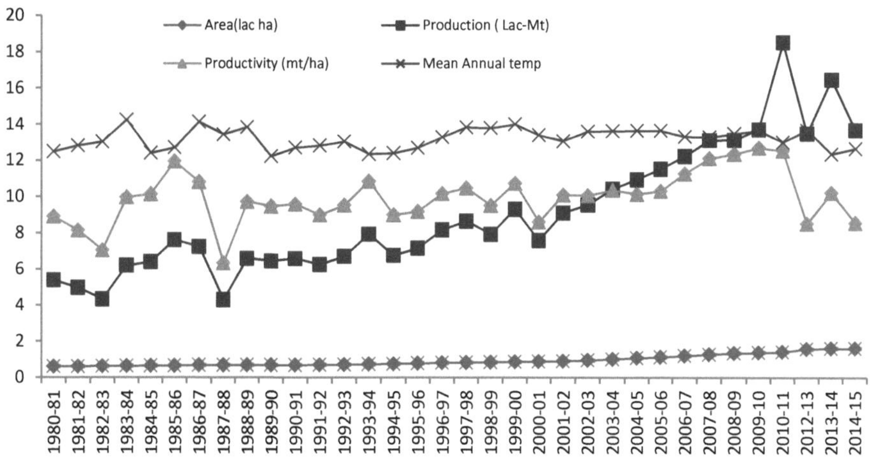

Fig. 20.2 Apple area, production, productivity and mean annual temperature in Kashmir valley from 1980–81 to 2014–15

Similarly the data on the area, production and productivity exclusively of the apple in three Himalayan states is given in Table 20.3. The relationship of the area, production and productivity of apple with average temperature is given in Fig. 20.2.

20.5 Impact of Rise in Temperature on Temperate Fruit Production and Productivity

In the past 30 years the average mean temperature in these hill states has risen between 0.44 to °C to 0.59 °C and is effecting vernalisation of high-chill fruits like apple, walnut, apricot, almond, cherry etc) and warming scenario exceeding 1.5 °C would expected to increase the risk of prolonged dormancy for both stone and pome-fruits. Over the last few years due to increase in temperature, change in season and moisture stress there has been distinct slow growth in production and productivity of temperate fruits and nuts especially in rain-fed areas where there is a clear trend of switching over from high chill varieties of fruit crops like apple, walnut, apricot, cherries and almond to growing of low chill requiring apple, peaches, apricots and other crops like Kiwi fruit. As a result of rise in the temperature and decline in overall precipitation apple in lower attitudes are shifting upwards replacing with low chilling crops like peach and apricots. Cherries are fast disappearing from some districts of Kashmir Valley due to erratic rains and water scarcity. With the unusual hailstorms and windstorms in summer months, all kinds of pome and stone fruits are getting heavily damaged. Due to increase in temperature quite often there is a coincidence of snowfall and flowering in most of the fruit and nut crops resulting in severe frost injury. The higher average temperature in winter leading to early bloom and maturity. The impact of temperature change is so much that most apple and almond trees are sprouting 2–3 weeks early in place of their normal sprouting in mid March and April respectively. For some varieties and crops warmer temperatures may be beneficial but higher temperatures have disadvantage because the warmth produces too many cells per fruit and that results in fruit that is too big, too soft and goes off quickly while fruit size is generally smaller, when low temperature occurs during spring which affects cell division in the post-bloom period.

20.6 Effect on Fruit Quality

For many apples their red colour is a trademark of quality. The low temperatures during the night in autumn, just before harvesting develop deep red colour. With the increase in temperature apples fail to change into their specific red shades. The other problem observed with increase in temperature is the increase in sunburn of apples. Sunburn in apples has existed earlier however, these days, frequency and intensity has increased significantly. Very easily about 10 to 20% of apples might get affected, particularly if there is hot afternoon sun without any protection. The warm temperature, heats the fruit flesh up and the cells break down resulting in sunburns.

20.7 Effect of Climate Change on Diseases and Insects

The climate change i.e. increases in temperature, droughts and CO_2 also effecting the three legs of the plant disease triangle. The diseases like *Alternaria* leaf spot and scab of apple, gummosis in stone fruits and nuts have become severe. The appearance of Sanjo Scale due to dry weather conditions in springs is another manifestation of climate change. The aphid attack is occurring approximately two weeks earlier for every 1 °C increase in average temperature. Many of the impacts of climate change on population dynamics of aphids will therefore have consequential effects on viral diseases. The red mite and white grub have emerged seriously in almost all crops in temperate areas including the scale insects are multiplying rapidly under congenial temperature and responsive host plant.

A survey of horticultural crop responses during hot dry summer has revealed higher incidence of powdery mildews, rusts and *Fusarium* diseases. Recent occurrence of stormy rains in Kashmir conditions are expected to increase bacterial gummosis of stone fruits as the pathogen mainly spreads by water splashes.

20.8 Management Strategies Under Changing Climatic Scenario

I. Crop improvement strategies

- Introduction of low chilling cultivars of pome, stone and nut fruits.
- Introduction and collection of gene source from plant and animal kingdom for future improvement.
- Diversification with other high value fruit crops like peach, apricot, walnut, kiwi etc. suiting to low chilling areas.
- Development of new genotypes which are efficient and high yielding at high temperature and CO_2 concentrations.
- Marker assisted selection and development of transgenics having resistance to biotic and abiotic resistance.
- Development of genotypes having resistance to heat and drought.

II. Development of agro-techniques

- Evaluation of the environmental impacts of apple production using Life Cycle Assessment (LCA) in order to evaluate alternative agricultural production methods that may reduce environmental impacts. This require assessment tools are required that measure the consequences of changing systems.
- Extreme events, such as late spring frost or windstorm, may cause crop failure. Future climate may also increase occurrence of extreme impacts on crops, e.g. weather conditions resulting in substantial reduction in yield and quality (for example severe drought or prolonged soil wetness).
- To develop a set of high resolution daily based climate change scenarios, suitable for analysis of agricultural extreme events
- To identify climatic thresholds having severe impacts on yield, quality and environment for representative crops and to assess the risks that these thresholds will be exceeded under climate change
- To analyze extreme weather impacts on reproductive and vegetative crop yields, using crop simulation models
- To identify knowledge gaps on physiological sensitivities, potential pest, disease and weed threats linked to extreme weather and to propose approaches to reduce the impact of extreme weather events
- Identify key inputs/outputs associated with crop production and determine their environmental impact.
- Determine the impact of horticulture in different geographical areas.

III. Plant protection strategies

- Strengthen surveillance of pest and diseases.

- Study the pattern of increasing climatic variability and observe the change that could lead to rapid movement of pathogens and insect pests.

IV. Post harvest management strategies

- Development of cost effective storage techniques.
- Development of varieties having longer shelf life.
- Studies on mitigation of post harvest spoilage

20.9 Introduction of New Climate Resilient Varieties by SKUAST-K

Apple Cultivars Released by SKUAST K To coup up with the changing weather patterns in the valley, SKUAST Kashmir through different conventional breeding programmes has developed different verities of apple which are mentioned below.

1. LalAmbri (2) Sunhari (2) Akbar (2) Firdous (2) Shiren (2) Shalimar apple-1, (2) Shalimar apple-2

Promising Exotic Varieties To full fill the growing demand of the apple in the market and to introduce climate resilient varieties many exotic varieties with high commercial value and early/fast fruit bearing potential and also suitable for effective cross pollination have been introduced. **Promising cultivars:** Summered, Mollie's Delicious, Gala Must, Imperial Gala Red Chief, Red Fuji, Jonica, Florina, Golden Delicious, Straking Delicious, Tydeman's Early Worcester, Starkrimson, Red Spur, Vance Delicious, Oregon Spur, Spartan, Red Gold etc.

Promising Clonal Root Stocks In order to overcome the weather vagaries, SKUAST K introduced many colonel root stocks viz.; M-7 (it is a high temperature and high soil moisture tolerant root stock), M-9 (it starts bearing 2–3 years after plantation, and is tolerant to high soil pH), MM-106 (it starts bearing in 3–5 years after plantation tolerant to high pH and suitable for high density plantation), MM-111 (it starts bearing 5–6 years after plantation, and is tolerant to drought and well adapted to all kinds of soils).

20.10 Strategies for Orchard Management Under Changing Climatic Scenario

Cultivation shallow cultivation in spring is advised in order to conserve soil moisture and facilitates incorporation of organic matter.

Mulching It is usually done in rain fed areas by using straw, hay, cut grass, FYM, rotten pine needles etc. Mulches conserve soil moisture and increase water holding capacity of the soil, improves aeration besides addition of organic matter and nutrients to the soil.

Irrigation In case there is a prolonged dry spell, 4–5 irrigations should be given at fortnightly intervals during fruit development period.

Drainage Areas that have excessive rainfall and are low lying should be provided with drainage channels so that excess of water is drained out to avoid damage caused by root and collar rot disease.

Pollination Management For the last couple of years farmers across the valley have been advised to introduce pollinizers in the apple orchards viz.; early season varieties (Tydeman's Early Worcester, Black Ben Devis, McIntosh, Allington Pippin), mid season varieties(Golden Delicious, Red Gold, Sunhhari, Yellow Newton etc), late season varieties (Rome Beauty, Granny Smith etc), spur type varieties (Gran Spur).

Nutrient Management The application of organic manures (fully decomposed FYM) to be preferably applied in February–March for conservation of moisture etc.

20.11 Conclusion

Climate change impacts are to be looked not in isolation but in conjunction with all the aspect of agriculture and allied sectors. The effects of climate change on temperate horticulture sector are due to variable climatic factors and the extreme weather conditions at critical periods of growth play a vital role in the yields and productivity of these crops. However more emphasis on development of heat-and drought resistance crops where crop architecture and physiology may be genetically altered to adapt to warmer environmental conditions besides developing such technologies which mitigates and makes full use of the effects of changing climate.

References

Annual Climate Summary (2014). National Climate Centre, India Meteorological Department, p 24. Available at http://imdpune.gov.in/Clim_Pred_LRF_New/Reports.html.

Bhutiyani, M. R., Kale, V. S., & Pawar, N. J. (2010). Climate change and the precipitation variations in the northwestern Himalaya: 1866-2006. *International Journal of Climatology, 30*(4), 535–548.

Chevuturi A, Dimri AP, Thayyen RJ (2016) Climate change over Leh (Ladakh), India. TheorApplClimatol. ISSN 1434-4483.

Choudhary, M. L., Patel, V. B., Siddiqui, M. W., & Mahdi, S. S. (2015). *Climate dynamics in horticultural science volume-I: Principles and applications*. Cambridge: Apple academic press.

Cruz, R. V., Harasawa, H., LalM Wu S, Anokhin, Y., Punsalmaa, B., Honda, Y., Safari, M., Li, C., & HuuNinh, N. (2007). Asia climate change 2007: Impacts, adaptation and vulnerability. In M. L. Parry, O. F. Canziani, J. P. Palutikof, P. J. Van Der Linden, & C. E. Hanson (Eds.), *Contribution of working group II to the fourth assessment report of the intergovernmental panel on climate change* (pp. 469–506). Cambridge: Cambridge University Press.

Dimri, A. P., & Dash, S. K. (2011). Wintertime climatic trends in the western Himalayas. *Climatic Change, 111*(3–4), 775–800. https://doi.org/10.1007/s10584–011–0201-y.

Fowler, H. J., & Archer, D. R. (2006). Conflicting signals of climatic change in the upper Indus Basin. *Journal of Climate, 19*(17), 4276–4293. https://doi.org/10.1007/s10708-016-9755-6.

Gujree, I., Ishfaq, W., Mohammad, M., & Gowhar, M. (2017). Evaluating the variability and trends in extreme climate events in the Kashmir Valley using PRECIS RCM simulations. *Modeling Earth Systems and Environment, 3*(4), 1647–1662. https://doi.org/10.1007/s40808-017-0370-4.

Hussain, M. (1987). *Geography of Jammu and Kashmir state* (pp. 11–18). New Delhi: Rajesh Publication.

IPCC. (2014). Climate change 2013: The physical science basis. In T. F. Stocker, D. Qin, G. K. Plattner, M. Tignor, S. K. Allen, J. Boschung, A. Nauels, Y. Xia, V. Bex, & P. M. Midgley (Eds.), *Contribution of working group I to the fifth assessment report of the intergovernmental panel on climate change* (p. 1535). Cambridge: Cambridge University Press.

Pielke, R. A., Sr., Stohlgren, T., Parton, W., Doesken, N., Money, J., & Schell, L. (2000). Spatial representativeness of temperature measurementsfrom a single site. *Bulletin of the American Meteorological Society, 81*, 826–830.

Qiang, F., Celeste, M. J., Stephen, G. W., & Dian, J. S. (2004). Contribution of stratospheric cooling to satellite inferred tropospheric temperature trends. *Nature, 429*, 55–57.

Rangwala, I., & Miller, J. R. (2012). Climate change in mountains: A review of elevation-dependent warming and its possible causes. *Climatic Change, 114*(3–4), 527–547.

Rasool, R., Shafiq, M., Ahmed, P., & Ahmad, P. (2016). An analysis of climatic and human induced determinants of agricultural land use changes in Shupiyan area of Jammu and Kashmir state, India. *GeoJournal.*

Shafiq, M. U., Bhat, M. S., Rasool, R., Ahmed, P., Singh, H., & Hassan, H. (2016). Variability of precipitation regime in Ladakh region of India from 1901–2000. *Journal of Climatol Weather Forecast, 4*(2), 165.

Shafiq, M., Rasool, R., Ahmed, P., & Dimri, A. P. (2018). Temperature and precipitation trends in Kashmir valley, North western Himalayas. *Theoretical and Applied Climatology*. https://doi.org/10.1007/s00704-018-2377-9.

Sharma, K. P., Moore, B., & Vorosmarty, C. J., III. (2000). Anthropogenic, climatic and hydrologic trends in the Kosi Basin, Himalaya. *Climatic Change, 47*, 141–165.

Shi, X. H., & Xu, X. D. (2008). Interdecadal trend turning of global terrestrial temperature and precipitation during1951–2002. *Progress in Natural Science, 18*, 1382–1393. https://doi.org/10.1016/j.pnsc.2008.06.002.

Xu, J., Grumbine, R. E., Shrestha, A., Eriksson, M., Yang, X., Wang, Y., & Wilkes, A. (2009). The melting Himalayas: Cascading effects of climate change on water, biodiversity, and livelihoods. *ConservBiol, 23*(3), 520–530.

Yadav, R. K., Park, W. K., Singh, J., & Dubey, B. (2004). Do the western Himalayas defy global warming? *Geophysical Research Letters, 31*, L17201.

Yao, J., & Chen, Y. (2015). Trend analysis of temperature and precipitation in the Syr Darya Basin in Central Asia. *Theoretical and Applied Climatology, 120*(3–4), 521–531. https://doi.org/10.1007/s00704-014-1187-y.

Correction to: Climate Change and Agriculture in India: Impact and Adaptation

Syed Sheraz Mahdi

Correction to:
S. Sheraz Mahdi (ed.), *Climate Change and Agriculture in India: Impact and Adaptation*, https://doi.org/10.1007/978-3-319-90086-5

The author group of the chapters 18 and 19 were inadvertently revised in the original publication. The author groups of these chapters have now been restored as in the original submission.

The updated versions of these chapters can be found at
https://doi.org/10.1007/978-3-319-90086-5_18
https://doi.org/10.1007/978-3-319-90086-5_19

S. Sheraz Mahdi (ed.), *Climate Change and Agriculture in India: Impact and Adaptation*, https://doi.org/10.1007/978-3-319-90086-5_21

MIX
Papier aus verantwortungsvollen Quellen
Paper from responsible sources
FSC® C105338

If you have any concerns about our products,
you can contact us on
ProductSafety@springernature.com

In case Publisher is established outside the EU,
the EU authorized representative is:
Springer Nature Customer Service Center GmbH
Europaplatz 3, 69115 Heidelberg, Germany

Printed by Libri Plureos GmbH
in Hamburg, Germany